GREAT BOTANICAL GARDENS OF THE WORLD

TEXT BY

EDWARD HYAMS

PHOTOGRAPHY BY

WILLIAM MACQUITTY

With a foreword by

SIR GEORGE TAYLOR

Director of the Royal Botanic Gardens, Kew

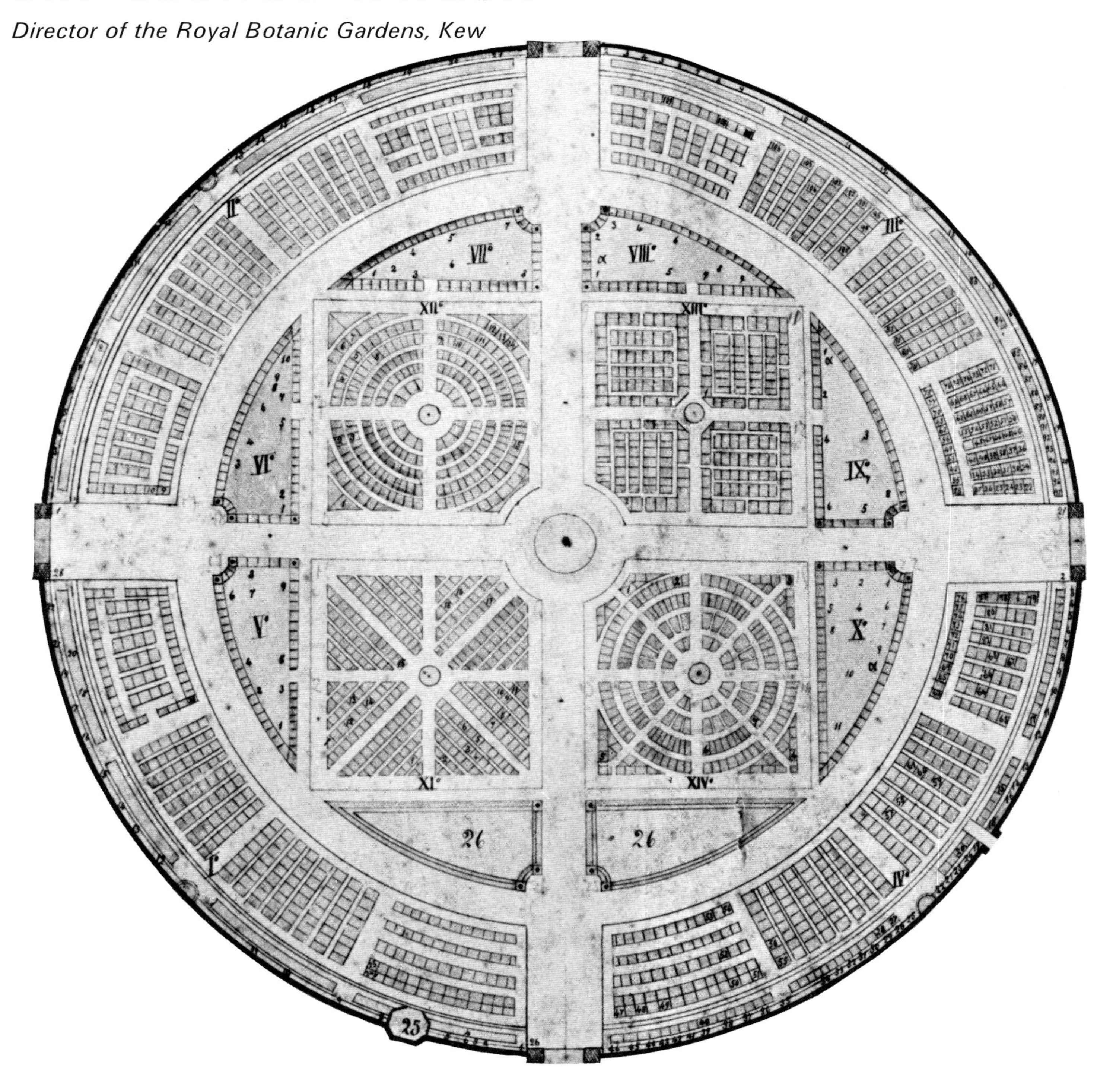

GREAT BOTANICAL GARDENS OF THE WORLD

THE MACMILLAN COMPANY

Library of Congress Catalog Card Number: 73-87880

First American Edition 1969

First published in Great Britain in 1969 by Thomas Nelson & Sons Ltd., London

The Macmillan Company

Printed in Italy

CONTENTS

TROPICS

SOUTHERN HEMISPHERE

FOREWORD

'The only other botanic garden which in my estimation approaches the grandeur of Kirstenbosch is that of Rio de Janeiro, with its magnificent amphitheatre of tooth-like mountain peaks'
—Sir George Taylor

By SIR GEORGE TAYLOR,
Director of the Royal Botanic Gardens, Kew

Botanical gardens should, in most civilized countries, aim to be sources of aesthetic and intellectual delight, and they have become a proud part of the national heritage, making the science of botany the handmaiden of horticulture.

Over the last forty years, and particularly in the last decade, I have had opportunities to visit botanical gardens in many parts of the world. One experience is exceptionally memorable. When I was a young graduate in 1927, before I had set foot on continental Europe, I went to Kirstenbosch when commencing my first plant-hunting expedition for the Royal Botanic Garden, Edinburgh. The impact of that visit has remained a cherished recollection over the years. With its superlative backcloth of mountain buttresses, for sheer grandeur the setting of Kirstenbosch is unsurpassed. At that time I made the acquaintance of General Smuts, who was a most knowledgeable botanist with a specialist interest in grasses. He spared time from his parliamentary duties to meet me at Kirstenbosch, and together we climbed Disa Gorge. He told me much about the plants, showed me the famous orchids and pointed out from the plateau the interesting vegetational and topographical features. It is perhaps understandable, then, why I have a special affection for Kirstenbosch in its dramatic situation and with its wonderful pageant of indigenous flowers. The only other botanic garden which in my estimation approaches the grandeur of Kirstenbosch is that of Rio de Janeiro, with its magnificent amphitheatre of tooth-like mountain peaks.

But setting is not everything and in one sense there is more merit in creating a lovely, scientifically and socially valuable garden in the ugly and difficult conditions of a city, than with all the advantages of the open country. In that respect, as in its magnificent teaching record, one would have to pick out the Brooklyn Botanical Garden, New York, for special praise. Another remarkable American achievement is the Missouri Botanical Garden which, originally planted in open country, has been overtaken by the growth of St Louis and yet retains all its beauty and freshness. But I must not give the impression that all the American botanic gardens are buried in cities: the Arnold Arboretum, though on the outskirts of Boston, might be in deep country remote from the town, and right across America on the west coast, the Strybing Arboretum is very pleasantly set in one corner of the great Golden Gate Park, itself a remarkable

Part of the formal garden at the Huntington Botanic. In the background is the mansion which houses the library, the art collection and the magnificent collection of manuscripts.

botanical achievement. Then, the Fairchild Tropical Garden, with its incomparable collection of palms, is in typical Florida water-scenery; and both the Huntington and Longwood Gardens are sufficiently 'far from the madding crowd' without being inaccessible.

For twelve years I have had the good fortune to direct the Royal Botanic Gardens at Kew, which are regarded by many as the supreme example of botanical and horticultural effort. It is for others to decide on the position of Kew, and Edward Hyams has been generous in his assessment, but tradition, which abounds at Kew, is a fickle jade, and cannot alone ensure the maintenance of a great garden. In my period as Director I have been helped by the understanding support of an enlightened government department and the loyal help of a succession of splendid colleagues. Kew, I am certain, will continue to play a leading role among the botanical gardens of the world and keep abreast of and contribute to the advances of the sciences which it practises.

Turning to the present book, much has been written in journals and elsewhere about the origin and functions of botanic gardens and informative accounts of individual gardens are common enough: but no work known to me deals with the major botanical gardens of the world with the scope and on the lavish scale of *Great Botanical Gardens of the World*. The author, Edward Hyams, and the photographer, William MacQuitty, have chosen a broad canvas on which to present their facts and impressions of their gardening odyssey. To most people the task of selecting, describing and depicting the great botanical gardens of the world would be much too formidable and indeed intimidating. Doubtless that is why no previous attempt has been made. The writer and the artist have given the fruits of their travels, which have taken them from the botanical gardens of Britain through Europe, and thence round the world. Edward Hyams absorbed as much as he could from the published histories and then visited the gardens to study their organization, meet their staffs and get to know at first hand the range of their collections and what is done with them.

William MacQuitty has captured the atmosphere of these widely differing gardens in his remarkable pictures: the formal elegance of the European planners, the magnificent sweep of the natural gardens, the cool vistas of the

temperate zones, the luxuriance of the tropics—everywhere the infinite variety of form and colour are revealed with skill and understanding. The result is a book which reveals the rewarding blend of science and beauty, and which will be valued for reference, and read with pleasure, by all who love gardens and are interested in their operation.

This, then, is not another treatise on gardening but a book dealing with representative botanical gardens throughout the world describing in detail and beautifully portraying the various features which give character to these famous places: their ancillary embellishments of sculptures and architecture; and something, also, of their work in assembling and distributing their great collections for the benefit of man.

This enterprise has clearly required most careful planning and the result is an extraordinary achievement and a splendid tribute to the most effective co-operation of writer and artist.

INTRODUCTION

It is generally supposed that the first botanic gardens which were something more than mere physic gardens were made and planted in Italy between 1545 and 1550. But the idea of scientific collections of plants may well have been suggested to Europeans by the remarkable 'gardens of plants' which so astonished the *Conquistadores* in Aztec Mexico. By the mid-sixteenth century there were three such gardens in Europe: by the middle of the twentieth century, over 500.

In the sixteenth century, botanical gardens were confined to Italy, Holland and Germany; today there are many such gardens in every economically advanced country in the world, and in a great many countries which cannot be so described. These gardens vary in extent from a fraction of an acre to over a thousand acres. Often the smaller gardens, for example those of Padua and Leyden, are aesthetically more pleasing and of greater historical interest than the large ones. In the small, early botanical gardens the number of species planted was a few hundred, rising to about a thousand, and late in the seventeenth century a few gardens had as many as 6,000 species. The Royal Botanic Gardens, Kew, today cultivates about 45,000 species, and all considerable gardens number their species in tens of thousands.

Botanical gardens have served, and continue to serve, three major and many minor purposes. In the first place, by the comparative study of the plants collected in them and the herbarium material collected by them, the modern sciences of taxonomic and experimental botany have been developed. Taxonomy is concerned with the system of classification and nomenclature, which gives clear expression to the kinship groups into which plants fall. In the case of green, flowering plants, for example, it has erected a grand and noble scheme establishing the natural relationships between over a quarter of a million species. Experimental botany is concerned with establishing the anatomy, cytology and metabolism of plants.

Secondly, the applied science of economic botany has been so practised in botanic gardens that they have served as acclimatization stations through which economically valuable plants such as rubber, coffee, tea, chocolate, cotton, hemp, vanilla and scores of others native to only one part of the world have been introduced to and established in others.

In the third place, there has been the specifically horticultural service of botanical gardens. Since the emphasis in such gardens has necessarily been on botany—a science—rather than on gardening—an art—this service has chiefly consisted in the trial, selection, hybridization and distribution into horticultural commerce of thousands of new or improved kinds of both useful and ornamental garden plants. In one or two great botanical gardens, however, the lessons of garden designing have been taught, too. Because two great garden designers have been concerned in laying it out, Kew is an object lesson in landscape gardening; so is Edinburgh, to which the best school of landscape gardening in the world is attached; and there are others in other lands.

An interesting by-product of botanical gardens has been the greenhouse. For if commercial rather than academic institutions have been the pioneers of the latest structural methods, of improved systems of heating, ventilation and irrigation, and finally of greenhouse automation, in an earlier age the academic gardens were the pioneers. Padua, Florence, Leyden and Leipzig were the first; but orangeries were developed by the French and Paxton's great hot-house at Chatsworth began that new era which at Kew was expressed in Decimus Burton's Temperate House and above all in Hooker and Burton's lovely Palm House, which became a model for every great botanical garden in Europe and America. And the work continues: the new greenhouses at Hamburg in Germany and at Meise in Belgium begin another new era, and these great new greenhouses will be models for first botanical, then commercial, and finally for private gardeners.

TECHNICAL DATA

CAMERAS: Two Nikon F Photomic Tn.
One Nikkormat F5.

LENSES: Nikkor Auto 28mm f 3.5
Nikkor Auto 35mm f 2.8
Nikkor Auto 50mm f 2
Micro Auto 55mm f 3.5
Micro Auto 105mm f 2.5
Micro Auto 200mm f 4

All lenses were fitted with lens hoods and protected with UV filters. Colour and neutral density filters were also used. Film stock used was Kodachrome 2 and Kodak Tri-X. They were given normal exposure at meter reading with shutter speeds of usually 125th of a second or faster. Flash was supplied by two Mecablitz 163 with mains charging units.

ACKNOWLEDGEMENTS

We are anxious to acknowledge the great kindness and generous help we have received all over the world in the writing and photographing of this book. At the planning stage, faced with the task of choosing representative gardens, we should have been very much at a loss without the help of Sir George Taylor, Director of the Royal Botanic Gardens, Kew; of Academician Tsintsin, Director of the great network of botanic gardens in the USSR; of Dr Russell J. Seibert of Longwood Gardens and Professor Richard Howard of the Arnold Arboretum in the United States.

In Italy, the staff of the *Istituto ed Orto Botanico dell' Università* were most helpful in Rome, and in Palermo Professor Francesco Bruno. Professor Carlo Cappelletti of the ancient *Orto Botanico dell'Università*, Padua, was most generous with his time and patient with our questions. In Germany we were most grateful for the advice and guidance of Professor Dr H. Merxmuller and Dr H. Chr. Freidrich of the *Botanischer Staatssammlung*, Munich, and Professor Dr Theo. Eckardt of the *Botanischer Garten und Museum*, Berlin-Dahlem. At the *Verwaltung der Bundesgarten* and the *Alpengarten im Belvedere*, Vienna Ing. Dip. Kargl gave up hours of his time to help us. Monsieur Maurice Fontaine, the Museum Archiviste-Palaeographe, and Monsieur Yves Laissus were most helpful at the *Jardin des Plantes*, Paris; and we are most anxious to thank Monsieur Marnier-Lapostolle for opening to us the *Jardin Botanique* of Les Cèdres, and Monsieur René Hebding, the Curator, for guiding us through that remarkable garden. In Belgium's new botanic garden at Meisse and in its great *Palais des Plantes*, Monsieur L. de Wolff helped us for several days with friendly patience, as did Professor W. K. H. Karstens in the historic *Hortus Academicus* of Leyden University in the Netherlands. From Dr Per Wendelbo in Göteborg, Sweden, and Dr Nils Hylander at Uppsala, home of Linnaeus, we had all the help and advice we needed and warm hospitality into the bargain. Dr T. J. Walsh of the National Botanic Gardens, Dublin and Dr H. R. Fletcher at the Royal Botanic Gardens, Edinburgh, were very helpful indeed. At Kew we had constant help and advice from Sir George Taylor and from Mr R. G. C. Desmond, the Librarian. We have also had valuable help from Mr P. Stageman, Librarian of the Lindley Library, Royal Horticultural Society.

Outside Europe our work began in South America and in Argentina our task was made not only easier but very much more pleasant by the help and warm hospitality of our friend Don Ricardo de Bary-Tornquist and Senora de Bary-Tornquist. In Australia our guide, philosopher and friend was Dr R. T. M. Pescott at the great South Yarra garden in Melbourne and the staff at the Sydney Botanic Garden, the new garden at Canberra and at Christchurch, New Zealand were very kind to us. Our work in Singapore was made pleasant and easy by the Director, Dr H. M. Burkill, by his deputy Dr Chew Wee-Lek and his Curator, Mr A. G. Alphonso. At the Kebun Raya, Indonesia (the erstwhile *Hortus Bogoriensis*), one of the half-dozen most beautiful gardens in the world, not only were we received and helped in the most friendly way by the Director, Dr Didin S. Sastrapradja, but Dr Sudjana Kassan, the horticultural curator, spent several days with us in the gardens and took a great deal of trouble to take us to the garden's lovely mountain satellite of Tjibodas, at Sindanglaja.

In Ceylon we had the patient and very friendly help of the Director of the famous Peradeniya garden, Dr D. M. A. Jayaweera, but we should have been much less at our ease in that gorgeous island without our guide and driver Charles de Silva. We should like to record our gratitude to Sir James and Lady Lindsay for their generous hospitality in Calcutta and to Dr S. N. Mitra for his guidance in the great botanic garden of that bewildering city.

At Kirstenbosch Dr Hedley Brian Rycroft took the greatest pains with us, especially in helping us to understand the plans and policy for an overall botanic gardens policy in South Africa; and Mrs Enid du Plessis made our task simple and agreeable in a very complex garden.

In the USSR we should like to thank Dr P. T. Lapin, Moscow; Dr N. V. Smolsky, who was so kind to us in Minsk and whose passion for rare lilacs was so catching; the several specialists who helped us in the great botanical garden at Yalta; and in Tashkent, Professor Dr Rusanov, that remarkable breeder of new Yuccas and new Hibiscus.

We have already mentioned the help we received from Dr Seibert and Professor Howard in the United States. Others who went out of their way to help us were Dr George S. Avery at the Brooklyn garden; Dr David Gates and Mr Ladislaus Cutak in St Louis; Dr P. H. Brydon and Mr Arthur Menzies at the Strybing Arboretum; Dr P. B. Tomlinson at the Fairchild Tropical Garden; and Dr Boutine at the lovely Huntington Botanical Garden, Pasadena, where Dr Myron Kimmach was also very kind to us.

Dr Yves Desmarais was most kind and helpful in the Montreal garden. And we are grateful to Dr Jiro Fumoto in Kyoto, and Dr Hurusawa in Tokyo.

One of the most troublesome and exacting tasks in connection with our work was the planning of the journeys; another, the right ordering of our manuscript, which was often in a state of some confusion. In both cases our thanks are due to Mrs Mary Bacon. Thanks are also due to Messrs Fairways and Swinford (Travel) Ltd, for their long-sighted and careful interpretation of our travel plans and to Messrs Kodak for their handling of the photographic material.

Edward Hyams
William MacQuitty

EUROPE

ITALY

Europe's first botanical gardens

Below A view of the garden at Rome.

Opposite Gates of the Padua Botanic Garden founded by a decree of the Venetian Senate in 1545. It is either the oldest botanic garden in Europe or two years younger than Pisa's, for which a foundation date of 1543 is claimed.

In botanical as in ornamental gardens, indeed in all the other arts and sciences, Italy was the teacher of us all. Early though some of the Dutch universities, notably Leyden, were in the botanical field, the Italians were earlier by between half and quarter of a century. Before coming to an account of the two gardens which dispute the honour of having been first, something should be said touching the lead of about a century given to them by the Vatican. For technical reasons, the fifteenth-century Vatican herb and medicinal or officinal gardens are not regarded as true botanical gardens; but they were the forerunners and the distinction is, perhaps, rather a fine one.

It was in 1447 that Pope Nicholas V set aside a part of the Vatican grounds for a garden of medicinal plants and, quite specifically, to forward the teaching of botany not, indeed, as a pure science, but as a branch of medicine. This Papal garden seems to have been the only one of its kind in Europe until the founding, about a century later, of the Paduan and Pisan Botanical Gardens. If the Vatican garden in question had been a true garden of plants, then there can be no question of Europe having taken the idea of such gardens from the pre-Conquest Mexicans. But it is no longer possible to establish what exactly was the nature of the garden planted by Nicholas V; and Italian historians of botany do not recognize the garden as botanical.

It became so, however, when, after two centuries, in 1660, by which time there were many gardens of plants in Italy, it was moved to a corner of the grounds of the monastery of S. Pietro in Montorio on the Janiculo. There it remained for another century and a half, but in 1823 it was abandoned, and Pope Leo XII started an entirely new botanical garden, as successor to the two older ones, in the grounds of the Villa Salviata on the Lungara. Nothing of this now remains but the stone on which the inauguration of the garden was recorded. In 1873 the principal plants of the Lungara garden were moved, systematically, in families, genera and species, to the garden of the old monastery of S. Lorenzo in the via Panisperna; but this site was, for some reason, not a success—perhaps it was required for some other purpose. At all events, ten years later, the Italian state and the Roman municipality received, jointly from Prince Tomasso Corsini, the *palazzo* Corsini and its grounds, on condition that they be used to house an Academy of Sciences; the gardens were to become the Official Botanical Garden of the City of Rome.

There had been a famous ornamental garden on this site: the Palazzo Corsini was the Roman residence of Queen Christina of Sweden from 1689, and she had made a garden in the grounds in which were performed a whole series of one-act plays written for her by the poet-noblemen of her circle. But after her time the garden fell into neglect, and all that Professor Romualdo Pirotta, first Director of the new Botanic Garden, found there was a wilderness overgrowing a few good surviving plants, a pleasing Triton fountain, a handsome stairway climbing the steep Janiculum slope, and very little else which he could usefully retain. However, what Pirotta did find of value for a true botanical garden was that some of the primal boskage of the hillside still remained, the true wild flora of the region. Among this stood fine Lebanon cedars. For the rest the palms, conifers, *Hovenia, Persea, Quercus* and other genera of trees and shrubs now grown into good specimens were all planted by Pirotta.

Those who visit the gardens today and who are accustomed to the impeccably maintained great Renaissance gardens of the Roman region or the Tuscan 'Humanist' gardens are apt to raise their eyebrows at the untidiness and apparent neglect of Rome's Orto Botanico. Such criticisms are only partly justified: during the war (when, incidentally, the head gardener and his family found the avocado pears of the *Persea indica* trees and the seeds of some of the edible pines a useful substitute for unobtainable bread at the time of the German occupation) it was impossible to maintain the gardens properly; for years thereafter there was neither the money nor the labour to restore them; and even now the seven gardeners employed cannot keep up with the work. But to some extent the unkempt look is intentional: writing in *Gli Orto Botanici Italiani* (Instituto di Technica e Propaganda Agraria), the late Professor Vincenzo Rivera explained:

> E' da notare che il nostro Giardino Botanico no se presenta con la toleta impeccabile di tanti giardini privati e pubblici di Roma. . . . La Villa non vuole rifatti rappresentare un campione di arti del giardinaggio, ma piutosto uno spettacolo di vita vegetale alquanto vicino a quello 'spontaneo' dell' ambiente romano. Invero i prati non sono 'in toleta' ed il bosco ed il sottobosco non sono sfoliti oltre lo stretto necessario . . .

This argument would, no doubt, be stronger if the vegetation included no planted exotics; but it is true that even these exotics look more 'natural' in woodland and boskage then they would in a formal setting and that they are just as useful for study so grown as if planted in a formal setting.

ITALY

PADUA

Opposite The original (1545) layout of the Padua Botanic Garden remains unchanged to this day, but some of the plants are of more recent introduction.

Below Original plan for the Paduan garden.

The botanical garden commonly distinguished as the oldest in Europe is the Orto Botanico of Padua. It was founded in May 1545 by a decree of the Senate of the Venetian Republic, and it is either slightly younger, slightly older than, or exactly contemporary with the Orto Botanico of Pisa.

I shall assume here that Padua is the older garden. The vote on its foundation was taken in the Venetian Senate on 29 May 1545: 137 ayes; 3 noes; and 17 *non sinceri*. The proposal became law one month later and on 7 July it was announced that the monastery of S. Giustina in Padua had ceded to the Republic and the University an area equal to about 20,000 square metres for the planting of a botanical garden. The architect Giovanni Moroni da Bergamo was commissioned to draw up a plan—he was working on S. Giustina at the time—and his plan was executed by Professor Pietro da Noale who, with the Venetian Patrician Daniele Babbaro, was in charge of construction. In 1546

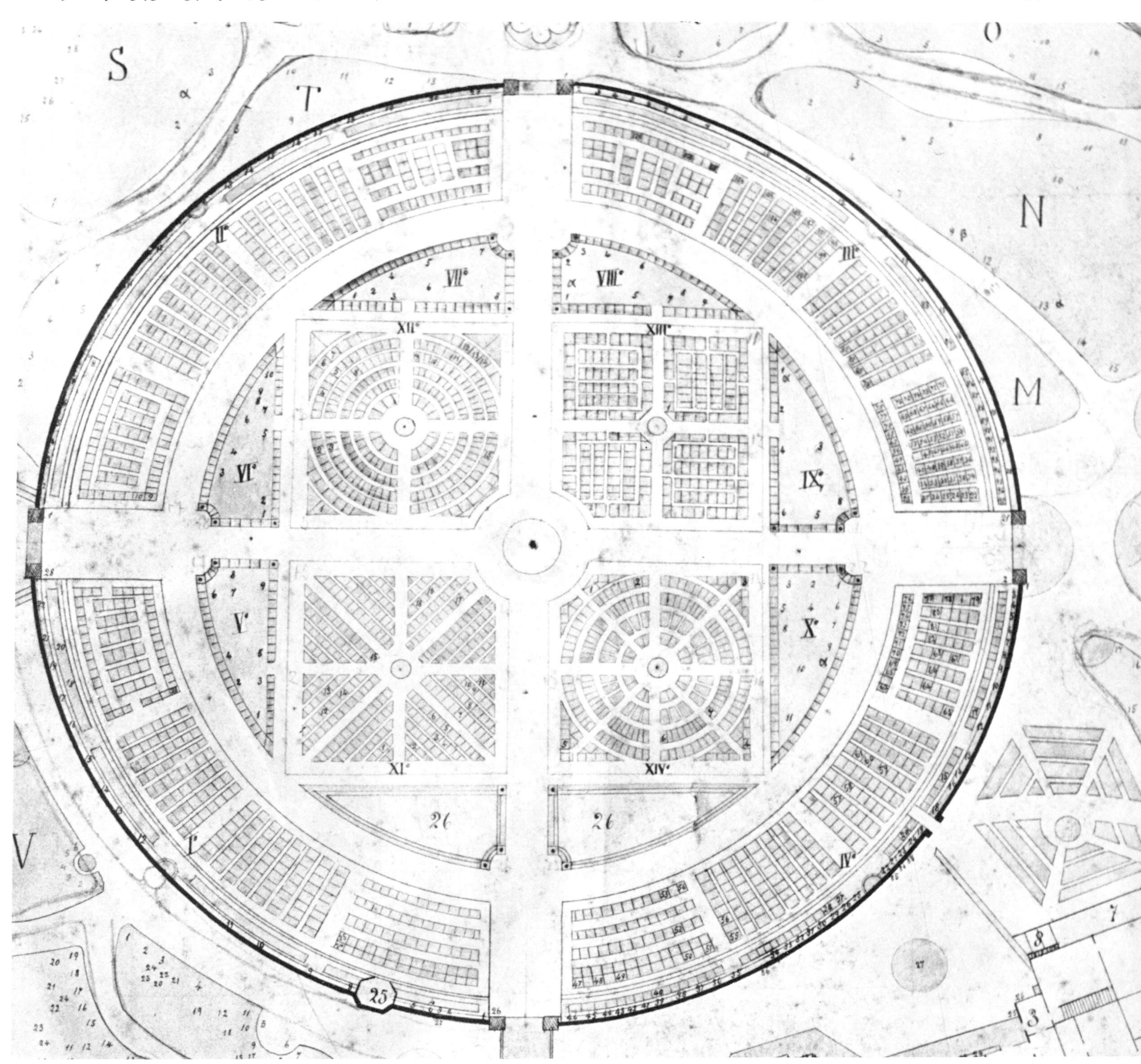

The greenhouse built in the Paduan garden to cover the bicentenarian *Chamaerops humilis*, known as 'Goethe's Palm'

Left and right Padua; seventeenth-century meteorological instruments in the garden.
Centre Herbarium sheet.

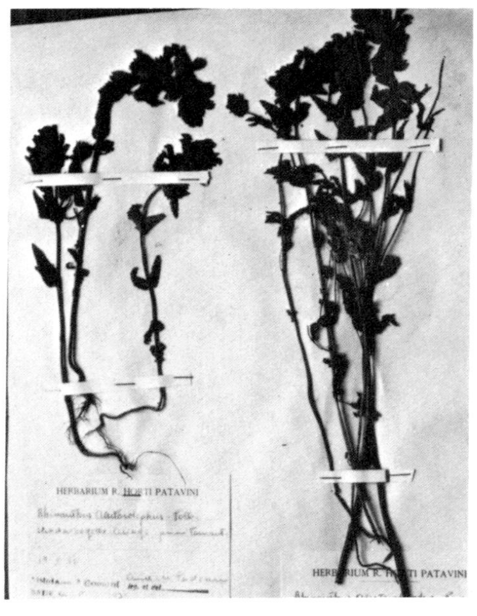

the garden was entrusted to the direction of Luigi Squalerno, called Anguillara (after the place of his birth), who retained his office until 1561. During that time the garden was still under construction and planting. In the choice of plants and their arrangement the Praefectus, as he and his successors were called, had the help not only of da Noale and Babbaro, but above all of another Venetian Senator, himself a great plantsman, botanist and gardener, Pier Antonio Michiel, author of a magnificent Herbal still to be seen in the Biblioteca Marciana in Venice.

As in all the Italian sixteenth-century gardens, elegance and taste in layout were brought to the service of science and combined with it in a fashion never again equalled until Kew Gardens were laid out two centuries later. The plan of the Paduan garden is formal, rectangular excepting for the charming central feature, a perfect circle 84 metres in diameter, defined by a handsome brick wall coped with a white marble balustrade. The original circular wall was built in 1551 to Giovanni Moroni's plan, but the present wall dates from the eighteenth century. The balustrade is decorated with busts of Prefects of the garden, Fabio Colonna, Saraceno, Prospero Alpino, Pontedara and Marsili, all of whom, especially Pontedara, were great botanists, fathers of botanical science. The interior of this circle—the area enclosed is approximately 5,555 square metres—is divided into sixteen divisions each subdivided and again subdivided, so that there is a perfectly symmetrical pattern of small beds. The main divisions are enclosed in iron railings of the early eighteenth century, the smaller ones in low stone edges. Each bed is devoted to a single species of plant. Until recently each species was labelled with its name according to Engler's classification, its number in the Della Torre catalogue, generic and specific names, and place of origin. However, the simple Linnaean binomial system has now been substituted.

The collections of grasses, of alliums, and iris are good. There are interesting collections of *Paeonia*; a pretty and representative collection of Mediterranean endemics planted as a rock garden—plants of the typical *maquis* or *macchia*; water lilies and other aquatics in and round the small pond. There is a tiny garden of medicinal and culinary aromatics. The iron railings are used to carry collections of *Vitaceae*, of *Lonicera* and of *Clematis*. Under glass are collections of cacti, including a very complete one of *Opuntia* species; aroids, orchids and other tropical exotics, but the Italian collections of these are nowhere impressive when they are compared with the German collections.

In the circular garden itself and in the pretty little Arboretum surrounding it there are some fine old trees; outstanding among them are the two-hundred-years-old *Ginkgo biloba*, an enormous specimen which has, unfortunately, lost a main branch; a *Platanus orientalis* planted in 1680, split, hollow but flourishing and having a young branch springing from near the base which has united itself by a sort of natural grafting to the old trunk in two places; and a fine *Cedrus deodara* which has attained a great stature and perfect form in the span of only one century.

The unique glasshouse in the Orto Botanico of Padua covers the plant known since Goethe's visit to the garden late in the eighteenth century as *La Palma di Goethe*. It is a specimen of *Chamaerops humilis* planted in 1585, and it was the basis of Prefect Giulio Pontadera's establishment of the genus *Chamaeroyphes* (subsequently changed by Linnaeus to *Chamaerops*) in 1720. It was Goethe's long and careful study of the morphology of this plant which led to the writing of his *Metamorphosis of Plants* published in 1790. Goethe gives an account of the impression which this plant made on him in his studies in natural history. Nor was this great palm the only plant which attracted his special attention when he visited the Paduan Orto Botanico. He records in his diary that

. . . on entering the botanic garden in Padua I was dazzled

Right Detail of Goethe's Palm, Padua.

Below Plan for a botanic garden showing influence of Renaissance ornamental garden design.

by the magical look of a *Bignonia radicans* which tapestried a long, high wall with its red bells, so that it seemed all aflame . . . This plant is still living and decorates a part of the garden wall.

The list of plants cultivated for the first time in Italy and often in Europe is a long and impressive one; it includes, by way of example merely, *Cedrus deodara* (1828); a number of bamboos; *Mesembryanthemum* (1713) —all of which have become of the first importance in Italian horticulture. There are also *Robinia pseudoacacia* (1662), now one of the commonest trees in Europe and thoroughly naturalized; *Pelargonium cucullatum* (1801), also since become an important element of European gardening; *Cyclamen persicum*, now one of the most important plants in florist commerce; *Jasminum nudiflorum* (1590), which, introduced from Italy to Britain, was lost, and reintroduced directly from China, by Fortune, nearly three centuries later;* the potato, *Solanum tuberosum* (1590) which reached Ireland, then

England, somewhat later or at about this time.

Although priority is commonly conceded to Padua,† there was much founding of Orti Botanici in the mid-sixteenth century, based on the recognition of the medicinal, agricultural and even industrial importance of the study of life; on the old foundation of monastic gardens of herbs; and perhaps on the introduction by Genoese in Spanish service of the Aztec or Toltec idea of gardens of plants.

*This conflicts with W. J. Bean's date for the first introduction to Britain but I believe it to be correct. It has certainly been in the Paduan garden since 1590 and it is very unlikely that British horticulturalists and botanists should not have known this.

†It has never been decided which of the two gardens, the Tuscan Grand Duke's Pisan or the Venetian Senate's Paduan, is really the older. First, the foundation of the Paduan garden is precisely documented; that of the Pisan is not. Secondly, it seems that the two States were using different calendars in the mid-sixteenth century with a difference of one year. The Pisans, however, have a letter in their archives which does seem to establish Pisan priority. But see, in this controversy, Giuseppe Martinoli in *Orti Botanici Italiani* ('Agricoltura', Rome, 1963).

PISA

and other sixteenth-century gardens

Reverting, for a moment, to the controversy between Padua and Pisa, Professor Giuseppe Martinole claims priority for the Pisan garden which he directed. His argument is founded on a letter written by Lucca Ghini, founder of the Pisan garden, dated Bologna 4 July 1545 and addressed to Piero Francesco Ricci, majordomo of Cosimo I de Medici, and which implies that the Pisan garden was then already in being. The date of that letter is only five days after the decrees of the Venetian Senate which founded the Paduan garden. The original Pisan garden was called the Giardino della Cittadella Vecchia, from its location, but it was moved to a new site in 1562. From the beginning Pisa, like Padua, contributed to the introduction and acclimatization of new plants to Europe. The full list would be a long one, but it includes the horse chestnut (*Aesculus hippocastanum*) in 1597, introduced by the Prefect Malocchi, presumably from Vienna where Clusius had introduced it by means of a supply of seeds sent to him from Constantinople; *Juglans nigra**, *Ailanthus glandulosa*, *Cinnamomium camphora*, *Chaenomeles japonica*, *Magnolia grandiflora*, *Photinia serrulata* and *Ginkgo biloba*†—a few names which would confer distinction on any botanical garden.The garden possesses what are probably the finest specimens of *Magnolia grandiflora* and *Liriodendron tulipfera* in Europe. Its Prefects, moreover, beginning with the first, Lucca Ghini, have made important contributions to botanical science: Ghini was the teacher of such great botanists as Aldrovandi and Mattioli. The Praefectus Andrea Cesalpino (1554–8), in his work on systemizing plant taxonomy, set the example for such great men as Tournefort, Ray, Morison and Linnaeus. The Prefect Georgio Santi (1782–1814) was the first botanist to publish a Tuscan flora. Giovanni Arcangelli (1881–1915) was the author of a valuable *Compendio della Flora Italiana*. Biagio Longo, expert in fig growing, did work of great economic value in his researches on 'caprification'. Another of the Pisan Prefects was the first botanist to apply palaeontological methods to the study of fossil flora.

The Orti Botanici of Florence, Ferrara and Sassari were the next in chronological order; all of them started from five to ten years later than the Paduan and Pisan gardens. Ferrara, notably, made important contributions to the science of plants. Next, in about 1568, the great botanist and physician Ulisse Aldrovandi, Lucca Ghini's pupil, persuaded the Senate of Bologna to give him and his colleagues an 'Orto dei Semplici'—a garden of herbs which soon developed into a botanical garden. The founder, Aldrovandi's, *Tavola di piante, fiori e frutta* is one of the most beautifully illustrated and sharply observed of the sixteenth-century botanical works.

Messina had its first botanical garden in 1638, but after the revolt of the Messinesi against their Spanish overlords the garden was spitefully destroyed by the Spaniards; it was refounded in 1678; and yet again, this time by the Bourbons, in 1838. The Sicilian climate made possible the cultivation of plants much too tender for northern Italy but, like every other activity of this great Sicilian city, the botanical work, and the garden itself, have suffered from severe earthquakes. Yet the work of its botanists on African and American plants has been valuable.

The next Italian city to have a botanical garden was Urbino. Then, in 1682, comes the Neapolitan garden.

If the kind of documentation which historians recognize—and there are no greater pedants than modern historians—were not wanting, Naples might have a claim to be the first Italian city to have had a botanical garden. The garden of 1682 was founded on a whole series of royal and noble herbal and ornamental gardens, going back to the thirteenth century, some of which might well have claimed to be 'botanical', that is, gardens of plants arranged in some kind of system for study. The immediate forerunner of the Orto Botanico proper was the Parco Reale, planted by Alfonso II of Aragon between 1450 and 1470. A French poem refers to this

> . . . parc tout clos ou sont maint herbes saines,
> Un bois plus grans que le bois de Vincennes . . .

as including among its delights olive, orange and pomegranate trees, figs, dates, pears and *allemandier* (almonds, presumably), apples, laurels and many sweet flowers.

Although this park was a forerunner of the Orto Botanico, the first garden devoted to botanical science and which was yet another of the ancestors of the garden of 1682 was Giovanni Vincenzo Pinelli's early sixteenth-century garden on the Collina dei Miracoli. Pinelli was a man of letters and a bibliophil, but his interest in botany was such that he corresponded with every leading botanist of his day, including, therefore, Clusius, Aldrovandi and Colonna, all of whom sent him seeds, herbarium material and information. Above all, he worked in close association and friendship with Ferrante Imperato, who, in his introduction to his own classical botanical work, referred to Pinelli as his principal collaborator, and praised his friend and fellow student as *huomo de elevata dottrina*.

Pinelli went to settle in Padua in 1558, perhaps to be near the new Orto Botanico, but Imperato had his own botanical garden and so did a number of other Neapolitan botanists, and this tradition of private or semi-private Orti Botanici continued until the foundation of the present garden by Michele Tenore in about 1810, who took over and 'institutionalized' the garden founded at the order of Joseph Bonaparte and continued by the Government under King Joachim Murat.

Urbino, as I have said, got its first Orto Botanico with its first university in 1672 from Pope Clement IX, the parent Academy of both being the 'Studio Urbinate', which had been founded by the Duke Guidobaldo I di Montefelto in 1506. Thereafter there was a pause in botanical foundations until in 1730 Turin, in 1762 Cagliari, in 1770 Pavia and in 1784 Siena planted Orti Botanici for the study of botanical science, Siena's tiny and very pleasing garden replacing the old municipal Herb Garden which had existed since 1588.

*Probably from Turkey or Greece in about 1600.
†About 1720.

PALERMO

and nineteenth-century gardens

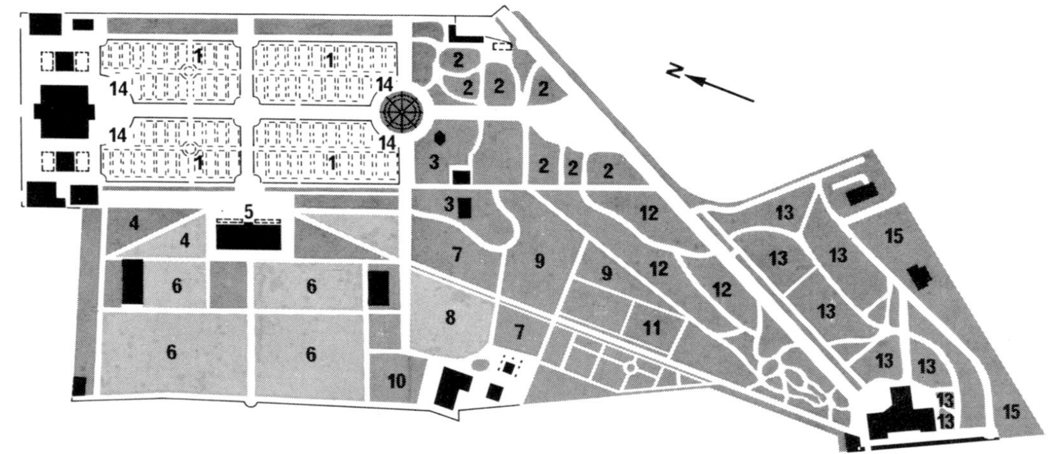

Key to Map

1 Irrigated and plunged pot parterres
2 Coppices of exotic plants; sub-tropicals
3 Xerophytic plant hillock
4 Citrus Gardens
5 Winter Garden
6 Experimental plantations
7 Fibre plants
8 Medicinal plants
9 Experimental plantings
10 Nursery
11 Essential Oils plants
12 Coppices of European plants
13 Plants arranged by Engler's system
14 Plants arranged by the Linnaean system
15 Experimental planting of exotic plants

Next in order of time was the magnificent Schola Regia Botanices and Orto Botanico founded at Palermo in 1795 by King Ferdinand II of Bourbon, and supported with money, and at least a show of enthusiasm, by the Prince di Caramanico, Viceroy of Sicily, by the Municipality of Palermo, a number of rich prelates in a city rich in rich prelates and, as the city records tell us, by *ricchi patrizi ed egregi cittadini*. The Palermitan garden, one of those we have chosen to illustrate, combines eighteenth-century taste with scientific system. Outstanding among its plants are the colossal specimens of *Ficus magnolioides* (now in their second century) and an extraordinary avenue of kapok trees. At the present time the garden is suffering from shortages of money and labour, but it is still impressive. There are some specimens of *Dracaena draco* probably larger than any (in Europe) outside the Canary Islands.

Genoa got its first Botanical Garden in 1803 following the foundation of the *Cattedra di Botanica e Storia Naturale*. There had, indeed, been an earlier garden, privately owned, maintained and directed by Domenico Viviani. Genoese botanists can hardly have been much better off after 1803, for the annual allotment of public money for the garden was L.500, about £25 gold. Viviani did good work on the *florae* of Liguria, Corsica and Libya.

Bari owes its Botanic Garden, at least in part, to Napoleon's suppression of the monasteries. There had been some mild interest in botany and in the flora of Puglia since the sixteenth century, and when Bari (Altamura) got its first small university in 1748 it also got a small Orto Botanico. In 1810 Joachim Murat decreed that the province must have an Agricultural Society; the planting of a larger Botanic Garden was mooted; and in 1813 the dispossession of the Capucin Friars left a suitable site vacant. That was not the last nineteenth-century foundation: in 1828 Camerino got its first botanical garden of particular value because it is sited 2,000 feet above sea-level; in 1858 came the Orto Botanico of Catania; and in 1872, that of Portici.

In a little more than four centuries Italy had founded botanical gardens in every province where the botanists who taught all Europe the elements of their beautiful science did their work.

Below left Tropical *Ficus* were planted in the Botanic Garden, Palermo, early in the nineteenth century.

Below The design, architecture and ornaments of the Botanic Garden, Palermo, were dominated by French taste, the garden fashion of the period.

VILLA TARANTO

The Villa Taranto Botanic Garden is a Scottish landscape garden on Lake Maggiore, stocked with tender exotics.

Overleaf The formal area of the Villa Taranto garden is used to display the best garden cultivars in season. All botanic gardens now do some purely horticultural work of this kind.

The latest botanical garden in Italy is a very beautiful curiosity on the grand scale, laid out, planted and maintained out of his own pocket and for his own pleasure by a Scottish gentleman who then presented it to the Italian nation. This is the great garden of the Villa Taranto at Pallanza on Lake Maggiore. It is one of the small class of botanical gardens which are of importance although created not by an academic or state institution but by a private collector of plants. Another important garden in the same class is that of Monsieur Marnier-Lapostolle in southern France (see pp. 87ff). There is nothing 'Italian' about the Villa Taranto garden; it is a great, romantic 'English' garden in looks and layout, 'botanical' in the range and diversity of species.

The maker of this garden was Captain Neil McEacharn, who, in 1910, inherited the great gardens of Galloway House, when his father died. Already interested in gardening, he set about modernizing the gardens; among his friends were two considerable Scottish gardeners, Sir Herbert Maxwell of Monteith; and the 4th Marquess of Headfort, who helped and guided him. Through him he came to know the Regius Keeper of the Botanic Gardens, Edinburgh, Professor Sir William Wright Smith (see p. 51).

Later, after World War I, Captain McEacharn set about replanting the Galloway House gardens in real earnest and was greatly helped when, in 1921, he formed a close relationship with the Assistant Curator of Kew Gardens, Mr C. P. Raffill.

But as well as being a gardener Captain McEacharn was a lover of Italy; he had been visiting that country since his boyhood, loved it and its people and realized that he could plant a greater range of plants in its climate than in that of Wigtownshire.

Key to Map
Only principal plantations are indicated.

1 Conifer walk
2 Tree ferns
3 *Paulownia*
4 Botanical species roses
5 Japanese *Acer* species
6 Herbaceous borders
7 *Rhododendron* wood
8 *Paeonia*
9 *Magnolia* wood
10 *Davidia involucrata*
11 Main lawn
12 *Fagus* species
13 *Magnolia*
14 Azaleas
15 *Cedrus*
16 *Camellia*
17 *Nelumbo* lake
18 Aquatics
19 Bog plants
20 Japanese cherries
21 *Erica* species
22 *Magnolia* specimen trees
23 Cotoneasters
24 The Villa
25 Greenhouses and Nursery
26 Office and Seed Rooms
27 Terraced gardens

In 1928 he started looking for a suitable property, found nothing to suit him, was returning by train from Venice one day in 1930, when he saw in *The Times* an advertisement for a villa for sale on Lake Maggiore. He got out of his sleeping berth, dressed, left the train at Pallanza and put up at an hotel. He found that the property for sale was La Crocetta, next door to the lovely garden of the Villa San' Remigio, and owned by the Contessa di St Elia, an Englishwoman. He found a hideous villa and a bad Italian garden full of bad statues; spent a day considering how this and the surrounding woods could be changed. He made up his mind to buy when he discovered that he could also buy the land adjoining that of the Villa, and would, therefore, have all the room he needed for the great garden he had in mind. While the various purchases involved were left in the hands of Captain McEacharn's Italian lawyers, he himself went to England and began to arrange for the exporting of the plants that he would need for his garden.

A man who knew *little* about plants, studying the weather and climatic statistics of the Italian lake district, would not have planted there a garden in which he planned to grow plants tenderer than those he could grow in Scotland. The climate of Scotland is milder and, on the face of it, the Scottish winter is less likely to kill subtropical plants than the winter in Pallanza.

But McEacharn knew better; he knew that if many tender trees and shrubs are killed in Scottish winters it is because Scottish summers are too mild to harden them, to cook them into that state which gives resistance to severe cold. The climate on Lake Maggiore is, by British standards, harsh: the temperature range, for example, is enormous with as much as 100 Fahrenheit degrees of difference between winter minimum and summer maximum; in winter it may fall to 10°F, in summer rise to and remain for weeks at 95°F. The 100-inch rainfall is badly distributed from the point of view of plants. But it is a fact that a number of tender Australian and other subtropical plants survive there and flourish which die in Scotland, simply because the hot summers 'ripen' the wood.

The plan reproduced opposite will give a better notion of the garden layout than a description would do; it should be borne in mind that the distance along the main drive from the lakeside main gate to the house is half a mile; that will give the scale. It was Captain McEacharn's intention when he gave his garden to Italy and its future and purpose were laid down in an act of parliament, that money he was leaving should endow a School of Horticulture and, in general, make his garden serve the usual purposes of a botanical garden.

But one way and another, what with heavy expenditure on the garden itself, inflation and other factors which erode large fortunes, the money was found not to be forthcoming at the testator's death. The garden is now financed and managed by the Ente Giardini Botanici di Villa Taranto, an organization of a kind peculiar to Italy,

The Villa Taranto collection includes hundreds of superlatively grown trees.

Below Villa Taranto; one of the fountains seen from the hall of the Director's house.

half private and part official. The private members of the Ente are various provincial organizations such as the Tourist Board, certain banks and other interests: they can, in theory, look to the State for help. Unfortunately the Italian Government finds itself unable to regard even so great a garden as worthy of much financial priority, and the Ente has been unable to obtain a grant of any kind towards the upkeep and new planting which are essential. About a third of the money comes from entrance fees and other expenditure by the one hundred thousand people who visit the garden every year. The balance is made up by private subscriptions paid by members of the Ente. In 1967 the Ente was coming to the end of its existence and the Director of the gardens, Consigliere Delegato Dr Antonio Cappelletto, has the heavy task of creating a new Ente to ensure the garden's future. Technical horticultural direction is in the hands of the Curator, Mr D. Barmes, a horticultural botanist trained in some of Holland's finest horticultural schools.

A few of the older trees in the garden were planted by the former owner and original builder of the Villa, Count Orsetto, or by his friends: for example, the mighty *Fagus sylvatica* behind the Villa; and some of the enormous specimens of *Magnolia kobus*. But for the most part the planting was done by Captain McEacharn after 1930; and the great stature of many of the trees and shrubs, attained in under forty years, is very remarkable, doubtless attributable to the humid warmth of the springs and summers. There can be no doubt that if the Italian State were more interested in the Villa Taranto than is, alas, the case, this could be one of the half dozen grandest botanical gardens in the world.

The collection of conifers is a good one: there are thirty-two species of *Abies*; specimens of *Metasequoia glyptostroboides* have made faster growth than in English and American gardens; fourteen species of *Picea* and ten varieties; *Pinus* was represented by thirty-one species, four hybrids and several varieties, but disease has killed a number of this genus and is attacking others. Both genera of cypress in numerous species have attained a fine size.

Nothing better demonstrates the importance of hot summers in making plants resistant to cold winters than the fact that Australasian trees and shrubs grow as well in the Villa Taranto garden as in Ireland. There are about fifty species and varieties of *Acacia* for example, and some very large trees. There are twenty species, varieties and cultivars of *Callistemon*. About seventeen species of *Eucalyptus* were originally planted and only *E. ficifolius* failed to survive. The present Curator has added thirty-three more species to the *Eucalyptus* collection, being convinced that the genus must be increasingly important in the nation's official reafforestation programme and that testing trees for reafforestation is part of the garden's proper work. Another Australian, *Grevillea*, is represented by twelve species. Yet other southern hemisphere genera have, on the whole, failed on trial, for example, *Nothofagus* and many of the genus *Olearia*.

Because the Villa Taranto garden was private, one man's creation, the emphasis on one genus rather then another depended on that man's taste rather than on some botanical purpose. For example, the garden is so rich in species, varieties and cultivars of *Viburnum* that a special study of that genus should be possible there. Apart from hybrids and cultivars there are no fewer than seventy-four species, fourteen of them with varieties. It is possible that viburnums grow too fast in the lake climate for their own good; at all events, some of the specimens are enormous and those over thirty years old are showing signs of senility.

The garden is very rich in *Magnolia*: thirty species, sixteen of them with varieties and a number of hybrid cultivars. Among the latter is a cross *M. campbelli* × *M. mollicomata* made at Kew by the late C. P. Raffill and which is about 30 feet tall. It has flowered since 1960 and the huge flowers are a deep, clear pink. All species of this genus reach a great size, notably *M. kobus* and plants of the × *M. soulangiana* complex, some of which are the size of forest trees. *M. tripetala* has reached 40 feet and a vast spread. If only as an arboretum of flowering trees the Villa Taranto would earn a place in this book.

Captain McEacharn planted a great many kinds of *Rhododendron*, for the soil is acid and the humidity generally high. Not all of this genus succeed, and the specimens would not impress botanical gardeners familiar with the rhododendron gardens of south-west Scotland, south-west England, and Ireland. For many rhododendrons the

Below Displays of garden flowers, such as tulips, of value to practical gardeners are a feature of the Villa Taranto.

Right Tulips, a lily pond, and a group of palms in the Villa Taranto garden.

winter climate of Pallanza, and the summer droughts, are unfavourable to this genus. (It is notable that the Villa Taranto tree-ferns, *Dicksonia antarctica*, which form a very romantic feature in summer, have to be moved under cover in winter, which is not the case in, for example, south Devonshire, while in parts of the west of Ireland, for example at Rossdohan, they are virtually naturalized.) However, *Rhododendron williamsianum* does quite well, *R. schlippenbachii* very well, and there is a good form of *R. augustinii* with almost blue flowers. Rhododendrons of the *grande* series have not made impressive growth but flower well. *R. preptum* and *R. falconeri* are the best in this series. The *arboreum* series is perfectly happy in the garden.

If the rhododendrons are not particularly impressive, although interesting as a large and diverse collection very attractively planted, the representatives of *Camellia* are, on the other hand, magnificent. There are eleven species in 171 varieties; specimens vary from 70-year-old giants round the villa itself, to the young but richly flowering *C.* × *williamsii* which Dr Capelletto and Mr Barmes have been planting. From the behaviour of the camellias at the Villa Taranto it is clear that *Camellia* in all but the tenderest of its forty-odd species, their varieties and cultivars, could be planted, studied and developed, that work of importance on that genus could very well be concentrated here. *C. japonica* in over a hundred of its cultivars is the principal species, but *C. sasanqua* and *C. reticulata* are well represented. Owing to forced neglect during the war and thereafter, labelling needs attention, as it does in the case of *Rhododendron*. Mr Barmes is doing all he can, but these two genera are such as to need experts in identification.

Other genera which are impressive at the Villa Taranto, for the diversity or range of representation and for size and quality of specimens are *Chaenomeles*, *Erica*, *Syringa*

Below Fountains—another Renaissance Italianate feature of this great garden of exotic plants.

and *Daphne*; *Berberis* (200 species and varieties or cultivars), *Buddleia* (40) and *Clematis* (64). *Daphne* behaves very oddly in one respect; some of the species usually quite easy to grow, fail; but *D. genkwa*, for example, the very difficult blue-flowered Chinese daphne, does wonderfully well.

Although Captain McEacharn's interests were chiefly in shrubs and trees, the gardens have important and charmingly planted collections of groundling plants as well: there are twelve species and varieties or cultivars of *Agapanthus*; 78 of *Allium*; 88 of *Campanula*; 90 of *Gentiana*; 50 of *Iris*; 14, plus many hybrids, of *Paeonia*; and seven of *Watsonia*. There are, of course, hundreds of other genera, but these figures give an idea of range and scale.*

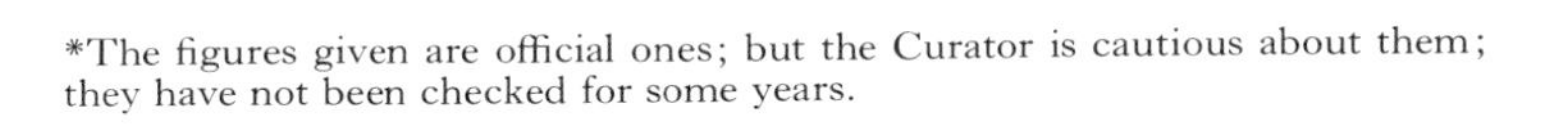

*The figures given are official ones; but the Curator is cautious about them; they have not been checked for some years.

NETHERLANDS

LEYDEN

Clusius and his successor

Appropriately the *Hortus Academicus* of Leyden University, created by Clusius, makes a fine display of tulips.

For its size, Holland has many botanical gardens, as one might expect of a country which has produced such excellent botanists and such great gardeners. The Leyden University garden is still of the greatest interest and this chapter is therefore devoted entirely to it and its long line of great Prefects and Directors. The other Dutch botanical gardens are the small but quite important, the *Cultuurtuin voor Technicsche Gewassen* at Delft; the *Poort-Bulten Arboretum* near De Lutte; the botanical gardens associated with the universities or privately owned of Doorn, Haren, Lunteren and Groningen; the two gardens in Rotterdam and the two at Utrecht; and the two at Wagneningen of which the garden of the Institute of Horticultural Plant Breeding is one of the most important centres of practical and experimental economic botany in the world. But Leyden remains outstanding for its four centuries of service to science and learning.

The Leyden garden is small; at its first planting in 1594—it was not planted for seven years after its official foundation—it was only about 130 feet by 100 feet. Even today it only covers about five acres; and the facsimile of the original (Clusius) garden beside it is under 10,000 square feet.

The University of Leyden was founded in 1575, and it was early realized that its medical faculty would need a *Hortus Medicus*; but by then, as a result of the Italian examples and that of Leipzig* it was also realized that a *Hortus Medicus* would be out of date, and that what was needed was a *Hortus Botanicus*. Nothing was done about it until twelve years after the foundation, when the Curatores of the University asked the Burgomasters of Leyden to yield them a plot of ground for a garden. This was granted. But as the Professors of Medicine, placed in charge of the garden were not very interested in botany outside their books, no planting was done.

Then, in July 1592, with what might seem impertinence in so young a school to so great a scholar, the Curatores approached Clusius with the offer of appointment as their *Praefectus Horti*. Old, crippled by a fall, and tired of moving about, the father of scientific botany refused.

Charles de l'Ecluse, known as Carolus Clusius, was a French Fleming, born at Artois in 1526. He read law at Louvain and Marburg, medicine at Wittenburg and while studying medicine became interested in plants far beyond the limits of medicinal botany. His interest hardened into the intellectual passion of a lifetime while he was again studying medicine, this time at Montpellier. An excellent linguist, his first literary work was in the translating of Dutch and Italian herbals into French. Meanwhile he was collecting, studying and describing plants and he spent the year 1564–5 plant-hunting in Spain and Portugal, where he described about two hundred species hitherto unknown to the new science. But his interest was not confined to the European flora. He translated Garcia di Orta's Portugese book on Indian plants into Latin, and Nicholas Monardes' work on the plants of the New World.

*The order of foundation is Pisa, Padua 1543–5; Florence 1550; Bologna 1567; Leipzig 1580; Leyden 1587

His work had already made him a European name when, in 1573, the emperor Maximilian II invited him to Vienna to found a *Hortus Medicus*, of which we know little excepting that in the hands of this great botanist it must, in practice, have been a botanical garden. It was while working in Vienna that he was able to introduce tulips and among other plants horse-chestnuts from Turkey. Clusius seized the chance to spend three years studying the flora of Austria and Hungary so that in 1583 he was able to publish his *Historia Stirpium per Pannonium*. From Austria he went to Germany, settled in Frankfurt and it was from there that he wrote to the Curatores of Leyden University saying that he did not want the responsibility of managing a public garden or of delivering lectures.

The intermediary who had persuaded the Leyden Curatores to apply to Clusius, Johan van Hoghelande, would not accept his refusal, and he persisted in persuading him until at last Clusius accepted on condition that they give him a *Hortulanus*, or Head Gardener. His salary was fixed at 300 *rijksdaalders* plus allowances. The post of the Hortulanus he had asked for was given to a pharmacist and herbalist named Dirk Outgaerszoon Cluyt, who was to be paid 500 florins a year. Cluyt, seeing no reason why he, too, should not Latinize his name, called himself 'Clutius', and the slightly absurd combination of Clusius and Clutius became an amiable joke in university circles, as was their friendly, if at times acrimonious quarrelling over the garden.

Together, in 1594, Clusius and Clutius undertook and completed the planting of the little *Hortus*, and within a year or two had a collection of a thousand species. It happens that the original plan and planting list, properly keyed, have survived and this enabled a much later *Praefectus Horti*, Professor Dr Baas Becking, in 1931 to reconstruct the Clusius garden exactly as it was, even to the plants, on a slightly reduced scale. In his account of this Dr Becking says:*

*See his *Hortus Academicus Lugduno-Bataaus* (Haarlem, 1938). It is a rare book but there is a copy in the Lindley Library of the RHS.

Key to history of Greenhouses Map

1 Tropical House (1878)
2 Tropical House (1866)
3 Orchid House (1861; extended 1872)
4 Tropical House (1856)
5 Tree Fern House completed 1877
6 Victoria House (1870)
7 Propagating House (1883)
8 Cactus House (1887)
9 Conservatory (1736)

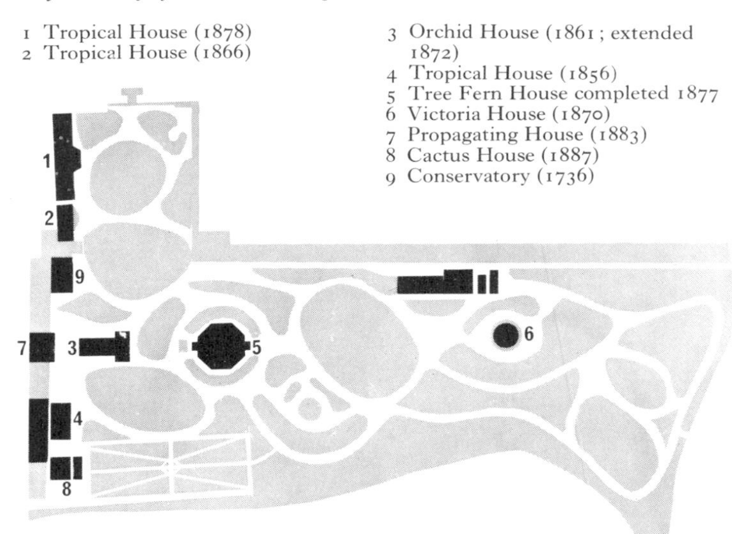

No primitive systematic arrangement was followed, nor were plants grouped according to their different virtues. . . . One point, however, is apparent: the *Index Stirpium* of 1594 is a catalogue of a *Hortus Botanicus* and not of a *Hortus Medicus*.

Cluyt, to whom the garden owed much and who was quite capable of arguing with his master about the identity or nature of a plant, died in 1598. When Clusius died there was a move to appoint the younger Cluyt, son of the first Hortulanus, in Clusius's place: the young man was a good Greek and Latin scholar, an excellent botanist and a fine gardener. But it did not seem fitting to the Senatus and Curatores of the University that a youth without a degree be offered a chair, and it was decided that the garden must be directed by the Professor of Medicine. Not until the last years of his life did Outgaers Cluyt get a footing in the Hortus he had helped his father and Clusius to make. Meanwhile he travelled, taught for two years at Montpellier University, and went plant-hunting in various parts including the Barbary Coast where he was beaten and robbed more than once. In 1636, as a sort of special honour, he was given the post, a sinecure, of Inspector of the Hortus at Leyden.

The interregnum following the death of Clusius was filled by the Professor of Medicine, Pieter Pauw. His successor was Everard Vorstius, whose principal contributions were the introduction of plants from the East Indies, accomplished through his friends in the Dutch East India Company; and of some species from North America, including the Stag's Horn Sumach (*Rhus typhina*) and some species of *Oenothera*, which became the parents of our garden Evening Primroses. He used the Hortus as a garden of acclimatization and distributed plants to other European countries. It was during his Praefecture that Robert Sibbald, founder of the Edinburgh Botanic Garden, studied medicine and botany in Leyden. Vorstius died in 1663 and was succeeded by Florens Schuyl, who was on probation at 1,000 guilders a year for some years, but confirmed as *Praefectus* at 1,500 guilders in 1667. He concentrated on enriching the garden, and European horticulture, with plants from South Africa —*Pelargoniums*, *Mesembryanthemums* and some species of *Gladiolus*. He was followed by an undistinguished *Praefectus* called Seyan, and Seyan by a Saxon physician in the service of the East India Company, Paul Hermann, who was in Ceylon when the offer of the chair and the garden reached him.

Left Garden design on the English Romantic model is a major interest of the Leyden Botanic Garden as it now is.

From the early fifteenth century the descriptive work of botanists using the new Italian, German and Dutch botanic gardens called for illustration by careful artists. The art of correct botanical drawing and painting was a by-product of botanical gardening; it is still practised at its best in the great botanic gardens.

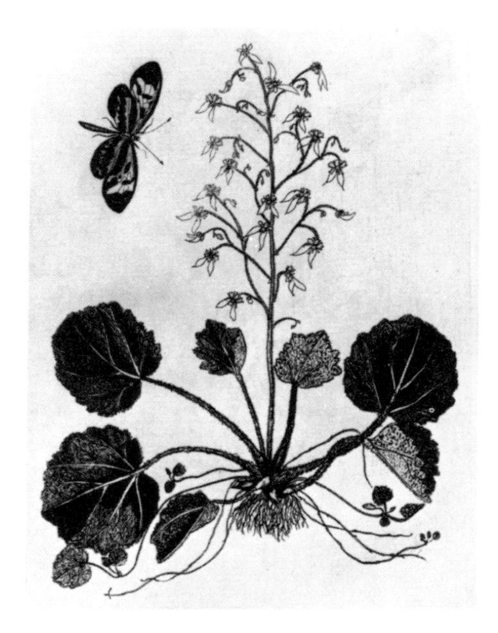

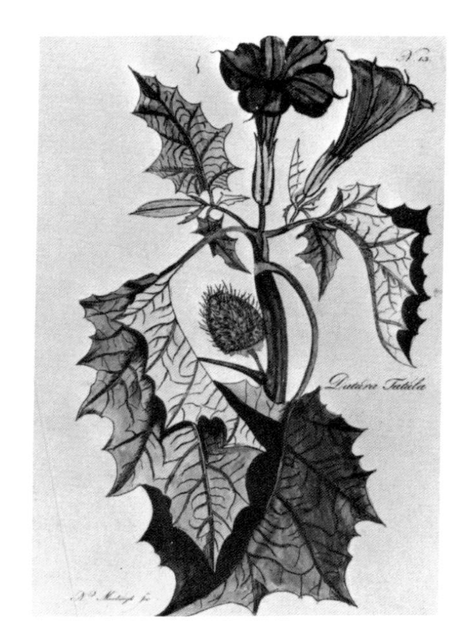

Paul Hermann was a man with many friends; too many, perhaps—the Curatores had to complain that he was cramming the garden conservatory and stove with plants belonging to his amateur gardener friends who wanted a safe place to winter their treasures. But among his friends were the Oxford Professor of Botany, the Aberdonian Robert Morison, and the staff of the Oxford Botanical Garden. Through these English colleagues he was able to introduce two hundred new species to the Leyden garden. Later he visited the Oxford Botanical Garden, and under the influence of what he saw there and of its Director, Robert Morison, Hermann reorganized the planting at Leyden on more systematic lines. Moreover, he raised the number of species cultivated there to 3,000. His Herbarium, from which Linnaeus, among others, worked, is in the British Museum.

Hermann's successor was a Leyden man, the physician Petrus Hotton, an undistinguished but conscientious *Praefectus* who died in 1709, making way for Leyden's greatest botanist since Clusius and one of the greatest naturalists in the history of science—Herman Boerhaave.

Boerhaave was the son of a poor clergyman of Voorhout, where he was born in 1668. He obtained a bursary to read theology and philosophy and got his Ph.D. in 1690. But he either had, or was suspected of, leanings towards the teachings of Spinoza, which barred him from holy orders, so he turned to medicine, studied it on his own for three years, and took his doctorate of medicine at the small University of Harderwijk in 1693. Thereafter, for seven years, he practised medicine in Leyden, meanwhile making a study of mathematics and the classics. He was appointed Reader in Medicine at Leyden. Three years later he was offered the chair of medicine at Groningen. Leyden, not wanting to lose a man as beloved for his humanity and sweet temper as for his learning, kept him by raising his salary by 50 per cent and promising him the first suitable chair to fall vacant. It was thus that in 1709 he became their Professor of Medicine and Botany.

A year later Boerhaave published his first catalogue of the Hortus—3,700 plants. The garden, though much enlarged, was becoming crowded and he started the policy of finding homes for new plants elsewhere in the town. In 1720 he published his most important work derived from the garden, a short history and descriptive catalogue in two volumes, the *Hortus brevis Historia & Index Alter*. By then the number of species and varieties grown had risen to 5,846, evidence of Boerhaave's activity as an introducer and describer of new plants. One distinction of the *Index* are the magnificent engravings of *Proteas* from South Africa. For this *Index* Boerhaave correctly separated Angiosperms from Gymnosperms,* used Ray's division into Monocotyledons and Dicotyledons, and Tournefort's classification by flower structure.†

Boerhaave kept a botanical diary throughout his term as *Praefectus* and notes on his correspondence with more than fifty botanists at home and abroad. These documents are now in the library of the Leyden Botanical Institute. So active was his policy of seed exchange that, for example, in the single year 1725 he received from abroad 1,416 packets of seed. Although he was overworked, delivering two separate courses of lectures and tutorials, as well as managing the *Hortus*, he raised hundreds of new kinds of plants himself in his own garden.

In 1729 Boerhaave began to feel too old and tired for his work. Retaining the Directorship of the garden, he gave up both his chairs, but a year later he retired to his country house and Adriaan van Royen became Director in his place.

The Van Royens, Adriaan and his nephew and successor, David, were both very good Directors and did the garden great service. Adriaan van Royen was very well

*Angiosperms are plants with their seeds enclosed in a pericarp. Gymnosperms have no such enclosure. The pericarp is the rind of shell of fruits; it is a development of the wall of the ovary.

†There was a second authorized edition of the *Index* in 1727. All others are worthless; the work was pirated in several countries and much corrupted in the process.

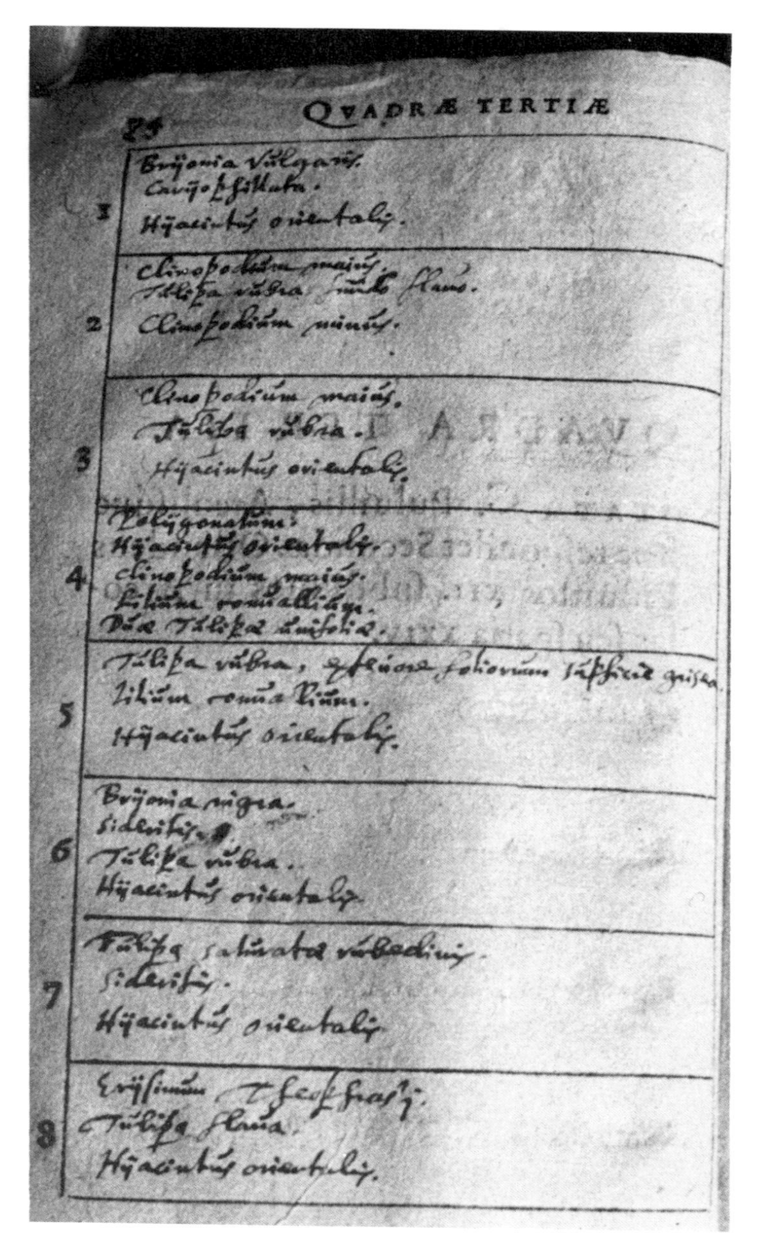
QVADRÆ TERTIÆ

1 Bryonia vulgaris. Caryophyllata. Hyacinthus orientalis.

2 Clinopodium majus. Tulipa rubra ... flava. Clinopodium minus.

3 Clinopodium majus. Tulipa rubra. Hyacinthus orientalis.

4 Polygonatum. Hyacinthus orientalis. Clinopodium majus. Lilium ... Tulipa ...

5 Tulipa rubra, ... Lilium ... Hyacinthus orientalis.

6 Bryonia nigra. Sideritis. Tulipa rubra. Hyacinthus orientalis.

7 Tulipa ... Sideritis. Hyacinthus orientalis.

8 Erysimum Theophrasti. Tulipa flava. Hyacinthus orientalis.

Below From Boerhaave's day-book in the archives of the *Hortus Academicus*.

Opposite The Clusius garden as reconstructed and planted by Dr Baas Becking at Leyden University.

connected; his elder brother David, father of Adriaan's successor, was Secretary to the Curatores of the University and his influence enabled Adriaan to do what a man of lesser social consequence could probably not have done, to get a large sum of public money for an extension to the garden.

There had already been three such extensions, the latest in 1685. But between 1685 and 1729 the number of species cultivated had risen from under 3,000 to over 8,000; 8,000 species, half of them trees, in a garden of 1,600 square metres. Boerhaave had been responsible for much of this crowding; he could not resist a new plant. In making his point for an extension to the Curatores David van Royen the elder pointed out that Utrecht was planting a *Hortus Botanicus* twice as large as Leyden's; that Amsterdam already had one three times as large; and that, so important had botany become, especially to medical schools, that the size of a university botanical garden influenced the number and quality of students attracted.

As a result of these representations in 1735 the University received a subsidy to enable it to buy and demolish three blocks of houses beside the garden, to wall in the whole garden, and to build a big new conservatory—it was over 100 feet long, 30 feet high and 22 feet deep. Nor was this the only great event for the Leyden garden in that year. For in June, Karl Linné (Linnaeus) had arrived in Amsterdam, on a visit to Holland which lasted several years.

Only a few days after his arrival the young Swedish botanist received the degree of B.A. from the little ducal University of Harderwijk,* and late in June he went to Leyden, where he was received and befriended by one of the great men of the University and city, the Regent, Senator, Burgomaster and Curator Jan Frederick Gronovius, himself a very considerable botanist. Through Gronovius Linnaeus came to know Boerhaave, and the old Dutchman and the young Swede became close friends. Linnaeus showed Gronovius and his friend Isaac Lawson the manuscript of his *Systema*. They made arrangements for it to be published at once; by the end of July it was in the printer's hands, and it was published in December 1735. Two years later, when his great *Genera Plantarum* was published, it was dedicated to Boerhaave who, in his letter thanking Linnaeus, wrote this:

Centuries shall praise it; the good will follow it; and all will derive profit from it.

Boerhaave offered to finance Linnaeus on a two-year plant-hunting expedition to South Africa. Linnaeus refused on the grounds that he could not stand hot climates, the truth being that he was becoming impatient to go home and marry the girl who was waiting for him in Sweden; all he would agree to do was to remain in Leyden long enough to help Adriaan van Royen to replant the enlarged *Hortus Academicus*, as it was now called, on the new systematic lines. It has since been suggested that the replanting plan was all Linnaeus's work for which van Royen took the credit. Both parts of this libel are untrue.

During Adriaan van Royen's term tropical orchids were flowered for the first time at Leyden, and by 1740 the garden had a collection of them, small, but unique in Europe at that time and including *Vanilla aromatica*. The quality of the collection was in part due to orders issued by the Prince-Stadtholder that captains of all East Indiamen were to bring back live plants for the Leyden Hortus, without making any charge.

A description of the gardens at this time is amusing: the translation is by Professor Dr Baas Becking (*op. cit.*):

The academic garden, or Hortus Medicus at Leyden is 600 roeden' in extent, inclusive of buildings. The Hortulanus is appointed by Curatores, and draws a salary of 300 guilders and 36 tons of peat, but no candles. (The candles necessary in the planthouses in winter are paid by the commonwealth.) At the discretion of the visitors, they may pay him [the Hortulanus] an entrance fee; a peasant, however, has to pay two groats. He [the Hortulanus] states that he would not miss the entrance fee for fl.800, —, but he has to keep a maid who has to serve as a guide for visitors, also his housewife should help showing visitors around. He—the Hortulanus and his gardeners—only rarely do this. All days and even on Sundays the garden may be visited, such at the discretion of the Hortulanus. Couples openly in love are on no account admitted. Herbwine and flowers for brides although seldom required, may be sold by the Hortulanus, but in case of the wine, the Professor should know in advance.

*Where he planted a *Ginkgo biloba*, still flourishing.

Redesigned in the early nineteenth century, the Leyden *Hortus* was replanted as an 'English' landscape garden to conform with the style which had conquered all Europe.

Plants are not sold but exchanged against others. The Hortulanus has to keep note of these exchanges and submit his correspondence to the Professor.

He [Hortulanus] provides students in medicine preparing for their candidature (baccalaureate) with plants, of which he may also derive some profit.

At the other hand, the Hortulanus has to work in the garden like a labourer, he has to get up at night in winter to cover plants and to take care of the fire etc., but also the gardeners living nearby are held to assist him of nights on changing weather—to this purpose they each have a door-key.

There are only two gardeners throughout the year, summer and winter, each making fl.250 a year—no helps are used for weeding and clipping.

In the past few years for the ordinary work of carpenter, mason, glazier and painter a sum of fl.180,—is granted. The tubs are made of common square wood.

Dr Becking continues:

The sticks are single square planed laths, made in winter by the gardeners and painted white, without extra compensation. The number labels are pointed sticks of double lath, worked and painted by the gardeners.

After the Professor's 'vidit' Curatores pay the money advanced by the Hortulanus.

The water was obtained from a reservoir, communicating with a city canal, in which reservoir it was pumped by means of a double pump. It was mixed with rainwater. The conservatory had two stoves, while the planthouses were heated by means of smoke-flues, the stoves being placed outside. The highest house, which held the 'arbor draconis' seems to have been about 28 feet high. The old *palm or date tree* was so high that a hole was dug in the conservatory floor in order to place the tub that held it.

'Muscovy-mats,' i.e. linden-bark mats used on the Baltic grainships for stowing grain were only used to make twine or bind plants (a custom persisting in our gardens up till 1931), and not for cover. In winter the planthouses were first covered with woollen Leyden blankets, on top of which a tarpaulin was placed. The frames, in which bulbs, seeds, and pineapples were grown, were covered with reed mats. The edges of the flowerbeds were not made of box, but of bricks, placed edgewise. Tender plants remaining in the open were first covered with leaves, over which mass a glass jar was placed.

There was no catalogue available for the public although the Hortulanus possessed one. Lessons were given in the gardens by the Professor from May to the summer holidays.

Professor Dr Karstens, Praefectus van Royen's present successor, has been unable to justify to us his predecessor's illiberal attitude to couples openly in love.

Adriaan van Royen resigned the chair of Botany to his nephew David in 1754. David van Royen, at various times Regent, Senator and Rector of the University, was inactive as a botanist but at least he supported the work of the garden's most remarkable Hortulanus, Nicholas Meerburgh. He had been head gardener for many years already when, in 1775, he published a beautiful five-volume book of his own engravings of the plants in the garden, hand-coloured—*Afbeeldingen van zeldzaame gewassen*. Meerburgh published other valuable works and he made a great impression on visiting foreign botanists.

David van Royen's successor was a naturalist who had already made a name for himself, Sebald Justinus Brugmans. A doctor of medicine and of philosophy at Gröningen, he was appointed to the chair of Botany at Leyden in 1799. Subsequently he became a sort of academic Pooh-Bah—at once Professor of Botany, Natural History, Medicine and Chemistry. He was a great diplomat, organizer and administrator. He served for two terms as Medical Inspector General to the armed forces, and did a magnificent job of evacuating thousands of wounded after Waterloo. Not until 1806 was he able to attend to the garden and he then reported to the Senatus that it was much neglected; he had it put in order and he managed to arrange for all the plants of the small botanical garden at Harderwijk University, which was being closed down, to be moved to Leyden. There followed more years of military duty; and not until 1816, released from his Medical Inspectorate General, did Brugmans again turn to the garden; but when he did so it was to some purpose, for he was able to get the money and land to quadruple its size. With this great extension went a radical change in the layout and planting style. English landscape gardening was all the rage: Brugmans got rid of nearly all the old formal plantations, abandoned the rectilinear plan, and had the garden landscaped, replanting the thousands of specimens 'after nature'.

The man appointed to follow Brugmans in 1819 was Caspar Georg Reinwardt; he was unable to take up the appointment immediately because he was in the Far East, doing important work there. One of Brugmans' pupils, a young physician named Gerard Sandifort, was appointed to fill his place until Reinwardt returned which, as it happened, was not for several years. Sandifort did excellent work, in part because he had an outstandingly good Hortulanus, of whom something should be said.

Meerburgh having died in 1814, the man appointed in his place was sixty-two years old but of quite exceptional ability, Herman Schuurmans Steckhoven. An experienced gardener and nurseryman he had worked as number two to the well-known garden designer-architect, Gerard van Swieten, and with him had helped not only to lay out and plant the *Hortus Botanicus* at Vienna, but had also helped to design and lay out the great imperial park of Schönbrunn; all this while he was still in his teens.

As Hortulanus of the Leyden Hortus, Schuurmans helped Sandifort to add considerably to the garden's collections. He also published a number of useful monographs, notably on the *Phanerogama*, the *Cryptogama* and, in 1825, a *Botanical Dictionary*. Sandifort and Schuurmans were so active in adding to the plants at Leyden that in 1822 a new catalogue became necessary: the *Elenchus Plantarum* of that year shows that the additions were chiefly of Australasian species. Professor Dr Baas Becking, writing of the work of these two men says: 'Reinwardt, on his return, found the gardens fit for instruction,

Below Oriental tulips, ancestors of the modern garden tulips, were introduced into horticulture by Leyden.

the ideal Brugmans had aimed at but did not reach.'

Although taxonomic botany had been advanced at Leyden, it was elsewhere that the experimental botany, phytology and plant physiology started in the seventeenth century had progressed, and at Leyden the science was falling behind the times: it was time that the *Hortus Academicus* was presided over by a new type of botanical scientist. The absentee professor whose place Sandifort was filling was to go some way towards giving botany at Leyden a new start, if only by being the first of its Directors to realize that the old botanists' dream of bringing together all plant species into a single collection was futile; that it was moreover impossible; Reinwardt, working in the Far East with the florae of the East Indies, Japan and Australasia, knew that the plant kingdom was immensely more populous than anyone had supposed.

Caspar George Carl Reinwardt was a Prussian, born in 1773, a pharmacist and chemist by training who had become Professor of Botany at Harderwijk, which miniature university made him Med.D. and Ph.D. *honoris causa*. From there he was called by King Louis Bonaparte in 1808 to be Director of the Imperial Botanical and Zoological Gardens first at Soesdijk and thereafter at Amsterdam. The restored legitimate government invited Reinwardt to serve on the Royal Commission for the Colonies in 1815; he went to Batavia, advised the government to found a botanical garden there, and was commissioned to carry out his own advice. He thus became the founder of the first great botanical garden in the Far East, the *Hortus Bogoriensis* at Buitenzorg, now Bogor, which was to play a very important part in introducing new economic plants to the East Indies and in sending East Indian plants to Europe. It was Reinwardt who began the enrichment of Leyden in Far Eastern and Australasian plants, although that increase, so noticeable when one compares the *Elenchus* of 1822 with Reinwardt's catalogue for the Hortus in 1831, was due less to Reinwardt and James Cooper, his English Hortulanus at the *Hortus Bogoriensis*, than to one of the half-dozen most remarkable horticultural botanists who ever lived, closely associated with the Leyden garden, but always as an outsider.

Jonkheer Dr Philip Franz Balthazar von Siebold was a Würzburger of aristocratic but hard-working scientific family who was appointed Court Physician to William I of Holland immediately upon taking his doctorate of medicine at Heidelberg in 1820, and, two years later, Surgeon-Major of the Army in the East Indies. His headquarters was in Java but hardly had he arrived there when he was sent to the Company's 'factory', Decima, in Japan, which had already been used as a base by two distinguished botanical plant collectors, the German Kämpfer and the Swede Thunberg.

Von Siebold soon endeared himself to the Japanese by giving them medical advice and treatment and by instructing whomsoever asked for it in European medicine; he was particularly successful in ophthalmology. He found time to study the flora of Japan and to make and send collections of herbarium material and live plants to Europe. This work was carried on steadily for nearly four years, but in 1826 he was detailed to accompany the Resident on that official's annual journey of homage to the Shogun; in the course of this journey it became known to the Japanese that von Siebold was making maps of the country; at the Shogun's court he was accused of high treason and ordered to leave the country and never to return. He was able to take with him an immense collection of herbarium material and a great many seeds of Japanese cultivars as well as live plants. He went to Leyden and there set about sorting and classifying his material and writing his study of Japan, *Nippon*. In order to subsidize the work the Netherlands government bought his collection but left it in his charge; and he was appointed Directing Medical Officer; the post seems to have been a sinecure, for he remained at Leyden. The *Hortus* there already had many of his plants.

Von Siebold bought land at Leyden, near the Lager Rijndijk outside the Zylpoort, and there planted a garden of acclimatization for Japanese plants, raising chiefly cultivars of *Paeonia*, *Chrysanthemum*, and *Lilium*, but also Japanese shrubs and trees. The *Hortus Academicus* obtained many fine specimens from von Siebold's garden, and about thirty of his plants still survived until World War II; they represented, of course, only a small fraction of the collection with which he had enriched the Hortus, and which enabled Leyden to distribute good Japanese plants all over Europe.

The pre-eminence of the commercial horticulturists of Ghent in Japanese azaleas and some other Japanese plants was due to von Siebold plus an accident of the Belgian war of independence against Holland to which it had been united at the Congress of Vienna: a shipload of plants from Japan, intended by von Siebold for Leyden, was captured by a Belgian warship, and those plants made Ghent's fortune as a centre of commercial horticulture. It is a singular fact that von Siebold himself, on the other hand, never made a penny out of the commercial potentialities

One of the many magnificent trees in the Leyden Hortus.

of his work. When, for example, he and C. L. Blume, erstwhile director of the *Hortus Bogorensis* following Reinhardt, and now director of the Herbarium at Leyden, formed a society for the introduction and distribution of Japanese plants through the *Hortus Bogoriensis*, although this society later became a 'Royal Society' it was a commercial failure and its activities led to a series of acrimonious rows involving von Siebold, the aggressive Director of the Batavian garden, Meyer, and even, though Reinwardt held aloof, the Leyden *Hortus*.*

Reinwardt, meanwhile, had been running the Hortus at Leyden on the old lines and yet preparing it for new ones. In 1840 he published a catalogue of the garden's collection of 260 species of medicinal plants. By way of the garden he introduced *Vanilla aromatica* from South America to Java. He retired in 1845 and died in 1854. His successor was a Brabanter born in 1806, enrolled as a student at Leyden in 1825, and received Doctor of Medicine in 1831—Willem Hendrik de Vriese. Perhaps his most important act as Director of the *Hortus Academicus* in Leyden was to appoint as Hortulanus first Heinrich Witte *père* and subsequently his son, also Heinrich. The younger Witte served as Hortulanus for forty-five years and it was he who placed the garden's accumulation of knowledge at the public's disposal in a series of excellent 'popularizing' botanical works, about fifty books in all.

The increase in the orchid collection during the De Vriese–Witte regime was remarkable, making the Leyden garden the centre for study of that family of plants. The Hortus had:

in 1818......3 species   1862....515
1857....294   1888....720

The 1888 collection is recorded in 362 fine water-colours preserved in the garden's archives. By 1930 the collection had declined to 300 species and it is not, at present, an important one.

*See pp. 194ff. for the history of the *Hortus Bogorensis* and an account of its present state as the Kebun Raya, Indonesia.

During De Vriese's directorship the Leyden garden began to shrink when it was forced, despite the fight he put up against the authorities, to yield some of its land to make room for the building of an observatory. However, the 1859–60 catalogue still has 7,400 species of plants. And De Vriese, although a descriptive rather than an experimental botanist, did serve the science in the new way by his work on such economic plants as *Vanilla*, *Cinchona* and *Camphor*. The number of *Cinchona* species and varieties sent to Batavia became, indeed, an embarassment to Meyer at the *Hortus Bogoriensis*. De Vriese was sent to the Indies to report on the state and prospects of tropical agriculture and he travelled so assiduously, worked so hard, and, despite warnings, with so little regard for climatic conditions, that, shortly after his return home he died of exhaustion.

The next Director was the man Leyden needed, not a descriptive or economic botanist, but an experimental scientist. Willem Reinier Suringar had collected plants in the East Indies, done important work in the new science of bacteriology, yet was the author of the first good *Pocket Flora of Holland*. He died, however, in 1898, in which year Hortulanus Witte at last retired, and it was left to Jacobus Marinus Janse to give Leyden a modern botanical laboratory—the first in the Netherlands. Janse retired in 1931, having set the botanical section of this ancient university on the modern, experimental lines followed by his successors, Dr Baas Becking and the present Director, Dr Karstens.

The garden we have been writing about may disappear. Other, more fashionable sciences and their faculties press for that five-acre plot which is so valuable in the heart of a modern city's university. And scientific, experimental botany no longer, perhaps, in the second half of Leyden University's fourth century, needs a garden. For the time being there it is, a charming, cool and peaceful oasis in a busy, noisy commercial and industrial town. The laburnum tree which Clusius planted is still there, battered and wizened but alive and flowering freely every spring. There are tulips, the tulips we owe to Clusius, about the little pond; there is a magnificent copper beech, a colossus of great beauty which those with a sense of this garden's antiquity call young, for it was not planted until 1820; the much older *Fagus sylvatica pendula* has lost one whole side but it is still a hemisphere as big as a large marquee. By the waterside, the lawns and shrubs and trees are charming. The rock-garden, the shrubberies, the trees and the walks every foot of which remind us of what we, gardeners as well as botanists, owe to Leyden, are meticulously cared for. Something of the pre-Brugmans formalism remains; and the Englishry of the post-Brugmans gardening is very apparent. The handsome bust of Jonkheer Franz von Siebold, to which every year Japanese visitors never fail to bow in reverence for a great teacher, is planted about with some of the plants he introduced to Europe.

But one is aware of the town closing in, of the garden being squeezed out, with the humanism which created it.

EDINBURGH

the Royal Botanic Garden

Edinburgh's botanic garden is world famous for *Rhododendrons*, including *Azaleas*.

The Royal Botanic Garden in Edinburgh is florally among the most pleasing in Europe. For reasons which will appear, it has specialized in certain Chinese genera, notably *Rhododendron, Primula* and *Meconopsis,* which happen to be among the most beautiful in the world and, in the case of *Rhododendron,* the most protean in variety of size, form and colour. But that is not the only reason for this garden's outstanding beauty; it had the good fortune to be served, late in the nineteenth century, by a Head Gardener (later Curator) James MacNab, who was a landscape gardening artist of distinction working under a Regius Keeper who let him reshape the garden. So that the beauty of the Edinburgh Botanic, as we shall call it for short, is not a by-product of its scientific work, but a product of art.

It will be recalled from p. 36 that in the late 1650s an Edinburgh Scot called Robert Sibbald was studying medicine at Leyden when Vorstius was Praefectus of the *Hortus Academicus* of that university. When Sibbald returned to Scotland in 1661 he was disturbed by the low standing of medicine in Edinburgh and by the ignorance of his fellow-physicians. Coming from Leyden, he was bound to think that the first step in their education should be the planting of a botanical, or at least of a medicinal, garden, so that these ignorant physicians and apothecaries could familiarize themselves with the herbs from which their drugs were derived. Sibbald was first in a position to do something about this in 1670, which happens to have been a very important year for botany: at Oxford Robert Morison, the university's first Professor of Botany, had just published his *Praeludia Botanica*, based on the work of an Italian botanist, whom we named above, Cesalpino; John Ray had just made his first experiments on the movement of sap in trees thus laying the foundation on which De Jussieu and De Candolle were to work at the Jardin des Plantes in Paris, and published his *Catalogus Plantarum Angliae*. Still in that year of botanical wonders, another Italian whom we have met with, Marcello Malpighi, submitted his first work on plant anatomy to the Royal Society in London on the day that the Society also received Nehemiah Grew's *Anatomy of Vegetables Begun*. It will be evident that while the Scottish contribution abroad was not inconsiderable (for instance Robert Morison was an Aberdonian), at home the Scots were behind the times.

Sibbald had a partner in his venture, Andrew Balfour, a physician who had studied under the great Harvey in London and who had become very interested in botany after visiting the garden which Robert Morison had made for the Duc d'Orléans at Blois. He had a small garden and even a botanical museum of his own. Sibbald and Balfour bought a small plot, 40 feet by 40 feet, planted a thousand specimen plants, and put them in charge of one James Sutherland, 'a youth who by his own industry had attained a great knowledge of plants and medals'.* As the garden was soon too small, in 1676 another, the Trinity Hospital garden, was leased from the Town Council and Sutherland given charge of that one too, as well as being appointed 'professor' (teacher) of Botany in the Town College, where he later planted a third garden. Finally, Sutherland was put in charge of the Royal gardens at Holyrood House.

These four gardens were the ancestors of the Edinburgh

**Memoirs of Sir Robert Sibbald.*

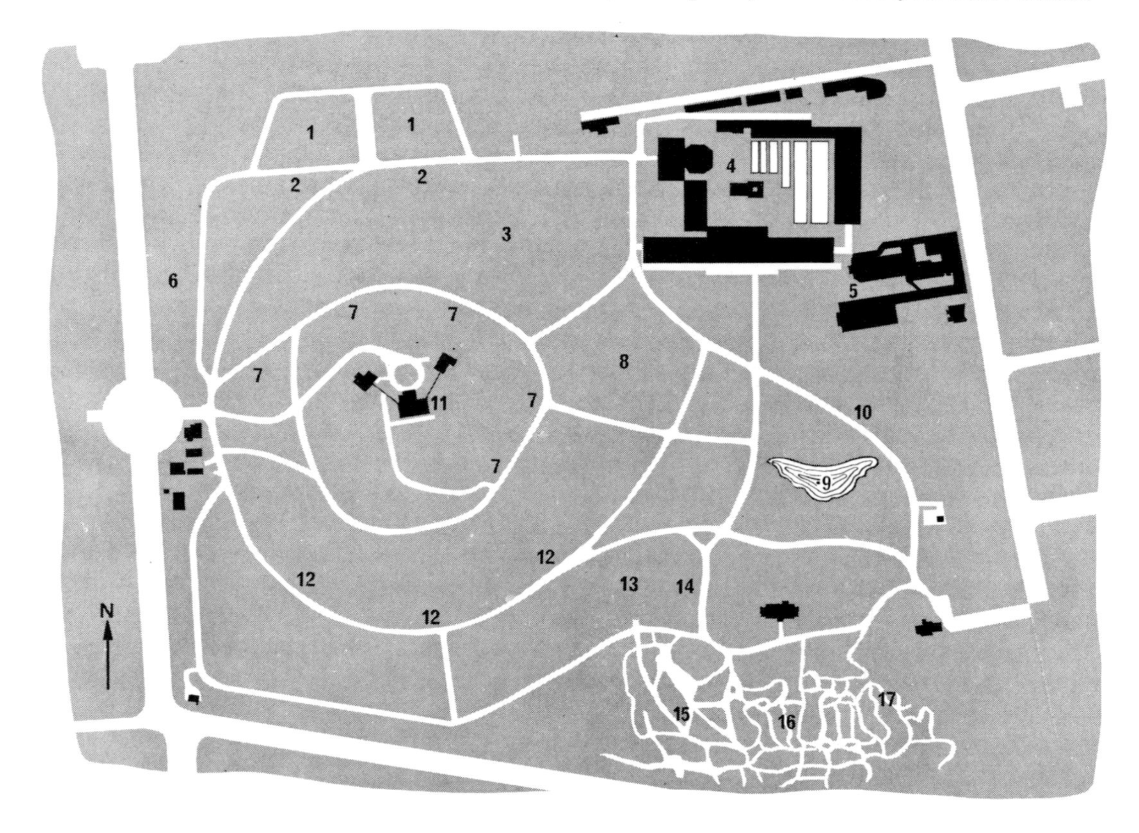

Key to Map

1 Demonstration garden
2 Herbaceous border
3 Copse
4 Glasshouse complex
5 Herbarium and Library complex
6 Conifers
7 Rhododendron walk
8 Azaleas
9 The Pond
10 Roses
11 Gallery of Modern Art
12 The Hill
13 Peat garden
14 The Glade
15 Woodland garden
16 Rock garden
17 Heath garden

Left Detail of the water garden.

Below left Detail of the Rock Garden. The Rock Garden is one of the finest in the world, remarkable for Himalayan alpine plants.

Left A page from James Sutherland's *Hortus Medicus Edinburghensis.*

Right A 'Plan of the Ground' for a Botanic Garden at Edinburgh, commissioned by John Hope in 1763.

EDINBURGH,

November 22. 1683.

THe Lords of His Majesties Privy Council having heard and considered a Petition presented by Mr. *James Sutherland,* Botanist and Overseer of the Physical Garden at *Edinburgh.* Do hereby Give and Grant to Him the sole Priviledge of causing Print and Publish the Catalogue of all the Plants he hath Collected, in Latin and English under the Name of *Hortus Medicus Edinburgensis,* and that for the space of seven years after the Date hereof; and do discharge all others to Print the same, or Import it from abroad and sell it here during the said space, under the pain of Confiscation of the Copies to the Petitioners use, and being otherwise Punished for their Contempt as the Council shall find Cause.

A Plan of the Ground Designed for A Botanick Garden at Edinburgh

Botanic. For years Sutherland added to them; meanwhile he published in 1683 his *Hortus Medicus Edinburghensis,* a catalogue of the plants in his care which shows that his gardens were still more medicinal than truly botanical. But Sutherland's second interest, coins and medals, had been taking more and more of his attention away from plants; there were serious complaints that he was neglecting his gardening and his teaching; and in 1706 he at last gave way to a new professor and gardener by resigning his several offices.

The first considerable academic type—albeit a poor botanist—to have charge of the gardens was Charles Alston who had studied under Boerhaave at Leyden. He became Keeper of the Gardens, Crown Professor of Botany, and University Professor of Botany, and his reign was long, 1716 to 1760. A good practical Keeper, he had the misfortune to take—and the pig-headedness to maintain—the wrong side in the European-wide controversy about whether plants had sex. Moreover he totally failed to grasp the importance of the progress being made in plant anatomy, physiology and taxonomy in England and elsewhere. Not, indeed, until he was succeeded by John Hope, who was Keeper of the Royal Gardens and incumbent of the places which went with it from 1760 to 1786, did Edinburgh get a really good botanist. Hope, who had studied under Bernard de Jussieu at the Jardin des Plants in Paris, was responsible for separating the professorships, and study, of Botany from those of *materia medica*; he initiated the study of plant physiology at Edinburgh and he encouraged his Principal Gardener, John Williamson, in his experiments on trees. As a systematic botanist he was a zealous exponent of the Linnaean system and helped by his students he initiated the study of Scottish flora. From our point of view, however, his principal achievement was the uniting of the Royal and the Town gardens into one, on a new site (now Haddington Place) and on a larger scale (five acres).

But there followed, after Hope's term as Regius Keeper and Professor, twenty-three years of stagnation at Edinburgh, under a Keeper who was indifferent to botany, while elsewhere in the world both experimental botany and horticulture were forging ahead. The garden itself went back a good deal until, in 1810 and at the recommendation of Sir Joseph Banks at Kew, a certain William MacNab was made Principal Gardener at a salary of £50 a year—he had been offered £40 at first and had refused the job. In a short time he had raised the number of genera in the garden to a thousand, of species to four thousand, and had restored the place to order. By 1812 he had the hardy perennials and other hardy herbaceous plants arranged in one part according to the Linnaean system, and in another part according to the system of de Jussieu. In the stove he had a fine dragon tree, a camphor tree, a date palm and five kinds of bamboo. So active and enterprising was MacNab that by 1820 when the garden got a new Regius Keeper, Robert Graham, who had been

Scotland

The art of rock gardening was pioneered in the Royal Botanic Gardens, Edinburgh, to accommodate the growing collection of alpine plants from all over the world in a 'natural' manner to conform with the taste for romantic gardening.

Professor of Botany at Glasgow and founded the Glasgow Botanic Society, he had to move the garden to a new site—the Inverleith site which it still occupies, though in Graham's time it was only fourteen acres and is now over fifty acres.

The moving of the garden resulted in the proving of MacNab's techniques for moving fully grown trees and shrubs and the special machinery he invented for the work which took three years to complete. Such innovations were not MacNab's only contribution to the progress of horticulture; as a consequence of his experience in re-planting the Royal Botanic, he published books on the planting and care of evergreen shrubs. And following up his special interest in the Cape Heaths in the cultivation of which he was remarkably successful, he published a book on them too. It was also during his term that the Palm House was built, at that time (1834) the largest in Britain.

John Hutton Balfour who became Keeper in 1845 had, like Graham, been Professor of Botany at Glasgow. His term lasted thirty-four years, and the most important botanical event during that time was the publication of Charles Darwin's *Origin of Species*. Darwin had studied at Edinburgh. Also during Balfour's reign the Herbarium and Library of the Royal Botanic were enormously enlarged and improved by the gift of the Herbarium and Library collected by the Botanical Society of Edinburgh. Hutton Balfour is chiefly remembered as a great teacher; although as a devout Church of Scotland man and the son of a clergyman, he was deeply shocked by Darwin's theories, he loyally included them in his teaching. He wrote a number of important textbooks, added laboratory to field botany on the practical side, and made some progress in the new science of plant ecology.

MacNab died in 1849 and was followed as Principal Gardener and later as Curator by his son, James MacNab. James was an even greater gardener than his famous father had been.

When, in 1875, the *Gardener's Chronicle* published a Supplement on the Edinburgh Botanic, its authors gave special praise to James MacNab's work on conifers and to the garden's Pinetum; but their highest praise was saved for the Rock Garden:

> This is the largest and most varied rock garden we have ever seen and the most fascinating . . . . the rock garden is charming beyond the power of expression. . . .

At that time, towards the end of Hutton Balfour's term as Keeper, the Rock Garden contained four thousand species of alpine plants. It is, even now, probably still the finest rock garden in the world, and what the MacNabs created has been steadily improved on. Writing of it in 1933 Sir George Taylor, subsequently Director of the Royal Botanic Gardens, Kew, said:

> To most gardeners, however, perhaps the chief feature is the magnificent rock garden, which contains one of the most comprehensive collections of alpine plants in the country, as well as many of the rarer herbaceous plants. On its north-facing slopes, which finally resolve themselves into an excel-

lently designed moraine furnished with a host of rare and tricky alpine treasures, are many of the gentians and their close cousins, the cyananthus, many of the exquisite nomocharis, in the cultivation of which the gardeners at Edinburgh excel, the beautiful *Ranunculus Lyalli*, several meconopsis, and numerous primulas, including the rare, deep plum-coloured *P. Calderiana*. Skilful construction, to afford a variety of planting aspects, combined with the provision of adequate soil pockets in vertical as well as horizontal positions, serves to meet the varied needs of a host of alpines gathered together from all parts of the world, and to provide every rock gardener with an excellent object lesson in rock garden design and planting. Meconopsis and primulas luxuriate in the partial shade of the woodland garden, perhaps one of the most lovely corners in the whole Garden, whose beauty has been enhanced in recent years under the improving hand of Mr Stewart, while on the more open grassy banks and under the trees, crocuses, scillas, grape hyacinths and daffodils provide charming drifts of spring colour.*

Today one would have simply to multiply the superlatives. In 1875 the actual rock work—made with the material of the demolished wall which had separated the Royal garden from the Experimental Garden of the Caledonian Horticultural Society which had been united to it—was not nearly so effective a work of art-after-nature as it is now. Contemporary photographs show it arranged in tiers of little boxes made by setting slabs of stone on edge, and there is very little attempt at making it a picture of a possible—if idealized—alpine scene.

The *Gardener's Chronicle* Supplement also praised the landscaping of the garden, as well they might. James MacNab was a very considerable artist in that field, and the first horticulturist to advocate, in published papers, the 'amenity' planting of wasteland, railway embankments and roadsides. It may perhaps be due to the spirit of fine art which he introduced into the gardening at the Edinburgh Botanic that Edinburgh University has, today, by far the best school of landscape gardening in the world.

When James MacNab died (1878) and Balfour retired (1879) they left the garden not only immensely improved, looking something like it does now, but three times as large as they found it. But the man who gave the garden the look and atmosphere it has today was the great Regius Keeper Isaac (later Sir Isaac) Bayley Balfour, who took office in 1887, when Dickson, Hutton Balfour's successor, died.

Isaac Bayley Balfour, Hutton Balfour's son, had come under the influence of the great Thomas Henry Huxley, and as a result he shifted the whole emphasis of botanical research and botanical teaching from the dead to the living plant, from mere morphology to function and structure. As a result, the garden became more important than ever to good botanical work. Balfour was, moreover, a widely travelled man and had, consequently, a broader view of his work than that of an academic botanist, and that too influenced the garden's development; he had been

**Country Life*, 23 September 1933.

An early photograph of Edinburgh's rock garden, before the art of building natural-looking rocky outcrops to accommodate alpine plants had been mastered. Note the then fashionable *Araucarias*—'Monkey Puzzles'—and the 'plum-pudding' effect of the mounds.

Above Statue by Reg Butler in the Royal Botanic Gardens, Edinburgh.

Below The Edinburgh garden is famous for the work of its specialists in alpine plants from all the world's mountain ranges, such as these *Celmisias* from New Zealand.

botanist to the 1874 Transit of Venus Expedition to Rodriguez. He had collected plants in Socotra, adding over two hundred new species to the total described by science and, incidentally, introducing *Begonia socotrana* to whose progeny we owe our winter-flowering begonia cultivars. Young, clever, active, a born administrator, Balfour transformed the Edinburgh Botanic; but despite his strength of character and personality, he was the executive who made the changes rather than the originator who inspired and made them necessary. The man of action whose instrument Balfour willingly made himself was, until 1904, an unimportant subordinate in the Royal Botanic's Herbarium, George Forrest, by name. It was in 1904 that Forrest made the first of his seven plant-hunting expeditions to west China and, by pouring an immense quantity of new botanical material into Edinburgh, almost forced Bayley Balfour to modify the work and appearance of the Royal Botanic Garden.

This is not, alas, the place to write even a brief life of that very remarkable man. All we can say is that, sponsored and financed by the Royal Botanic Garden and by a syndicate of amateur plant-lovers, at the instance of Isaac Bayley Balfour, Forrest sent back from China such a wealth of perfectly prepared herbarium material as provided a life work for more than one botanist; and such a wealth of seeds of new plants that in due course it transformed not only the Edinburgh garden, but the gardens first of Britain and in due course of the whole Western world. Forrest sent home at least 309 new species of *Rhododendron*, including such magnificent trees as *R. sinogrande*, *R. giganteum* (which may attain 80 feet) and *R. griersonianum*; 154 species and sub-species of *Primula*; scores of new *Acers; Aconitum* and *Adenophora* in new species; 20 species of *Allium* and 11 of *Arisaema*; 18 species of *Buddleia ; Camellia saluensis* and *C. tsaii*; new *Clematis* and new *Codonopsis* species; a number of new firs of which the silver *Abies forrestii* is very beautiful; about 10 *Cotoneasters*; 22 species and varieties of *Cyananthus ; Daphnes, Delphiniums* and *Desmodiums*; a number of *Deutzias*; more than a score of *Gentiana* species new to science including the lovely *Gentiana sino-ornata*; new *Hemerocallis*, new *Hypericums* and many new *Incarvilleas*; 14 species of *Iris*, all new; a few new jasmines; 12 species and varieties of *Lilium*; 38 species and varieties of *Lonicera,* several of them new to science; 8 new *Magnolias*; 12 species of *Meconopsis*; 2 *Meliosmas* and 10 *Nomocharis*; species belonging to no less than 50 genera of orchids; 3 *Paeonies*; 8 *Pieris* species, including the lovely *P. forrestii*; 50 *Polyganums*, 8 of them new; a new *Pyrocantha* and a new *Rheum*; 3 *Sorbus* species; a new lilac; 5 species of *Trollius*; and a great many representatives of other genera. The number of actual specimens, whether as herbarium material or live seeds, sent home was, of course, ten times as great as this. It was material for a botanical and horticultural institution's work for a generation—and that was what Isaac Bayley Balfour thought when he applied himself and the Edinburgh Botanic to ordering and exploiting it. For Balfour grasped the magnitude of the task which his sponsorship of Forrest had created for himself, and thereafter devoted his whole life to it. He described hundreds of new genera and species; he became the leading authority on *Rhododendron*, creating a special system of classification of its species; and on *Primula*. And, by raising Forrest's seed, turned the Edinburgh Botanic into the world's primary rhododendron and primula garden. Writing about this in the *Country Life* article which we have already quoted, Sir George Taylor said:

If considerable success has attended the efforts of the authorities at Edinburgh in the growing of the vast majority of new plants that have reached our gardens through the last thirty years and more of intensive botanical discovery in China and its borderlands, it is only because they have made full use of the resources at their command in the shape of a situation about a mile from the sea and whose highest point is only a little more than about 100 ft. above sea level, with the ground falling away on all sides, thus offering a variety of aspects for planting, and a soil which is for the most part alluvial sand overlying clay. In its plant furnishing the garden is a mirror of the last fifty years of horticultural exploration, and if rhododendrons play the most important role in the display—largely owing to the enormous number of species and hybrids that are grown—they are well supported by a full cast of trees and shrubs, herbaceous and rock plants and bulbs, where each member is of established reputation.

On the eastern slopes of the hill, from which some of the most lovely views of the town are to be obtained, rhododendron species find a comfortable and satisfying home; and here, from April until June, the visitor can feast his eyes on a glorious display of colour and blossoms, as well as compare the relative merits of the many different species of this magnificent race. If the members belonging to the triflorum series—such as *RR. augustinii, yunnanense*, and *davidsonianum*—seldom fail to smother themselves in bloom every spring, others—such as the fine yellow *R. campylocarpum*, the blood red *R. thomsoni* and its hybrids, the handsome leaved *R. falconeri*, the rosy lilac *R. hodgonsi, R. fictolacteum*, the exquisite yellow *R. lacteum* itself, and the beautiful *R. schlippenbachii* and *R. argyrophyllum*—if less dependable, perhaps, are no less beautiful on their day. It is impossible to enumerate the hundreds of species that are represented in the garden and contribute to the rhododendron pageant on the hill; in the rock garden, where all the dwarfs, such as *R. racemosum, hippophaeoides, repens, scintillans, orthocladum, keleticum, calostrotum* and *radicans*, which flourishes amazingly, find a place; on the rootery—a mound, cleverly constructed of old tree stumps, ingeniously devised by Mr Stewart as a means of establishing some of the more tricky of the dwarf species as well as numerous primulas such as *P. sonchifolia* and *P. winteri*; and in the wild and woodland garden on the western side of the rock garden, where such species as the charming yellow *R. wardii* to name but one finds a comfortable home.

In the botanical part of the great work which Forrest's material imposed on him, Bayley Balfour had the help of his Deputy and subsequent successor, William (later Sir William) Wright Smith; and in the horticultural part which gave our gardens hundreds of new species, hybrids and other cultivars of *Rhododendron* and other flowering shrubs, *Primula* and *Meconopsis* and other lovely perennials and annuals, he was helped by his Curator, Robert Harrow.

In part because the new Chinese plants which transformed the Edinburgh Botanic are of such spectacular beauty, and in part because, although taxonomy and economic botany still have a use for botanic gardens,

'Reclining Woman' by Henry Moore. Royal Botanic Garden, Edinburgh.

modern experimental botany can do without them, the Royal Botanic Gardens at Edinburgh have become more horticultural than botanical in spirit. Some botanical garden workers still feel defensive about such a change; Mr de Wolf, at the Rijksplantentuin, Meise, complained to us that having been impressed by a lecture delivered at an international congress by a famous British botanist who urged that the orchid collections in botanical gardens should be confined to species, he was disconcerted to find, on his next visit to Britain, large numbers of cultivars in the British collections. I do not think that the authorities at the Edinburgh Botanic would deny my judgement, but if they did I should point, in the first case, to the Rose Garden. True, many botanical species are cultivated there, showing the material from which our garden roses have been made; but there are also collections—the planting in that part is formal—of both the old and the modern garden roses. This same spirit will be found in some, at least, of the other principal divisions of the garden. Here are some of them, more or less in the order which is suggested by the official guide to the gardens as a good one to follow.

The Pond, an artificial small lake fed by natural springs, accommodates hardy aquatic plants, bog plants, and, further back from the water, such moisture-lovers as some of the *Primula* species for which Forrest made Edinburgh famous, hardy *Lobelia* species, hardy ferns; and the Willow collection. The willows, both about the pond and

Dwarf and prostrate conifers are another of Edinburgh's special interests.

in the Rock Garden, are outstandingly good at Edinburgh. Above the pond are some fine plantings of trees, including several genera remarkable for vivid autumn colour, and the swamp cypress, *Taxodium distichum*.

Next comes the Rose Garden already mentioned, and then the complex of glasshouses. Among these are houses devoted to temperate-zone plants including alpines which are best under cover; to tender species of *Rhododendron* with which are planted some other genera, notably the spectacular *Clianthus*; to succulents; to economic plants; to insectivorous plants, always a great draw at botanical gardens; to orchids; to tropical shrubs and herbaceous plants; to the nepenthes and aroids which have become so important in commerce as house plants; to ferns and their kindred such as the ground-covering selaginellas; to Bromeliads, again plants of great interest to indoor gardeners; to tropical aquatics. There are also the Palm Houses, and a central area, with its glass corridors, devoted to ornamental plants.

The Arboretum of trees and hardy shrubs, although planted after nature, is botanically arranged. As it includes many flowering kinds, barberries, for example, and viburnums, it is in beauty throughout the growing seasons.

The oldest plant in the garden is Sutherland's Yew, the only living link between the modern garden and the Sibbald–Balfour garden of the late seventeenth century. On the rising ground above it is the feature called the Azalea Lawn, a mass-planting of azalea cultivars underplanted with early spring bulbs. Here, in June, the colour is overwhelming. On the hill above are lilacs, ligustrums, forsythias and buddleias in variety.

We have quoted Sir George Taylor on the subject of the Edinburgh rhododendrons. But many of the most interesting and beautiful species are not where he writes of their hardier kindred, at the point where the garden commands that tremendous view of the city, but are planted in the Rhododendron Walk which encircles Inverleith House and is sheltered by a yew hedge. Further on there is a remarkable hedge, 50 yards long, of *Rhododendron* × 'Praecox', which flowers in February, usually so freely that the evergreen foliage of this remarkable cultivar is completely hidden by the flowers. And there are still more species and cultivars of *Rhododendron* in the copse, growing in the light shade of Scots pine, birch and oak.

Bearing in mind my claim that this garden favours horticulture rather than botany to an extent much greater than, for example, the great botanical gardens of the Continent, the Herbaceous Borders, two hundred yards long between magnificient beech hedges, and the West (Shrub) Border, although some botanical species have a place in them, seem to be dominated by cultivars of more direct interest to gardeners than to botanists. Again, in the Woodland Garden and in the Peat Garden, although in these botanical species and their varieties predominate, Edinburgh offers the visitor a superb example of woodland, or wild, landscape gardening in the tradition which was started by James MacNab. In the Woodland Garden the most beautiful flowers are the blue yellow *Meconopsis* species. In the Peat Garden the primulas, nomocharis and dwarf rhododendrons for which Edinburgh is famous, have their place. The gardens are also very rich in *Lilium* and *Rodgersia*.

It is impossible to describe effectively the Rock Garden which has been world-famous among gardeners and botanists for more than a century. The present one is not, however, the Rock Garden so highly praised in the *Gardener's Chronicle* of 18 September 1875 which we quoted above, but a reconstruction on the same site carried out in 1908 by Sir Isaac Bayley Balfour. It is a noble work of art, the rock, turf, water and dwarf shrubs which seem to unite with the rocks they half-cover married and blended so perfectly that the art used in creating the work is concealed—as of course it should be—by the very perfection of the creation itself. It is, beyond question, Edinburgh's masterpiece, for the gardener a place of rare delights from early spring into late autumn, and for the botanist one of the world's half-dozen greatest collections of alpine plants from all five continents and an astonishingly wide range of latitudes.

It is difficult to be certain that the Royal Botanic Garden, Edinburgh, has a future in the service of botanical science; but there can be no doubt whatever about its future in the service of horticulture botany.

NOTE: Scotland's other botanical gardens are: Aberdeen: The Cruikshank Botanic Garden (7 acres); Benmore: Younger Botanic Garden (100 acres); Glasgow (42 acres); and St Andrews (5 acres).

The New House at Edinburgh, designed and built by the Ministry of Works, compares with the Climatron at St Louis and the Palace of Plants at Meisse, as a new answer to the problem of big greenhouses without internal supports.

GLASNEVIN

The National Botanic Garden, Glasnevin, Dublin, an eighteenth-century foundation, pioneered in the construction of large conservatories which has become a centre for the advancement of horticulture as well as botany. It is remarkable for its good displays of garden plants as well as botanical species.

One of the most pleasing of the smaller European botanical gardens which have made useful contributions to the progress of botany and horticulture is the National Botanic Garden of the Irish Republic at Glasnevin in Dublin. Although not one of the oldest, it is of respectable antiquity: in or about the year 1790 the 'Members of the Right Honourable and Honourable Dublin Society resolved to form a Botanical Garden for the promoting of scientific knowledge in the various branches of agriculture and planting as well as to form a taste for practical and scientific botany'.* The Society in question must have been an influential body, because their resolution was soon embodied in an Act of the Irish Parliament which moreover endowed the foundation with £300, later increased to £1,700 by further grants. One can safely multiply by ten to bring this up to modern values. Nevertheless, the management of the garden remained in the hands of the Rt Honourable Society, which appointed a committee to buy a site and begin planting.

Who was the prime mover? The *Dublin Magazine* for July 1800 carried the following economium:

> In the planning and execution of this garden it has been uncommonly fortunate that the abilities and assistance of first-rate character which this nation or any other can boast of were most condescendingly and arduously devoted to further this great national object, and while the name of Foster remains respected and beloved by every Irishman, so long will this garden perpetuate the taste and abilities of this great and good man.

The gentleman thus praised was the Speaker of the Irish House of Commons, later Lord Oriel. The Committee which put Parliament's decision into execution included the Dublin Society's lecturer in Botany, Dr Wade; Professor Dr Hill (Botany) of Trinity College; the Secretary of the Royal Irish Academy, Dr Percival; and the Bishop of Kilmore. In 1795 they bought the demesne called Glasnevin, two miles from Dublin Castle, 27 acres of thin loam over Calpe limestone—a thoroughly bad choice which proves that no one on the Committee knew any gardening. Nevertheless the difficulties of the poor soil were overcome and the garden planted with an economic botanical purpose in view—the advancement of agriculture and horticulture by the study of botany. Thus although the layout was systematic, the plants chosen were such as were useful, for example the *Graminae*, which were the subject of a monograph by the first Curator's assistant, John White ('Indigenous grasses of Ireland and their use in Agriculture').

Guide to the Botanic Gardens (1885).

Key to Map

1 Poplars, Hollies, Thorns and Elms
2 Horse chestnuts, Alders and Birches
3 Arboretum
4 Maples
5 Ashes
6 Yews
7 Walled garden
8 Oak collections
9 Economic plants
10 Native Irish plants
11 Vegetable garden
12 Alpine House
13 Hardy herbaceous plants in natural orders
14 Conifer collection
15 Rock garden
16 Bamboos
17 Rhododendrons
18 Greenhouses

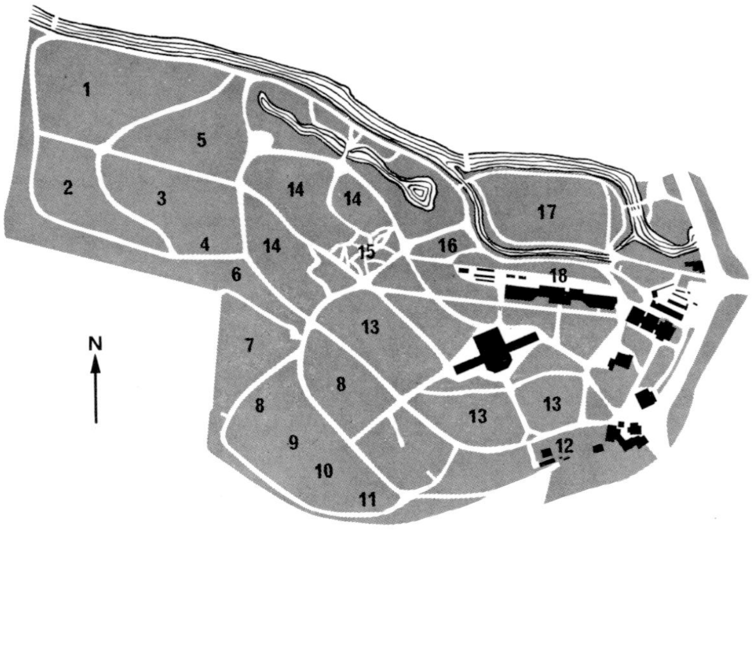

However, there were also a *Hortus Linnaeensis*, in which both native and exotic plants were used to demonstrate the Linnaean system of classification; a medicinal garden; a Hibernian garden and some other, purely botanical, sections, to which a Hardy Fruit orchard was added in 1836. Before then greenhouses for tender plants had been built and stocked.

In the garden's first three decades the original economic purpose of the formation became increasingly less important and pure science more so. In 1834, when a skilful landscape gardener, Ninian Niven, became first Curator and later Director, the gardens were remodelled and much of his work survives; the stoves and greenhouses were rebuilt and enlarged and a course of horticultural training—continued ever since†—added to the more strictly botanical curriculum. Niven's policies of improvement and his horticultural bias were continued by his successor, David Moore.

During the first half-century of the garden's history, the botanical lecturing had been done, in the gardens, by the Dublin Society's botanists (until 1845, Dr Litton). But in 1854 the lectures were transferred to the College of Sciences and became the Professor of Botany's responsibility, though he continued to lecture in and draw his material from the Botanic Garden which also supplied material to the School of Art and which, today, still does so as well as providing plant material to other Dublin educational institutions.

Nothing, in the world of botanic gardens, is more curious than the way in which, following Paxton's example at Chatsworth and thereafter the example of Kew Gardens, every self-respecting botanical garden in north Europe had to have a *Victoria regia* (now *Victoria amazonica*) house. It becomes a sort of status symbol, justified by the spectacular quality of the giant water-lily. Glasnevin got its 'Victoria regia' House in 1855 and new conservatories and a Palm House were built in the following decade. Throughout all this time and until an Act of Parliament put the garden under the Science and Arts Department of the government in 1877, the gardens were still administered by the Dublin Society.

†About forty such students pass out with a diploma every year.

Left The Rock Garden, Glasnevin, has a large collection of alpine species and cultivars.

Below left Tulips in the foreground and in the background the big Glasnevin Palm House, which Turner, the Dublin engineer, pioneered.

Below Round Tower seen from the National Botanic Garden, Glasnevin. These towers date from the tenth or eleventh centuries.

It has surprisingly often happened that a botanical garden has enjoyed a sort of Golden Age when a son has succeeded his father as Praefectus, Keeper, Director or Curator, their joint reigns being long: one thinks of the Cluyts, the Wittes and the van Royens at Leyden (in the last case uncle and nephew); of the Balfours at Edinburgh; of the de Jussieus in Paris and the Hookers at Kew. Glasnevin had such a Golden Age during the joint reigns of David and F. W. (later Sir Frederick) Moore who succeeded him; David Moore had become Curator in 1838; Sir Frederick retired in 1922, a joint reign of eighty-four years.

It was during F. W. Moore's long term that the title 'Curator' was changed to 'Keeper'; in 1966 it was changed to Director after the old title of Curator was revived for the incumbent of a new post in 1964.

At David Moore's death the gardens covered 31 acres; his son added to them the 12-acre West Arboretum and was so active in bringing the collections up to date, in getting specimens of all important new introductions for the garden, and in encouraging and helping the makers of such great private Irish gardens as Mount Usher, that he made Glasnevin one of the most important botanical gardens in Europe.

It became—and it remains today—particularly famous for its magnificent *Cycad* collection and its splendid *Orchid* collection.

For the state of the gardens under the present Director, Dr T. J. Walsh, we quote from our own *Irish Gardens*:*

The first division of the Glasnevin Gardens immediately inside the gates is a shrubbery, separated from the path by a border used for the growing of early flowering perennials. On the other side of the path is a good collection of herbaceous perennials. This path leads to the Tree Fern House, which has a fine collection of dicksonias, cyatheas and other tree ferns, as well as many species of ground ferns and a very interesting display of Filmy Ferns including the famous 'Killarney Fern'. The next house in the range is the Victoria Regia House, where the giant water-lily from the Amazon flowers every year in the central tank, lesser tropical water lilies in the side tanks. After that, and in sharp contrast, comes a house full of arid-soil succulents and cacti, one of the best collections of them I have seen anywhere.

The next range of greenhouses is called the Curvilinear Range, and is the one which was built in 1843. It consists of a Cool House, where many plants which are only just not hardy, and which often grow perfectly well out of doors in the south-west of Ireland, are housed; the tall central section, with a collection of tender conifers and some banksias; and the Stove which, among other tropical plants, contains a good collection of bromeliads and specimens of such economic plants of the tropics as coffee, cocoa, and sugar cane.

The Orchid Houses are separate from the other ranges of greenhouses, and the collection of orchids is a good one; it

*Edward Hyams and William MacQuitty, *Irish Gardens* (London, 1967).

Below Turner, the constructional engineer who built the principal greenhouses at Glasnevin, was later called in to build the great Palm House at Kew. The display houses are of horticultural rather than botanical interest.

was, at one time, among the best in the world; whether this is still the case I am not competent to judge. One curious fact connected with the Orchid Houses: in their entrance porch was housed a small collection of sarracenias, the North American Pitcher Plants, and as a consequence, apparently of their semi-exposure, some of the species have established themselves and become naturalized in a bog in Roscommon.

There are two important greenhouses: the Palm House which is seventy feet tall and contains a very fine collection of plants both economic and otherwise, some giant bamboos, cycads and giant ferns. And then, the Camellia House, less important for camellias which are in any case all hardy in Ireland, than for its display of flowering plants in summer and notably for its display of many species of begonias.

There would be no point in describing divisions of the garden whose name describes them: there are, for example, the Herbaceous Walk, the Rock Garden, the Bamboo Walk, the River Walk. It should be said, however, that even where the principal purpose of the authorities is to display a particular genus for botanical and educational reasons, ornamental gardening is practised by, for example, underplanting with daffodils and snowdrops, so that even for the non-technical visitor the gardens are very pleasant.

The area of the gardens known as the Mill Race, once a willow collection, but ruined by flooding, was recovered and raised during the Keepership of Mr Besant, and planted with conifers and rhododendrons. The banks of the river at this point have been used for a good collection of primulas, meconopsis and other moisture-loving plants. The River Walk itself lies between the tolka and the pond. The pond is rich in aquatic plants in great variety, from the spectacular water-lilies to the more modest species, and its margins have been made into a very well-stocked Bog Garden.

Glasnevin includes a good Pinetum and, in the West Arboretum, an interesting and well-grown collection of maples and of many other genera of broad-leaved trees; there are also collections of ivies, a relatively new collection of oaks, and a collection of all the different kinds of yews.

The Vine Border Walk is a division of the garden which I found rich in useful lessons for the ordinary gardener. There is a wall with a good collection of climbing and scandent plants; a Crocus collection; an Iris collection of the May and June flowering kinds; a *Paeonia* collection confined to the botanical species, and for autumn visitors, a Michaelmas Daisy collection. Other collections beyond this area and on the way back to the Curvilinear Range of greenhouses, are the Chinese Shrubs, the Leguminous Shrubs; the Magnolias; the species roses; the *Prunus*, the *Pyrus*; and the *Syringa*.

While the Glasnevin Garden remains an important teaching garden and does valuable work in economic botany as its founders intended, its bias, like those of Edinburgh, Copenhagen, Meise and Göteborg, has become horticultural, and it seems likely to continue in that way as more and more of the economic work passes into the hands of specialized research institutions or departments and more and more academic work into specialized laboratories.

BELGIUM

MEISE

Belgium's new botanic garden at Meise. The castle in the centre of this great landscape was formerly inhabited by the Empress Carlotta, wife of Maximilian von Hapsburg, who went mad when her husband, Emperor of Mexico, was captured and shot by the revolutionary patriot Juarez.

The French style and formal parterres of Belgium's old botanic garden in the heart of Brussels are in sharp contrast with the new landscape site at Meise, to which the Herbarium, Library and living collections have been moved.

Belgium has five botanical gardens, if we count the Kalmthout Arboretum as one of them, which indeed it is. All have done and continue to do useful work, less perhaps in pure science than in horticultural botany, at which the Belgians have always excelled; and in the economic botany of tropical plants. At the present stage of the world's political history there is a tendency to think and write that the formerly imperial powers, of which Belgium was one, took from but did not give to their tropical provinces. This, of course, is not true, and not the least of what European-applied science gave to the tropical countries were the new and improved plants which are often the basis of their economy. There was no rubber in the Far East until European economic botanists planted it there (with the aid of the greatest botanic garden in the world, Kew). European botanists were responsible for the distribution and acclimatization beyond their habitat of cotton, bananas, tomatoes, potatoes, chocolate, tobacco, citrus fruits, and avocado pear, to name only a few, and there was no coffee in the Congo, to mention but a single example, until Belgian botanists had 'made' a plant which would produce good coffee in the conditions of that country.

But of Belgium's botanic gardens, there is only one which has, at the time of writing, the kind of distinction that makes it interesting in the context of this book; the distinction may be historical, aesthetic, scientific, technical; or, as in the case here in question, the garden may represent a growing-point in the story of botany and botanical gardens. The great distinction of Belgium in our field, at the present time, is the vast Palace of Plants in the new Nationale Rijksplantentuin van België at Meise, near Brussels.

Until very recently Belgium's National Botanic Garden, the Herbarium, Laboratories and other services associated with it, have been in the heart of Brussels; for that matter the garden is still there and this old Botanic Garden in the Rue Royale is one of the prettiest, small, formal botanic gardens in the world. But following the nation's acquisition of the lovely Domaine de Bouchout park, things have changed. The park, with its beautiful castle beside the lake, has a history; it was there that the unhappy Empress Carlotta of Mexico, Maximilian von Habsburg's wife, spent some of her last and maddest years. Into the park has now been built the magnificent Palace of Plants conceived, designed and pushed to completion by Dr Demaret and the Curator of the Palace of Plants, Mr L. de Wolf, one of the most enthusiastic plant-lovers we have ever met, much of whose working life has been devoted to making the superb collection at Meise, and to seeing it nobly housed. Meanwhile the Herbarium, library and laboratories have been in process of transference from Brussels to Meise.

It will necessarily be some years before the 225-acre Bouchout estate—and surely no botanical institution in the world has a more beautiful site for its work—becomes a real botanical garden. At the time of writing it is still a gentleman's park in the manner of the English landscape garden of the late eighteenth century, and much care has been taken, for example, in screening the handsome modern building which houses the million and a half herbarium specimens, the 150,000 volumes of the library, and the laboratories, not to spoil the picture. Nor is there any reason why the specimens of the Botanical Garden should not be planted into this park without overmuch disturbance of its lovely sweeping lawns, magnificent trees and pretty little lake.

Dr Demaret and Mr de Wolf have a clear idea of what, in the 1970s, a Belgian botanical garden and its associated services should be doing. First comes research in systematic botany and phytogeography concentrated on the florae of Belgium and of tropical Africa. In the second place it should provide and maintain living plant-material for study; then, it should do all in its power to help native and foreign research workers in economic and other kinds of botany. Dr Demaret lays much stress on the importance

Right The botanical park at Meise has a good collection of cypresses in their most graceful forms.

of botanical gardens in another function, that of conserving wild species and especially for their genetical potential in a world in which they are increasingly likely to be destroyed by human activity and by the population and industrial 'explosions'. Finally, there is the duty of educating the public, concerning which Dr Demaret has written:

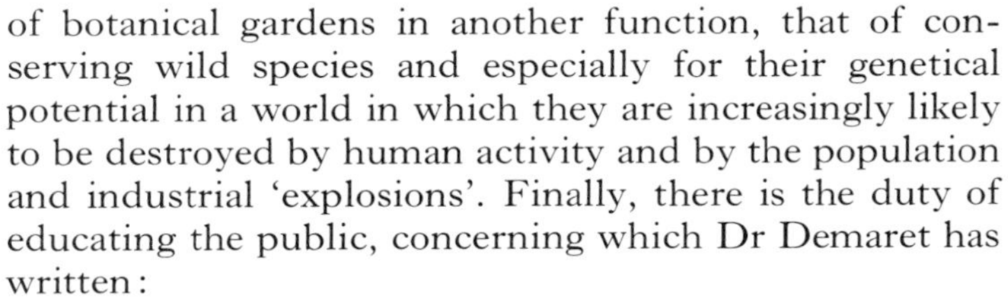

This mission of education is extremely important in our epoch, in which the problem of preserving natural resources confronts us more than ever before. Even economists are studying this problem and realizing that the time to safeguard our natural heritage has come indeed.

It is in that constatation that he brings together the botanical garden's twin missions of conservation and education, education not of specialists but of the public in general. For, 'If,' he says, 'we can inculcate a love and respect for plants in the public, it will become more receptive to the idea of respecting Nature Reserves and natural resources [in plant life].'

Certainly the great new glasshouse complex called the Palace of Plants, covering two-and-a-half acres of ground and comprising thirty-five halls or rooms full of plants, is well designed and calculated to inculcate that love and respect for plant-life of which Dr Demaret and M. de Wolf make such a point. But in attracting the public interest to the garden they have a task which is curiously different from that which faces British botanical gardeners. The English and the Scots can always be drawn by a good show of flowers; not so the Belgians. Far less of their leisure than of ours is given over to gardening and garden-visiting: one academic but flower-loving botanist we met there, bewailing the difficulty of interesting his fellow-Belgians in plants, spoke of them bitterly as *abrutis par les sports, m'sieur*. A thrifty and practical people, however, they like to see and to hear about economic plants.

The new Palace of Plants at Meise is the largest greenhouse in the world, and pioneers a new type of tropical house construction and heating. Contrast its practical severity with the pretty little Royal Crown Conservatory left over from the nineteenth-century garden of the castle.

Fortunately, the Palace of Plants collection is very rich in coffee, cocoa, vanilla, banana and other tropical fruits, cotton, kapok, oil-bearing plants and so forth, and it is these which, according to the Curator, attract the people, and especially schools, to the collection.

Before coming to the plan of this vast and newest of conservatories, and to the planting plan, a few details touching its building and the unexpected difficulties which have held up the completion of planting should be of interest. The subterranean galleries containing the heating, irrigation, humidification, lighting and automatic (pneumatic) ventilation machinery and the foundations of the building are of concrete, 5,000 cubic metres of it. The *centrale thermique*, energized by electricity, but heating water, supplies pipes which are sunk in very deep concrete trenches all round the inside periphery of the complex. Rising hot air travels upwards along the glass, cooling as it rises, to the highest point, and, as it cools, sinks again through the houses, to be drawn outwards and back into the trenches to replace the hot air which continues to rise. In short, this simple but most effective device provides a continuous vertical circulation of warm air, the motion providing those conditions which are best for plant health. There are about 17 miles of heating pipe.

The superstructure, its strength founded on 350 metric tons of steel standards and rafters, is of Mulmein teak, 300 cubic metres of it. An absurd rule of the Office of Works (a fair translation of the name of the responsible government department), has resulted, despite M. de Wolf's protests, in the teak being painted, which is not only a waste of public money but spoils the appearance of the woodwork. The Palace of Plants is completed by 14,000 square metres of double glazing, the two panes of each panel being sealed together so that algae cannot grow in the space between them. As will be clear from our illustration, this is the first glasshouse of anything like its size to break away completely from the plan originated by Paxton at Chatsworth and perfected by Hooker at Kew.

The principal and quite unexpected difficulty encountered in planting this huge conservatory has been with the irrigation water. The soil in the Palace of Plants is entirely artificial, that is, it was filled in to a depth of two and a half metres (to accommodate deep-rooted tropical trees, giant bamboos and palms) with many thousands of tons of balanced compost made of sterilized loam, peat and grit with added nutrient salts. No trouble was expected from the water of an acid-soil region and an artesian well was sunk, the water being pumped electrically. The water proved to be badly contaminated with calcium in some form, which was not only bad for plants requiring a low *p*H soil, but which left a chalky deposit on the leaves. Deeper and deeper drilling for water at much lower levels was very expensive and resulted in no improvement whatsoever. About a million francs were spent before that well was given up and another drilling tried in a different part of the grounds. The water from this second well at the higher levels was contaminated with chlorine in such quantity as to be toxic for plants;

deeper drillings brought in water contaminated with salt. Only then did the National Botanic department and its financial masters face the fact that a demineralization plant would be necessary and remineralization of the purified water with suitable nutrient salts.

Thirteen of the halls or rooms of this glass palace are devoted to the display of plants to the public; twenty-two others to collections for scientific work and these are not usually open to the public. Each is devoted to a single family of plants—Orchidaceae, Cactaceae, ferns and mosses, Bromeliads and so forth.

In the thirteen houses of the complex devoted to the permanent public display, the arrangement is geographical, and the planting 'natural'—that is, the plants are not simply put on show as specimens, but planted as they would be in a garden designed 'after nature'. But there are exceptions to the geographical rule: there is, for example, one large conservatory devoted entirely to the

Opposite Meise has fine old avenues of elms and other species of deciduous trees.

'Knees' of *Taxodium distichum* beside the lake, Meise. There are magnificent specimens of this Swamp or Bald Cypress in the garden, as there are of many other tree genera, often planted as avenues.

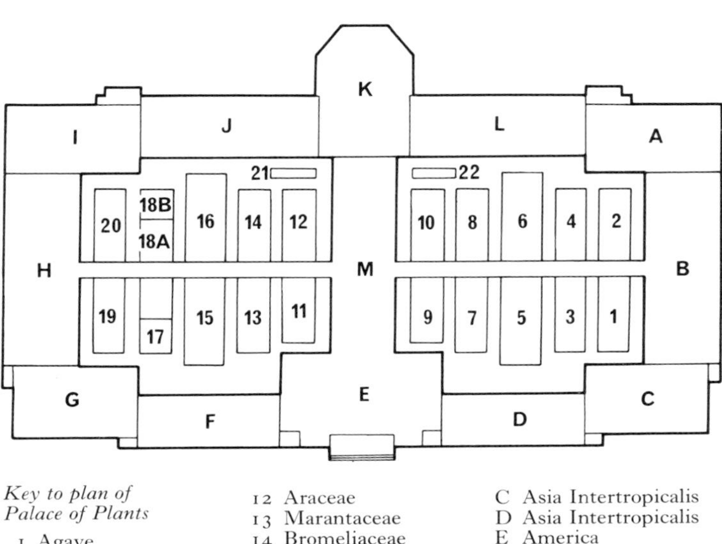

Key to plan of Palace of Plants

1 Agave
2 Cactaceae
3 Succulentae
4 Cactaceae
5 Succulente – Gerontogaeae
6 Succulente – Americanae
7 Begonia
8 Plant. Hort. Subtrop.
9 Amer. Intertrop.
10 Plant. Hort. Intertrop.
11 Araceae
12 Araceae
13 Marantaceae
14 Bromeliaceae
15 Oecologia Intertrop.
16 Cycadaceae
17 Orchidaceae
18A Orchidaceae
18B Orchidaceae
19 Pteridophyta
20 Pteridophyta
21 Insectivorae
22 Bryphyta
A Himalaya. China. Japan. Nova-Zelandia
B Australia
C Asia Intertropicalis
D Asia Intertropicalis
E America Intertropicalis
F Africa Tropicalis
G Africa Aequatorialis
H Africa Subtropicalis
I America Subtropicalis
J Plantae Utiles Intertropicales
K Victoria
L Plantae Utiles Subtropicales
M America Intertropicalis

economic plants of the tropics. Among them coffee varieties are very well represented, because the National Botanic department was responsible for the improvement, by selection and propagation, of the coffee grown in the erstwhile Belgian Congo. Also the object of important practical work, there is a wide range of *Musa* species, from very small dwarf banana cultivars to very tall species which, although bearing fruit superior to any in commerce, have yet to be exploited commercially because of some fault or complex of faults from the farmer's point of view. There are good *Theobroma* (cocoa) specimens in the collection, and there are both species and cultivars of the pineapple plant. Cotton, rubber and other economic plants are equally well represented, and the planting of all these is as at Munich (see below) but on a larger scale, such as to exploit their ornamental qualities. A similar house is devoted to the economic plants of the subtropics. A third exception to the geographical plan is the *Serre à Victoria*, with a handsome pool planted to aquatics regardless of provenance and including the giant water-lilies of the Amazon, whose huge floating leaves were used by Joseph Paxton as the model for the structure of the first really large glass conservatory ever built. The pool is closely planted round with tropical plants, including a fine collection of huge Aroids. During our own visit of several days a *Dracontium gigas*, of that family, came into flower, the spathe 2 feet tall, thick and fleshy, a dark purple in colour and stinking of overripe Camembert.

The range of species and of economic cultivars grown is very large and widely representative. Their presentation is as aesthetically pleasing as it is intellectually stimulating.

We have said that if Belgian botany has a bias it is one in favour of horticulture rather than of pure science.

In the Palace of Plants there are some displays, chiefly in the communication corridors, of garden cultivars, including some from the tropics. But it should be made clear that if the National Botanic contributes to the progress of horticulture, other than food-and-raw-materials horticulture, it is only incidentally; it is not trying to compete with commercial nurseries. Its workers are interested in, but do not exploit commercially, the cross-breeding of *Vriesias* and other Bromeliads which is going forward in the nurseries of Ghent and elsewhere; and they were only incidentally, as it were, responsible for the progress which led to the commercialization of the spectacular *Medinella magnifica*, for long and still the glory of the fantastic display of 'Edwardian' *de luxe* gardening in the *Serres Royales* built by the first King Leopold of the Belgians.

Systematic, experimental and economic botany, with horticultural progress as a by-product, these, and above all teaching, are the purposes served by the Palace of Plants.

MUNICH

Opposite Most botanic gardens seem to have, in their site or architecture or ornaments, some feature which instantly identifies them—Kew's Palm House, Glasnevin's Round Tower, St Louis's Climatron, Sydney's Opera House, Edinburgh's view of Auld Reekie, Calcutta's gigantic Banyan. The Meissen China ornaments in Munich's botanic garden are unique.

The *International Directory of Botanical Gardens*, published by the International Bureau of Plant Taxonomy and Nomenclature in Utrecht in 1963, lists fifty-five botanical gardens in Germany, that is, in the Federal Republic and the D.D.R. together.

Some of these gardens are large, Geisenheim for example, with its 350 hectares (875 acres), Dortmund with 65 hectares (162 acres) and Bremen with 34 hectares (77 acres). Most, affiliated to universities and used principally for teaching and research, are much smaller. Size is not a criterion of importance in the case of such gardens, for it is a curious fact that of the botanical gardens of large extent only Kew can rival such tiny ones as those of Pisa and Leyden and the Paris Jardin des Plantes in the importance of the work which has been done in them. And there are other distinctions: Leipzig was the first botanical garden in Europe outside Italy, founded in 1580; Göttingen is very remarkable for its unique collection of insectivorous plants; and Hamburg for the bold design and execution of its new Palm House.

To represent Germany in this field we have chosen two gardens which seem to be of outstanding interest and superior merit.

MUNICH'S BOTANISCHE STAATSAMMLUNG

The garden of the Botanische Staaatsammlung in Munich is superlative; the great complex of greenhouses, stoves and palm house for example, while it cannot compare with the Palm House at Kew for external beauty, is better and more beautifully planted. Under the direction of Professor Dr H. Merxmüller, botanists and gardeners have combined to serve the purposes of science while paying a proper respect to art. There is an important point here, of which Dr Merxmüller and his colleagues are keenly aware: botanical gardens exist chiefly for the teaching and forwarding of botanical science, but some botanists are too apt to forget the 'garden' aspect. If the public is to be attracted and so instructed—and it is on the public after all that the gardens depend for funds—display and presentation are important. At the Munich Botanic Garden the display and presentation of plants are impeccable.

The garden is not an old one. The first botanical garden in the city was founded under the aegis of the Einrichtung der Bayerischen Akademie der Wissenschaften by Franz von Paula von Schrank in the year 1812. So far as I can determine, while it served a useful purpose in conveying to the Germans the original work which was being done elsewhere, it did not play an important part, internationally, in botanical science.

About a century after the first foundation came the second, that of the present garden, in 1914 by Karl von Goebel, who moved it to the site it now occupies in what was formerly a part of the Nymphenburg park. Its extent is about 20 hectares (50 acres).

The big greenhouse complex, started in 1910 and finished, if one counts subsequent additions, in 1958, consists of a main *Palmenhaus* more or less on the plan originally established by Paxton at Chatsworth, refined by Hooker and Decimus Burton, and copied ever since until the innovations at Hamburg, Meise and St Louis, with eleven side houses opening off it on both sides. In these temperatures vary from cool to full tropical, and humidities from arid to rain-forest humidity. There are micro-climates, within that single complex of houses, for plants from every climate in the world.

As I have said, the most striking quality of this house, to the non-botanist, is that planting for science has not been allowed to exclude planting for art. The botanists have decided what is to be planted; the gardener has controlled the manner of planting, and each of the collections is grown in the kind of order which, although not strictly 'natural' makes a picture after nature in the way that an

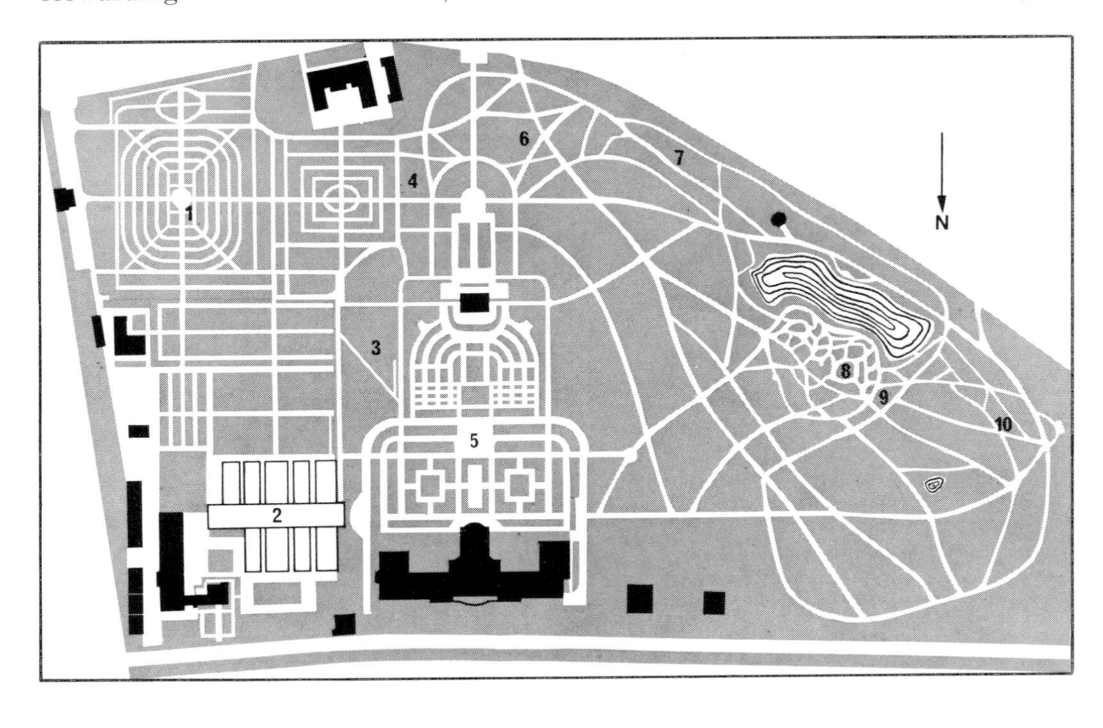

Key to map of open garden

1 System garden
2 Work-rooms
3 Frühlingsweg (Spring Walk)
4 Bavarian flora
5 Ornamental garden
6 Rhododendron grove
7 Fern Gorge
8 Alpine garden
9 Sand-dune plants
10 Heathers and junipers

The *Frühlingsweg* in Munich's botanic garden is planted entirely with botanical species and in all the 'wild' parts of the garden cultivars are *verboten*.

Opposite From the Munich hot-house collections. *Left* a *Bromeliad* in fruit; the collection of these chiefly epiphytic plants is a fine one.
Far right Cattleya cultivar (*above*) and *Masdevallia* species (*below*) from the Orchid Houses.
Overleaf: left Part of the formal horticulture display at Munich, in which garden flowers succeed each other in their seasons.
Overleaf: right A detail from a botanical drawing from the Library and Herbarium of Munich's botanic garden. The subject is *Cocos nucifera*, and the meticulous detail shows the superiority of the artist over the photographer for scientific and educational purposes.

English garden, wholly unnatural in the juxtaposition of naturally remote genera, in the trimness of lawns and the cleanness of walks, is nevertheless an idealization of the natural scene. As we have seen, this policy is being modified in an interesting way in the huge Palace of Plants at Meise; there geographical as well as botanical order are being respected, yet display and presentation of plants follow the best rules of horticultural art for the woodland garden.

I do not make this comparison with the 'English' garden without having a very clear justification in mind: it is, for example, particularly valid in House 12, which is devoted to cycads and ferns: in much the same way as the great English gardens of the eighteenth century were landscape paintings rather than landscapes, the interior of this house is like a romantic painter's conception of a tropical gorge; the U-shaped walk passes between steep, rocky banks densely planted with many representatives of the genera in question. The specimens are just as numerous and diverse as they could be if the space were filled with rows of plants in pots, and therefore as valuable for study. But the overall effect is very much more pleasing.

Precisely the same is true of Hall B, the big *Palmenhaus* which is the nucleus of the block. Walks lead through a sufficiently open and well-ordered idealized jungle of great palms and bamboos; the soil beneath these tall subjects and the bushier forms among them is densely covered with such plants as *Selaginella martensii*, the trees are grown over by climbing aroids and other scandent plants. The 'atmosphere' created, the feeling of being inside a vivid three-dimensional picture of a tropical scene, is very strong. It was amusing and pleasing to note how visitors tended to speak in lowered voices. However, none of this impairs the purely botanical value of the collections.

Again, in Hall C given over to large succulents, and Houses 8 and 9 given over respectively to African succulents and small cacti, American succulents and large cacti, the same spirit in the planting has produced the same happy result.

Obviously, there is in nature no such scene, no such diverse assembly of genera and species in one small space, as that of cacti and other succulents in House 9. But the 'picture' of an arid Central American landscape is true in the sense that any work of art is true.

There is another instance of the same spirit at work in a small house of tropical flora, mostly aroids, at the far end of the orchid houses, again a 'picture' of tropical 'jungle' about a small, dark pool; and where, however, the specimens serve the Botanic Garden's purpose of teaching and research.

The collection of orchids, which is a fairly good one, cannot, in the nature of the plants, be presented quite in the same way (although an attempt at it is made with some species at, for example, the *Estufa Fria* in Lisbon). And in Munich the nepenthes and other insectivorous plants are not as well presented as at the Berlin-Dahlem Botanic

Below left Flowering trees and walks planted with thousands of herbaceous species and bulb flowers make Munich's garden in spring one of the most colourful in Europe. Although every species is present for a scientific or practical reason, the quality of the actual gardening is superlative.

Below and opposite More examples of Meissen ornaments in the Botanic Garden, Munich.

Garden. But the beautifully planted little moss-collection house is surely the best thing of its kind in Europe. The bog plant displays are good and the collection of under-water plants in aquaria excellent.

The *Victoriahaus* is broad, spacious and light, covering an elegantly shaped pool of water-lilies and other aquatics, the star exhibits being the *Nymphaea lotus*, and the two huge water-lilies *Victoria cruciana* and *V. amazonica*. A display of this kind has become a sort of botanic garden *cliché* since Chatsworth and Kew began it, but the pool at Munich is planted round, as nearly after nature as possible, one interesting species of this planting being *Mimosa pudica*. And in May the house was prettily decorated with a show of *Malpighia* with rose and silver leaves.

The open garden is divided into two main areas; a central area on strictly formal lines, as rectilinear as any sixteenth-century Italian or Dutch botanic garden; and an outer area, planted on woodland garden or wild garden lines, which is at once an arboretum and a series of collections of hardy plants arranged geographically. Apart from the fact that there is always, throughout the growing seasons, a part of the Gardens devoted entirely to garden plants, for the benefit of the local amateur gardeners, the policy of the Munich garden is very strictly 'botanical', which means that only botanical species are planted, cultivars being wholly excluded; that correct associations of plants are respected as far as possible; and that this means that the geographical arrangement is firmly adhered to. And this is as it should be for, as will appear, botanical geography is a science which we owe primarily to German botanists.

In the wild part of the garden, then, each geographical region is represented by a sort of wilderness of herbaceous, annual and bulb plants, of shrubs and trees, native to the region in question. This needs not only stricter, but more skilful work on the part of the gardeners than would appear at first sight to be the case, for it is all too easy for seeds to be transported by one agency or another from, say, the European to the North American region, or from the North American to the Japanese. Moreover labelling presents a great difficulty, not to mention identification of plants which have lost their labels; and in any teaching garden, labelling is, of course, very important. Labelling at Munich is very well maintained and a member of the scientific staff, Dr Kress, is chiefly engaged, all the time, in the correct identification of plants.

In the Arboretum proper the trees stand in grass which is kept down, and they are arranged botanically, not geographically. Considering that most of these trees are young their stature is remarkable; some of the specimens of *Picea* and *Abies*, notably, have attained a fine size in the few decades of their lifetime. This is the more remarkable when it is considered that so poor is the soil of the vicinity of Munich that much of the garden soil had to be brought from other regions and is, in a sense, artificial.

The heather and juniper garden is not a great success; difficulties have been experienced with both genera, and

the junipers particularly are poor. On the other hand, the collections of sand-dune plants and of the flora of the Eurasian steppe are botanically interesting and socially valuable. Botanical gardens and the research laboratories and studies associated with them have worked on the problem of sand-dune planting to hold the dunes from blowing and shifting for at least three centuries, and no complete answer to the questions posed has yet been found.

Several of the collections in the open garden are so charmingly planted and their ground so effectively land-scaped that they are of quite as much interest to the gardener as to the botanist. The *Farnschlucht*, for example, the Fern Gorge, is a little sunken walk beside a small stream whose steep and rocky banks are planted with hardy ferns and with those flowering plants properly associated with them and with those conditions. As the walk broadens from point to point there are some fine groups of *Trillium* and of *Erythronium* in variety, of *Hylomecon japonica* and of *Primula*, a genus of which Dr Kress, the garden's taxonomist, is making a special study, although as a botanist of course, not as a gardener. Another very pleasing feature is a bank-sided, winding walk, again all on a small scale, called the *Frühlingsweg*, the Spring Walk, which is planted with all the herbaceous and bulb flowers native to Germany and flowering in May.

The Alpine Garden is a very important feature of this

botanical garden. Made on a group of artificial rocky hills, it is again divided into geographical regions and strictly confined to botanical species, cultivars being excluded at the expense of colour but to the benefit of botanical science. To a gardener it is strikingly different from a gardener's (as distinct from a botanist's) rock-garden; one realizes how much, after all, we owe to the plant selectors and breeders who have heightened the floriferousness and colour of our garden plants. The comparison was particularly instructive because we had come almost directly from the enchanting little Alpengarten im Belvedere in Vienna, to the Munich garden.

But botanically the very large collections of alpines, each in its geographical place, at Munich are of the greatest interest and all the art of advanced rock-gardening has been used in the display and presentation of the plants. An Alpine House, with the two benches landscaped in miniature and planted as a picture of nature idealized, completes the Alpine collections with a large number of exquisite and difficult rarities.

THE FORMAL SYSTEM GARDEN

The formally laid-out parts of the garden, the large rectilinear area which is surrounded on three sides by the wild garden, arboretum and alpine garden, is distinguished by some unique decorations, large coloured figures of parrots with fruit, and of boys, in the famous Nymphenburg pottery. They are used to decorate the pillars at the heads of stairways and edges of terraces. The plan of the 'system' garden is very well ordered for teaching, with the most primitive plants at the centre, the most evolved at the periphery, and the connecting link plants between them. It is, as it were, possible to stand and survey the process of plant evolution, each stage being represented by a few hardy genera from each epoch for as far back as there are surviving species.

Left The Palm House and range of connected hot-houses in the Botanic Garden, Munich. German greenhouse building tends to massive rather than graceful, but the plantings are the most imaginative and meticulously tended in Europe.

Left A fine crop of insectivorous *Sarracenias*.
Centre Tillandsia, superbly grown at Munich.
Right Dicksonia tree ferns, palms and giant bromeliads in one of the houses.

The rectangular gardens devoted to special purposes, such as these demonstrations of systematic taxonomy and of evolution, are separated by hedges and stone pergolas overgrown with a good collection of climbers. As well as the system garden, there is one devoted to each of the Botanical Institute's two special subjects of research and study, genetics and ecology, and one to hardy economic plants. Although all these parts of the garden are strictly scientific, even so, art has been called on in the planting; for example, the rectangular beds are edged with clipped box in the traditional knot-garden style; or, by way of a change, with clipped, dwarf *Berberis thunbergii atropurpurea*—incidentally a most successful dwarf hedge.

Because Dr Merxmüller holds that it is very important for a botanical garden to attract public interest and sympathy and so public support, one area of the garden is always devoted to collections, prettily planted for display, of good garden plants, each group of cultivars being labelled so that visiting amateurs can make notes on the plants they would like to grow in their own gardens. The garden thus does on a small scale what Wisley does for British gardeners on a big one. This may be why the small private gardens of Munich look better planted and kept than those in most of the German towns one passed through. Munich's Botanical Garden, with its ten or twelve thousand species planted with art and cared for with love, is a model of what a botanical garden associated with a scientific institute but also serving the people of a great industrial city for their recreation, should be.

In many countries the development of botanic gardens into gardens of recreaction has meant the running down of scientific work and teaching. Like Kew, the German gardens have been successful in maintaining their scientific and teaching work in gardens which nevertheless are planted, at least in part, entirely to please the taste of the ordinary citizen who wants a colourful park.

BERLIN-DAHLEM

The Berlin–Dahlem Botanical Museum, as it is now called, was founded as such by Heinrich Friederich Link in 1815. But Link was building on the magnificent beginning made for him by his predecessor, one of the greatest men in the history of botany, Carl Ludwig Willdenow. And Willdenow, although not the first, was the first effective maker of the Berlin Botanical Garden.

There were some flower collections in the royal garden of the Hohenzollerns, the Lustgarten, as early as 1646. And from 1679 there was what seems to have been an old-fashioned physic garden on the site of the present Kleistpark.* But that garden does not qualify as 'botanical': it included a hop-garden for the royal brewery, an orchard, a kitchen-garden and, during the reign of Frederick I, a pleasure garden. For two years, 1713 to 1715, it had an active director, Dr Gundellsheimer, who tried to give the garden some scientific and educational quality; thereafter it was little more than a plantation of medicinal herbs for the use of the Court apothecary, and so neglected at last that the greenhouse began to fall down and wild pigs broke through its fences. Johann Gottlieb Gleditsch, a distinguished physician, made an effort to restore the gardens, but as he could get 'neither obedience from the gardeners nor the money due from the Academy of Sciences' (Eckhart), he did not accomplish much. So that when, in 1801, Carl Ludwig Willdenow was elected director of the gardens he found a wilderness of weeds and only twelve hundred species of plants including those weeds. In the following eleven years he restored the garden to order, raised the collections to 7,700 species and created what was soon to be and has since remained, both on the old garden-sites and in Dahlem, one of the most important botanic gardens in Europe.

*The authorities for the dates, and the sources for most of what appears in this chapter other than the parts derived from my own notes taken in the gardens, are Ignatius Urban's *Geschichte Des Königlichen Botanischen Museums Zu Berlin-Dahlem 1815–1915* (Dresden, 1916); and Dr Theo. Eckhart's lecture, reprinted in the journal *Willdenowia, 150 Jahre Botanisches Museum Berlin*, 1 December 1966.

Willdenow was born in Berlin in 1765 and he lived there all his life. His father was an apothecary and also studied medicine and became qualified as a physician; but he never practised medicine; he inherited his father's shop, the Red Eagle Apotheke, and drew an income from it until, in 1810, he sold the business. Meanwhile under the tuition of his uncle, Gottlieb Gleditsch, he had become a skilled botanist. He began to teach medicine and botany, giving private lessons, and in 1798 was appointed teacher —one can hardly say professor—of Natural History at the Berlin College of Medicine and Surgery. This appointment must have been due, at least in part, to his publication in 1792 of an *Outline of Botany* which was frequently reprinted, and translated into English, Danish and Russian. In 1794 he had been elected a Fellow of the Berlin Academy of Sciences, the body responsible for the Botanic Gardens; and thus it was that he became their director in 1801.

Nine years later Berlin University was founded and Willdenow was made its first Professor of Botany. A number of his pupils became men of consideration, but one of them was among the greatest scientists of all time, Alexander von Humboldt; Humboldt's teacher, Willdenow may be called the grandfather of botanical geography.

As Director of the Botanical Gardens Willdenow was active and ingenious: it was he who first had pools made for the cultivation of water-lilies and other aquatics; who built new glasshouses for tropical plants; who first undertook the cultivation of a collection of Alpine plants. He formed connections with twenty-four European botanical or natural history academies and institutions and contributed to the transactions of most of them. He made the Gardens of use in the fields of pharmaceutical botany, agricultural botany and forestry. In his day the sexuality of plants was still in dispute; he proved it by a series of classic experiments with dioecious plants, and notably by successfully fertilizing a female *Chamaerops humilis* in Berlin—curious how important that palm has been in

Below left The Berlin–Dahlem Botanic Garden is remarkable not only for its fine living collections but for landscape gardening. A view of its 'Italian Garden'.

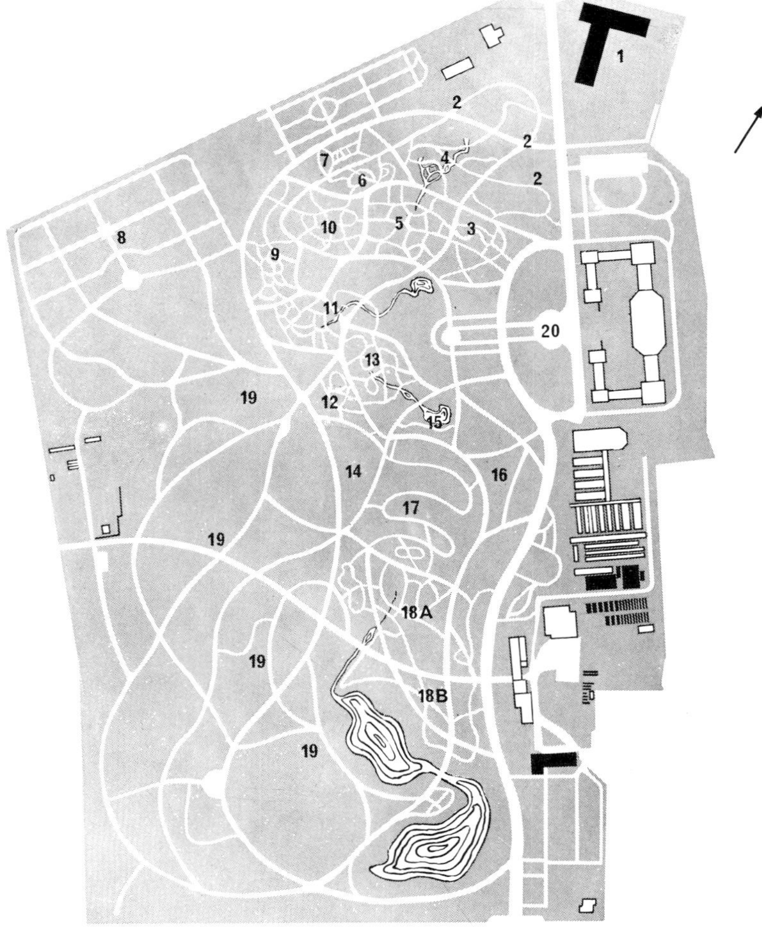

Key to geo-botanical and other divisions

1 Botanical Museum
2 German Forest flora
3 Pyrennean flora
4 Alpine foothill flora
5 Alpine flora
6 Sudetenland flora
7 Scandinavian flora
8 Systematic garden
9 Balkan flora
10 Carpathian mountain flora
11 Caucasian mountain flora
12 Altai flora
13 Himalayan flora
14 Siberian flora
15 Chinese plants
16 Cape plants
17 Japanese flora
18A Pacific North American plants
18B Atlantic North American plants
19 Arboretum
20 Italian garden

fundamental botany—with pollen from a male one at the old Leipzig Botanic Garden. His Herbarium collection of over 20,000 specimens was of the greatest value to systematic botanists and so were his written works, notably his *Hortus Berlinensis* and the even more magnificent *Enumeratio plantarum horti regi botanici Berlinensis.**

In his own science Willdenow's name is commemorated in the genus *Willdenowia*, so called for him by Carl Thunberg. He had described three of its species in 1790. Thus he had his monument; he died in 1812.

Willdenow's place was not filled at once, but in 1815 Heinrich Friederich Link, founder of the *Schaumuseum* as such, became Director of the Botanical Gardens. During his term—1815 to 1851, the longest in the Museum's history—the Botanical Gardens, the Royal Herbarium, and the Library were combined into a single institution, although the site of the gardens was twice moved (as it was a third time, into the old Schönberg gardens, in 1880; and finally to Dahlem).

*During World War II the Berlin-Dahlem Museum and the Garden were almost totally destroyed by bombing and infantry fighting. Those who subsequently helped to clear up the mess and bury the dead, including many Russians, which lay about the grounds, will not soon forget it. By a sort of miracle of luck most of Willenow's Herbarium material was saved. It included prototypes used in the beginning of the Eichler–Engler system of classification. Prototypes—the dried specimens from which plants have been described for the first time (in this case by Willdenow in his *Species Planatarum Linnes*, his *Hortus* and his *Enumeratio*) are important because description without the specimen described is very often too slight for correct identification. Among Willdenow's Herbarium material are over 3,000 specimens sent to him by Alexander von Humboldt from his great tropical expedition of 1799–1804.

Link was a very versatile scientist, researching and writing not only in botany, his principal subject, but in physics and chemistry, in pharmacology and *materia medica*, and in pure philosophy. Within the limits of botany he was equally versatile, being at once a taxonomist, an anatomist and a physiologist. He established about a hundred new genera and still more species, but many of these have disappeared, for he had the vice, common to so many botanists, of excessive genus-making and species-making. But many of his genera survive, notably *Penicillium*. Link was no mere scientist, and if the Museum and Gardens prospered under him as a great institution, it was as much because he was an admirable man of business and, in his social life, a man of wit and fashion whose sharp tongue and sardonic, very Berlinish, humour were relished.

Link was also a good judge of men, and many of his colleagues, chosen by him, were distinguished in some field or other. Diederich von Schlechtandal, Curator of the Herbarium, founded the journal *Linnaea*; the poet and satirist Adalbert von Chamisso was one of the Museum's best plant-collectors abroad, being, like Goethe, a botanist; Carl Sigismund Kunth, a pupil of Willdenow and Link's vice-director of the Museum, was remarkable for a colossal work, the seven-volume *Nova genera et species plantarum*, with seven hundred fine plates, all based on the material collected by the Humboldt–Baupland expedition. Huge though it be, the book is not

Berlin–Dahlem, through the 'Italian Garden' towards the Palm House. Jets from the Templehof Airport soar up over the big conservatory every few minutes. The greenhouses contain fine collections of *Sarracenias*, *Nepenthes* and other carnivorous plants, and good representative collections of tropical succulents including cacti. As at Munich, arrangement and association of plants are imaginative.

Carl Ludwig Willdenow, who made Berlin-Dahlem one of the most important gardens in Europe.

Below A garden pavilion.

Right Drawings of fruits of *Colutea, Paeonia* and *Hyoscyamus* ssp. made in the Berlin–Dahlem botanical laboratories.

Kunth's only work. His own Herbarium of 44,500 specimens, bequeathed to the Museum, was destroyed by British bombs in 1943.

Link was succeeded as Director by Alexander Braun, an original taxonomist and a good gardener; he extended and improved the gardens and at the same time worked at a system of plant classification which, published after his death by his pupil Ascherson, became the basis of the later Eichler–Engler system. When Braun died in 1877 he had been projecting a new Museum building; it was carried out by his successor August Eichler who, co-author of the Eichler–Engler system, published a work which is still of value to botanical taxonomists and morphologists, his *Blüthendiagramme*. But Eichler died prematurely of leukaemia and was succeeded by one of the most interesting men in the history of botany.

Adolf Engler was one of those German *savants* of the nineteenth century who undertook and accomplished the colossal task of bringing good order out of the chaos of new knowledge which had been accumulating faster and faster for three centuries. It is not clear that he made any particular impression on the botanical gardens in his charge; but his *Naturlichen Pflanzenfamilien* and his *Das Pflanzenreich* are great standards in the science of botany and so is his *Der Vegetation der Erde*. Almost as important as a service to science was Engler's appointment, as vice-director, of a man as remarkable as himself, Ignatius Urban. Urban put the flora of the West Indies in order and, working with Martius, did a like service for the South American flora in Martius's *Flora Brasiliensis*. Moreover it was Urban, rather than his Director, who, through his friendship with a Minister, Friederich Althoff, had the gardens moved from Schönberg to their present site in Dahlem.

Many who worked at the Museum under Engler contributed importantly to the advancement of botany: Schweinfurth, the explorer, in his work on African plants; Paul Hemmings, the mycologist; Georg Hieronymus, with his work on the Argentine flora; and Otto Warburg with his three-volume *World of Plants*. Two of Engler's lieutenants succeeded him as Directors, Ludwig Diels, who had to keep the gardens going during the terrible years of the inflation, only to see his work destroyed by the bombs of World War II; and Robert Pilger, who stepped into the task of rebuilding and replanting which was continued by Johannes Mattfeld and Eric Werdenmann, and is now being completed by Dr Theo. Eckhart.

THE GARDENS NOW

As in Munich, the Berlin policy is to serve scientific botany first but to attract the general public to the gardens as well and to be of use to amateur gardeners in manner, although not on the scale, of the British Royal Horticultural Society's Wisley garden. In the Museum associated with the gardens, too, it is the director Dr Theo. Eckhardt's policy to make the study of economic botany easy and pleasant for schoolchildren and enquiring adults as well. A part of the gardens is devoted to displays of garden plants of all kinds and at all seasons, so that Berlin's amateur gardeners have a permanent exhibition from which to draw ideas for their own gardens.

The plan for the purely botanical part of the garden was drawn up by Engler (see above), and it has been so strictly adhered to that very few botanical gardens, perhaps only Munich's, have so perfectly reliable a geographical layout. As at Munich, outside the special area devoted to garden plants and horticulture, only botanical species and varieties are planted, cultivars being excluded. The climate of Prussia is harsh, and therefore only those regions whose winters are as cold or colder and whose summers as hot and dry, can be represented in the open garden. These are northern Europe, notably Scandinavia and Germany itself, with a miniature *buchenwald*; Asia, notably the colder climates of Japan, northern China and the Eurasian steppe; and North America. Small sections, however, are planted to represent the florae of Central America, South America and Australasia with such few species as can stand the cold of a Berlin winter or can easily be taken under glass during the cold weather.

As at Munich, again, the five geographical regions represented are planted in the manner of a woodland or wild garden (English influence at work here), threaded by a maze of small walks so that every part can be seen from close-to. The rock-garden of alpine plants is different from Munich's in that the genera are grouped according to the kind of soil they prefer, and not geographically. The collection of alpines is a very large one and pleasingly planted on an artificial rocky hill with a mountain stream beside which semi-aquatics native to the various European mountain ranges are grown.

At the centre of the garden, maintained purely for recreation and repose, is a so-called 'Italian' garden which, however, is not really very Italian. True there are clipped yews and clipped box, but the lawns are more English than Italian; and still more German in being prettily decorated with geometrical patterns in massed pansies, etc.

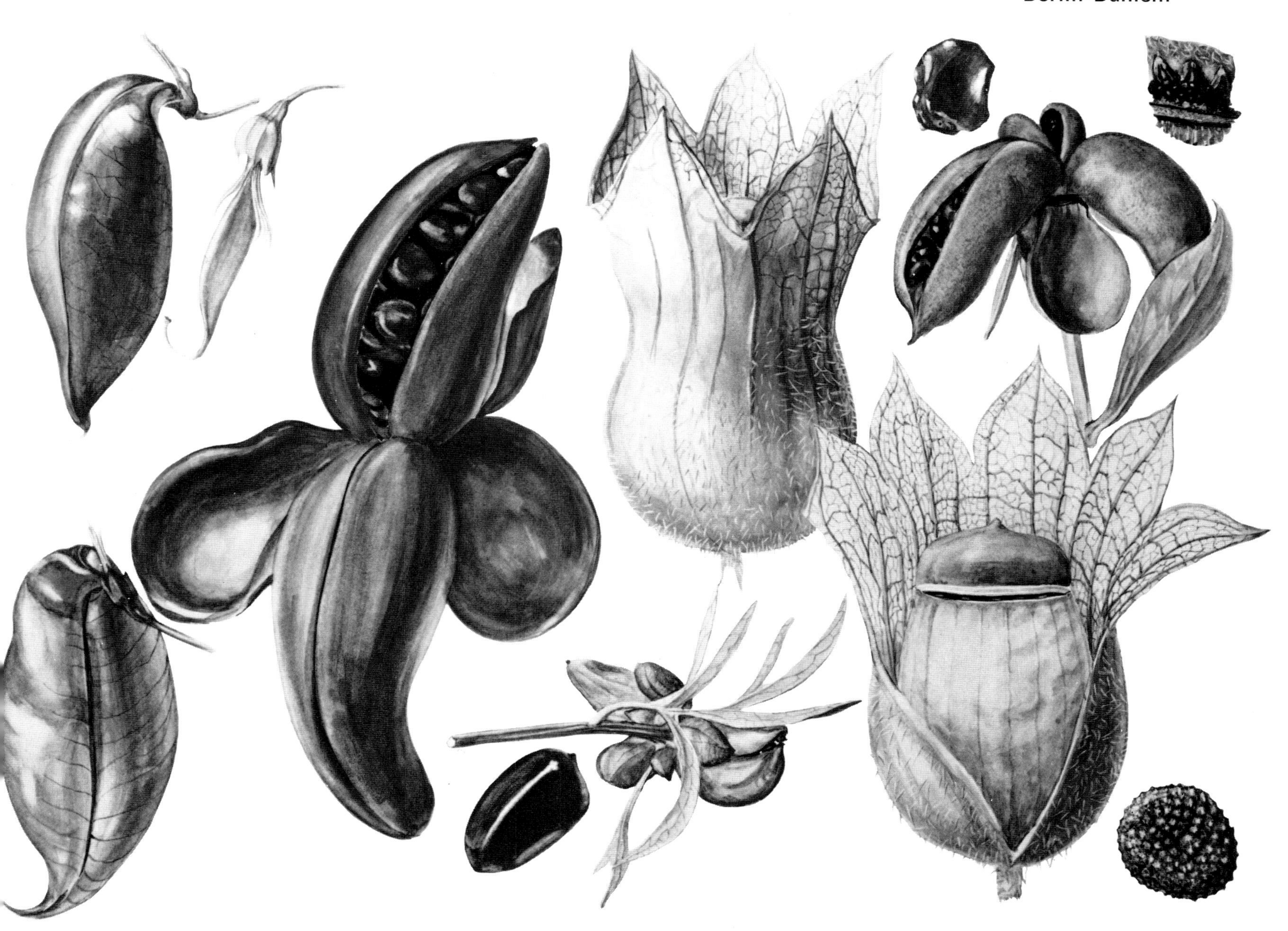

The geographical divisions described above are themselves subdivided, and each small division is devoted to the plants proper to that sub-region or micro-climate. Again, I doubt whether any garden in the world is better designed than the Berlin–Dahlem one for the study of the botanical geography which its botanists 'invented'.

The gardens have a large and well-planted complex of greenhouses, stoves and conservatories but the rebuilding of the colossal *Palmenhaus*, which was destroyed by bombing, was only just completed in the spring of 1967, and planting was just beginning. This palm house is one of the largest glasshouses in the world but certainly not one of the most beautiful: the metal work is too massive for beauty, and could, one would have thought, have been considerably lightened in the rebuilding since not glass but one of the clear plastics was being used for the glazing. In those houses already planted the Munich plan is followed but not, I think, quite so pleasingly as in that impeccable garden. But with the Munich house and the mighty Palace of Plants at Meise as models, something even finer may be accomplished at Berlin–Dahlem when the *Palmenhaus* replanting is mature.

In some respects the under-glass collections at Berlin–Dahlem are better than those of Munich; the range, diversity and presentation of succulents and cacti are superlative, but above all, the collection and display of insectivorous plants—*Sarracenia*, *Dionaea*, *Darlingtonia*, etc., and especially *Nepenthe*—are outstandingly good.

From what we have said of its history it will be clear that the Museum part of the museum-and-garden, the *Schausmuseum* complex, has always been important at Berlin–Dahlem. What is now admirable there is that the same policy of instruction for the general public, and above all for schools, is followed in the Museum as in the Gardens themselves. The exhibition includes a series of set scenes, skilfully illuminated, three-dimensional pictures let into the walls, each representing a particular flora in a natural scene, Northern Tundra, for example, or Alpine, or Arid Tropical, and, among the most fascinating, some submarine, and sublacustrine florae. Very attractive, and stimulating to the imagination, is again, the representation of the life of a giant sequoia tree depicted in historical scenes occurring at various points in its life: at one point Alexander is invading India; at others great wars are being fought, great inventions made, tremendous ideas expounded. The visitor is made to see with his own eyes and to realize vividly that there are trees alive whose youth was passed when the pyramids of Egypt were being built and which will yet outlive our present civilization.* Prehistoric florae as reconstructed by palaeobotanists, and the nature and processing of economic plants are demonstrated in admirably conceived *tableaux*.

*There is probably no living sequoia more than 3,000 years old; but some *Pinus aristata* in southern California, ring-counted from microscopic cores taken by boring, are as much as 4,700 years old.

JARDIN DES PLANTES

According to Monsieur M. Y. Laissus, Archivist, Librarian and Palaeographer of France's Muséum National de l'Histoire Naturelle, the oldest botanical garden in France is Montpellier's, which, he says, was founded by Henri IV in 1593.* The same author notes some older but private foundations in Paris and elsewhere, probably dating from the 1570's, by which decade Pisa and Padua had set a fashion for planting such gardens of plants. Coming to our own time, the *International Directory of Botanical Gardens* lists, under FRANCE, small, highly specialized gardens at Alford (Vetinerary School); Antibes (Vegetable Pathology); Besançon (University); Caen (Municipal and University); Col du Lauteret (Alpine Trees); Dijon (Local Flora); and thirteen other gardens all with a special purpose and all under ten acres, some much smaller. In addition to these there are the Jardin des Plantes in Paris, associated with the Muséum Nationale de l'Histoire Naturelle, one of the world's greatest scientific institutions; and the very remarkable garden of Monsieur Julien Marnier-Lapostolle at Saint Jean-Cap Ferrat.

*See his monograph in *Revue de l'Enseignement Supérieure*, No. 2, 1962. The Ecole de Medicine of Paris did, however, have a Physic Garden of some kind as early as this, in charge of Jean Robin, the celebrated gardener, who also had a private botanical garden on the point of l'Ile Nôtre-Dame in the Seine.

THE JARDIN ROYAL DES PLANTES MÉDICINALES

The garden which we now call the Jardin des Plantes dates from the year 1635 and is, therefore, about a decade younger than the Oxford Botanic Garden (see below). The prime maker of this world-famous garden was Louis XIII's physician-in-ordinary, Guy de la Brosse, who, by skilful and persistent lobbying at last extracted an edict from the government, in January 1626, for the planting of a state Physic Garden. After another seven years of lobbying and persuading, de la Brosse contrived to get the site purchased, and in 1635 obtained another edict

Left From a seventeenth-century print. The Jardin du Roi, which after the Revolution became the Jardin des Plantes, was originally a Physic Garden laid out in formal parterres.

Right Bernard de Jussieu, one of a score of great botanists, naturalists and biologists who were trained or worked at the Jardin des Plantes.

which made it possible to start building and planting. De la Brosse received, of course, the appointment as Intendant of the Garden, which was to be known as the *Jardin royal des plantes médicinales*; and his head gardener, or curator, was Vespasien Robin, Robert Morison's mentor in horticulture and son of the great Jean Robin, whose Ile Nôtre-Dame garden supplied the new royal garden with some of its first plants, including a specimen of the genus which bears his name, *Robinia* (*pseudo-acacia*). That tree, in its third century, is still flourishing.

From its foundation until about 1718 the Jardin royal was primarily, if not exclusively, a Physic Garden. Its Intendant was, automatically, whoever happened to be *Premier Médecin* at the Court. Other officers of the institution were likewise all physicians. As a teaching garden, all its alumni were medical or pharmaceutical students; in short, it was much like its Edinburgh contemporary. In one respect the policy of the Garden was liberal, the lectures delivered by its officers were free, open to the public, and—what is very remarkable for the time—not in Latin but in French. This liberalism was due to de la Brosse, and as a result of its tremendous success it roused the bitter hostility of the Faculty of Medicine of Paris. The Faculty did all in its power to hamper the growth of the garden as a medical school. It raised objection after objection to the Jardin royal in the *Parlement* of Paris—all of which were in vain because every time the Faculty forced through a resolution against the Jardin royal, it was immediately vetoed by the Crown.*

The first outstandingly able administrator of the Jardin royal, after de la Brosse, was the founder's great-nephew, Guy-Crescent Fagon. Not only was he able, as a Paris Faculty member and *Premier Médecin* of the Dauphine and the royal children, to force the Faculty to recognize the Jardin royal's *jubilé* (Licence), but he made the Jardin royal into a real scientific centre. It was Fagon who found and installed Tournefort, one of the greatest taxonomic botanists; Vaillant, one of the pioneers of experimental botany and the earliest exponent of sexuality in plants; and Antoine de Jussieu, first of an illustrious family of botanists. Fagon's activity enriched the Jardin's herbarium with all Tournefort's personal collection and several others; and raised the number of species cultivated to 4,000.

A notable distinction of the Jardin royal is the collection of what are called *vélins*, now in the care of the Muséum Nationale de l'Histoire Naturelle. These are exquisitely executed flower paintings on parchment (see p. 85). The nucleus of the collection had been inherited by Louis XIV from Gaston d'Orléans (Robert Morison's patron at Blois; see below), for whom the flower portraits had been painted by Nicholas Robert, the King's miniaturist-in-ordinary. When the collection was placed in the care of the Jardin royal, Nicholas Robert continued to work there, adding to the *collection des vélins* portraits of all the rarest and most beautiful of the plants of the Jardin royal. When Robert died in 1685 the work was carried on by a succession of other flower painters, Jean Joubert, Claude Aubriet, Magdeleine Basseport, Gerard van Spaendonck; none of them attained the perfection of Robert's work.

From early in its history the Jardin royal and the schools which grew up in association with it were more concerned with other sciences—chemistry, anatomy, zoology, mineralogy—than botany. In our context we shall have to ignore this part of the Institution's work.

An important development on the botanical side is reflected in the change of name in 1718 from *Jardin royal des plantes médecinales* to *Jardin royal des plantes*. From that time neither Intendant nor staff had to be medical men—Antoine-Laurent de Jussieu, for instance, was a pure botanist, and neither Lamarck, Lacapède, Buffon or du Fay were physicians.

Du Fay, Intendant in 1732, was influenced by the de Jussieus; he built the garden's first big hot-houses and he made the Jardin royal into a real botanical garden by throwing it open to all species of plants whatsoever. But probably du Fay's most important work for the garden and for scientific botany was that of nominating, to be his successor as Intendant, one of the greatest natural scientists of all time, Buffon.

Buffon remained Intendant for fifty years, so that not even the Hookers at Kew broke his record. Under his very authoritarian and *ancien régime* style rule† the botanists Lemonnier and A-L de Jussieu and the *professeurs de culture*‡ Thouin, father and son, continued to improve and

*Laissus, *op. cit.* This remarkably bitter conflict was aggravated by the fact that no 'foreign' (i.e. provincial) physician had the right to practise in Paris excepting by royal appointment. De la Brosse was a Montpellier graduate and hence anathema; and later Intendants of the royal garden were often 'foreign', for the Montpellier medical school was a famous one. The Faculty's active hostility seriously hampered, though it failed to ruin completely, the Garden.

†The man we know as Buffon was Seigneur de Montbard, marquis de Rougemont, vicomte de Quincy, seigneur de la Mairie, Les Harens et Les Berges et autres lieux . . . l'un des Quarante de l'Academie française . . . (Laissus) . . . perpetual treasurer of the Academie royale des sciences, Fellow of the Royal Society, member of the Academies of Berlin, St Petersburg, Bologna, Florence, Edinburgh and Philadelphia.

‡These teacher-gardeners would have been called curators in England, *Hortulani* in Italy and Holland.

A *vélin* by Pierre Redouté (1759–1840) of *Cactus speciosissimus*. from the former Royal Collection at the Jardin des Plantes.

embellish the Garden, whose area of 20 acres Buffon struggled for thirty years to enlarge. In 1771 he succeeded in getting his progressive policy accepted; by 1787 the area of the Jardin royal had grown to 40 acres and although part of the new territory was covered by museums, herbaria and schools, the plants too received a share of the extra space. De Jussieu and Thouin were able to treble the number of living species in their charge and to reorganize the whole Garden, at the same time going over to the Linnaean system of nomenclature and to Bernard de Jussieu's natural system of classification.

The *ancien régime*, in the Garden as well as in the nation, more or less ended with Buffon's death. His two successors, both noblemen, did nothing and the second one's term ended, very much in the spirit of the times, with an (albeit bloodless) revolution, in the Jardin royal: it seems to have been led by the *professeur de culture* (Head Gardener), Thouin, and Daubenton, author of the great *Prodromnus systematis naturalis regni*, but they were supported by the whole staff. The epithet 'royal' was dropped from the garden's name so that it became simply the Jardin des Plantes; a democratic government of the Jardin by a committee of all its professors and departmental heads was substituted for Buffon's monarchy; the Intendant became a sort of chairman and his new title was Director. This revolution was confirmed in the decree of 10 June 1793 which established the Muséum Nationale de l'Histoire Naturelle. Twelve professorships were established by the same decree, but in practice this simply confirmed the twelve Jardin royal professors in their respective chairs.

Thus A-L. de Jussieu remained professor of Botany, Thouin, professor of Gardening. Their colleagues were a brilliant set, including the botanist and minerologist Daubenton, the great Lacapède, and the still greater Lamarck, first of the evolutionists. Moreover the men who were to work in the Museum and Jardin des Plantes in the first half of the nineteenth century were, virtually, the founding fathers of the modern natural sciences: Cuvier, for example, and Saint-Hillaire; Brongniart founded palaeobotany in the Jardin, and Naudin, genetics. Claude Bernard, the physiologist, and Becquerel, the physicist, were both Muséum Nationale men, and in the arts the Jardin des Plantes fostered such botanical painters as Redouté and Huet.

The Revolution enriched the Jardin des Plantes not only with men and a better constitution, but with books and material: most of the looted private collections ended up there, thus enormously enlarging the Library and the Herbarium. Moreover, even the collections of living plants benefited largely:

André Thouin, professor of cultivation, busied himself seeking rare trees in the gardens of prisoners and proscripts, and having them moved and transplanted. He himself, and his successors, distributed seeds of esculent and ornamental plants *gratis* to the four quarters of France . . .*

*Laissus, *op. cit.*

Kew's first great period of overseas plant-collecting initiated by Banks had its equivalent at the Jardin des Plantes: the voyages of discovery made by Freycinet (1817–20), by Duperrey (1822–5), by Dumont d'Urville (1826–9), by Vaillant and by Dupetit-Thouars resulted in its botanical enrichment. And there was a second phase of such activity, equivalent to the Hooker phase at Kew, in the *Traveilleur* and *Talisman* voyages of 1880 and 1883. But whatever may have been the French results in discovery and the advancement of knowledge and science, they were not at all equal horticulturally to those of Kew. Nor does the Jardin des Plantes seem to have played so important a part in economic botany as did Kew under, for example, Thistleton-Dyer: there is really nothing equivalent to the introductions of rubber and chocolate from West to East, or Forrest's introduction of tea from China to India.

The reason, of course, is obvious: France had nothing like Britain's overseas empire and it was that empire which so vastly enlarged the ideas of English and Scottish economic botanists.†

There is, in the botanical work of the Jardin des Plantes, an elegance, a freedom from practical considerations which were always wanting at Kew. But it was Kew that did more for both farmers and gardeners.

But the greatest difference between these two great botanical gardens, or rather between their respective achievments, is in horticulture rather than in botany. While botany progressed at Paris, ornamental horticulture did not. At Kew, while botany progressed, so, strikingly, did gardening, until Kew became what is now one of the world's most beautiful and richly planted gardens whereas the Jardin des Plantes is, today, a museum-piece, pretty and interesting of course, but, simply as a garden, without distinction. It is no doubt significant that the art of making gardens is the only one in which the English surpass the French.

But the past must not be forgotten; they were Frenchmen at the Jardin des Plantes rather than Britons at Oxford, Chelsea and Kew, who founded modern botany. To quote M. Laissus once again:

It is now more than three centuries since the Jardin royal was founded on the very spot where the Muséum Nationale de l'Histoire Naturelle continues to live and work, three centuries rich in event and episode, in research and controversy, in learning and triumph, rich above all in men, scientists whether illustrious or relatively obscure whom time, little by little, now carries into its long darkness. And yet, on the very ground which formerly they trod, there still quivers the broken shade of Robin's *Robinia* which in the spring of each new year burgeons greenly yet again.

It is doubtful whether any similar institution in the history of European civilization contributed more to the forwarding of the natural sciences, led by botany, during the period of their foundation, than did the Jardin des Plantes.

†Though the Brussels Jardin Botanique was responsible, for example, for introducing better coffee plants to the Congo.

LES CEDRES

Les Cèdres, the greatest garden of succulent plants in the Old World, commands some of the loveliest views in Europe. Monsieur Marnier-Lapostolle's collection of *Cactus* is rivalled only by that of the Huntington Botanic Garden, near Los Angeles, California.

Below The Villa built for King Leopold I of the Belgians at Les Cèdres.

I think it could be shown that if, in terms of pure science, the Jardin des Plantes has been by far the most important of France's botanical gardens, the economic work done by the Jardin de Botanique of the University of Montpellier since its foundation in 1593 has been of much greater importance. The part played, for example, in introducing viticulture to the Americas, South Africa and above all Australia has had tremendous consequences for the world's trade in wine. It was from Montpellier that Busby took cuttings of over three hundred varieties of grape vines to Australia. But when we come to pure botanical horticulture, to collections of rare and beautiful plants, there is a private garden which far surpasses any institutional garden in France and most institutional gardens elsewhere. This is the botanical garden of Les Cèdres at Saint Jean-Cap Ferrat, the very remarkable creation of Monsieur Marnier-Lapostolle, latest in the great line of amateurs who have, from time to time, outdone the professional botanical gardeners.

Monsieur Marnier-Lapostolle's father bought Les Cèdres, a handsome Italianate villa built in the third decade of the nineteenth century, in 1922. It had about 15 hectares of land on a site commanding a magnificent view of one of the world's most beautiful coasts, views of vividly blue sea, of wooded, hilly capes, of pretty little bays and ancient towers and churches—views, in short, of high civilization's most pleasing achievement on one of nature's loveliest sites. The soil of this land is thin but fertile and it responds well to improving cultivation. The mean annual rainfall is about 36 inches. The sun shines all the year round. In midsummer the temperature may well rise to the tropical level; we experienced day after day well up in the eighties Fahrenheit and passing into the low nineties on occasion. In midwinter there may be an occasional slight touch of frost, and a thin and ephemeral sprinkling of snow is not unknown. Bananas are grown commercially by at least one cultivator near Menton, and from there to Nice in one direction and beyond Imperia in the other, the climate is the best in Europe.

Monsieur Marnier-Lapostolle began to plan his garden before his father died; planting actually began in 1923. Les Cèdres shows two distinct interests, the purely botanical and the horticultural. But the botanical clearly predominates, and although there are many acres of 'romantic' woodland-garden plantings with rare and beautiful material, even there and in the beautiful little lake garden the botanical requirement is allowed to predominate over the horticultural—for example in the very prominent labelling of the aquatic plants in the lake.

Les Cèdres, then, must be considered under two heads, that of the purely botanical collections; and that of what we would call the garden and what the French call the *parc*. The two sectors, comprising 15,000 species of plants, are cared for by twenty-seven gardeners under the general direction of the proprietor Monsieur Marnier-Lapostolle, and the immediate supervision of the curator, Monsieur René Hebding, whose own specialities are Orchids and Bromeliads.

A fine cast-iron garden ornament? No; a living plant in the Les Cèdres cactus garden, St Jean-Cap Ferrat.

THE COLLECTIONS

In two families of plants Les Cèdres is world-famous—*Bromeliaceae* and *Euphorbiaceae*, and some other succulents. In the cases of both families the specimens are displayed in two ways; simply as collections—rows and rows of thousands of pots in greenhouses, frames, slatted houses and other kinds of covered or semi-covered quarters; and as material in schemes of ornamental planting in the open garden.

Not being professional botanists, we can say little about the pure collections except that the collection of Bromeliads is the most complete in the world and the collection of succulents probably the best. Botanists from all over the world come to Les Cèdres to study material which they cannot find, alive, anywhere else.

But enormous numbers of these plants are used ornamantally in parts of the *parc* and in such conditions that they can be more easily admired and appreciated by gardeners. There is, for example, a quite extraordinary garden of *Cactus* which is called the Mexican garden because its plants are predominantly Mexican, although not exclusively so. The fantastic forms assumed by succulent plants, notably in the genus *Euphorbia* but also in, for example, *Monadenium arborescens*, a cactus with an umbrella-like head of leaves, in *Dorstenia gigas*, an immense green coral, in the extraordinary branchless *Euphorbia columnaris*, in the *Stapelias*, in the fantastic *Hodias*, in the *Echinocacti*, even in the *Aeoniums* and *Sempervivums*—these fantastic forms make possible compositions which, sometimes vaguely reminiscent of the work of Burle Marx in Caracas and elsewhere, realize a kind of gardening which is proper to the spirit of the other arts in our time.

The Mexican garden at Les Cèdres is infinitely richer in species, and hence in texture, than the Monagesque *Jardin Exotique* and there is nothing planted in the open anywhere in Europe which can possibly be compared with it. But that is by no means all that has been achieved with succulents at Les Cèdres: to take only one of a hundred surprises, chosen at random, the spectacle of night-flowering *Cereus* species climbing to the full height and completely embracing the trunks of giant palms as grown at Les Cèdres is surely a unique sight in European gardens.

But the marvels among the succulents are endless—the extraordinary *Euphorbia abdelkawi* from Socotra; *Adenia globosa*, a vast and amphorous lump of tissue with a tangled mane of cactus-like tentacles—one could go on for ever about these succulents.

It is more surprising to see Bromeliads from the special collections being used ornamentally in the *parc*, than succulents. Even some terrestrial and epiphytic orchids are so used; *Cypripediums*, for example, are very successful, hardy down to freezing point, and experiments are being made in establishing epiphytic orchids.* Bromeliads, epi-

*It is worth noting that M. René Hebding is emphatic about this degree of hardiness since it is generally accepted that about 45°F is the low limit for these orchids.

Below A few of the hundreds of cactus species in one of the greenhouses at Les Cèdres.

Below *Sempervivum* species at Les Cèdres. The collection of Old World succulents is as remarkable as the Cactus collection.

Bottom right One of the Palm Avenues at Les Cèdres. Epiphytic plants are cultivated on the Palm trunks and many Bromeliads and Orchids survive the Mediterranean winter.

phytic in nature, have been established on trees, palms and tree-ferns in the *parc*. Among the genera so represented are *Vriesia*, *Guzmania*, *Aechmea* and *Tillandsia*. Some of the curtains of *Tillandsia* species hanging from poles or branches or even stretched wires are a magnificent sight when vividly in flower. The *Tillandsia* collection includes very fine specimens of the rare *T. xerographica*. Another extraordinary Bromeliad is *Abromeitella lorenziana*, which is a great dense cushion of short, stout, dagger-shaped leaves.

Before coming to a general impression of the botanical *parc*, here are a few of the strikingly remarkable plants which we starred in our notes during our visits: *Echinoccacti*, 4 feet in diameter*; a coppice of *Dendrocalamus giganteus*—growing from the ground each season these colossal bamboos, the largest of the grass family, attain a

*Not, however, the largest in cultivation on this coast where there is a nursery with some centenarian specimens which are even larger and are perhaps the first introduced into Europe.

Les Cèdres is one of the very few European botanic gardens in which the Lotus (*Nelumbo* ssp.) and the great Amazon water-lilies (*Victoria* ssp.) flourish in the open.
Below Fruits of *Nelumbo*: seeds of this plant known to be 3,000 years old have been germinated by botanists.

height of 27 metres (90 feet) in three months, a growth rate of about a foot a day or half an inch an hour. Then, *Jacobina* in flower, *Aloes suzianiae* in flower—this aloe flowered for the first time in Europe at Les Cèdres in 1963, an event made the more important by the fact that there are only thirteen plants left of it in the wild. Then, again, *Camphoro-carpus fruticosa;* and the climbing Aroid *Rhadophora decursiva,* whose gigantic flower, like that of some related giant Aroids, becomes hot as it opens: the spathes of some species reach 50°C (122°F) but this Indo-Chinese *Rhadophora*, which is not unlike a *Monstera* to look at, had its spathes at 27°C when its temperature was taken.

We were much struck with the size attained by *Ficus hookeri*; by the fact that *Impatiens marianae*, from Assam, a plant which is supposed to need hot-house conditions, here flourishes better in the open. The asparagus fern we are familiar with as a florist's plant reaches 20 or more feet at Les Cèdres, clambering up trees and palms in the *parc*. The *Encephalartos* and other *Cycads* we know in the Kew greenhouses are used at Les Cèdres in ornamental plantings. There are fine trees of the wonderfully decorative *Sterculia discolor* from Australia, up to 40 feet and, when we saw them, covered with their rose-pink flowers. Among climbers, the passion-flowers are very fine. But for our own taste perhaps the two most beautiful trees in the *parc* are *Eucalyptus ficifolia* in vivid, scarlet flower; and a magnificent specimen of *Cupressus cashmiriana*, lovliest of cypresses.

THE LAKE

That part of the *parc* which is closely associated with the villa is formal and Italianate, as is the architecture of the house itself. There is a complex of grand stairways, marble ornamented terraces, water stairways and small ponds, rising from the low level of the carriage drive to the forecourt of the house. And associated with this is a pretty and richly planted little lake in a stone-edged basin, of fanciful shape and sufficient extent to be impressive. It is encircled by a footpath and the surrounds are well planted with a fine collection of exotic shrubs and herbaceous plants.

The water itself carries a large collection of water lilies and other aquatic plants, all strictly 'botanical', and these include *Victoria amazonica* and *Victoria cruciata*, which grow and flower very well at Les Cèdres, as well as the lesser water-lilies in cream, yellow, red, rose-pink and blue. There are many other aquatic plants including, for example, *Taro*, and the dreaded water-hyacinth which is so beautiful in flower. But above all there are magnificent clumps of the incomparably lovely *Nelumbo indica*, both the rose-pink and the white.

THE WILD

A considerable area at Les Cèdres is planted as woodland garden in the manner, but with totally different and very exotic material, of an English 'wild' garden. It is planted after nature, but idealized, since the maze of small paths is scrupulously maintained and the 'jungle' they thread

thoroughly gardened. Here are sub-tropical and even tropical trees, their trunks bearing many epiphytes, of tropical and sub-tropical shrubs, of giant Aroids, some of them climbing the trees, of huge palms and tall tree-ferns, *Dicksonia* and *Cyathea*. The ground is covered with shade-loving species, from *Aspidistra* and *Tradescantia* to several kinds of Orchids, *Begonia, Impatiens*, and many lovely ferns.

In other parts the planting of shrubs and trees has been done in wide borders beside broad, straight drives; and the associations of plants there are more formally botanical than horticultural. There are, to give a few examples, complete collections of *Acacia* and some species of true *Mimosa*;* of the *Myrtaceae*; and of the *Proteaceae*. Some of the sub-tropical species which one might, from one's knowledge of the Abbey Gardens at Tresco, and of the Irish gardens, expect to find at Les Cèdres, do not grow well there simply because the atmosphere is too dry for them; *Embothrium*, for example, and *Telopea* and even *Eucryphia*. But other southern hemisphere genera flourish such as *Callistemon, Melaleuca, Banksia* and *Protea*.

It has been possible to give only the sketchiest idea of this very remarkable private botanical garden. It is a grand work of both science and art which Monsieur Marnier-Lapostolle has accomplished; and it is perhaps appropriate that France, which, despite advancement in social legislation, has remained, in its higher classes, a nation of fiercely independent individualists, should have what is frankly a deficiency in good institutional and state gardens made so very good by a private one. Les Cèdres makes to practical botanical gardening in the present a contribution by no means unworthy of the Jardin des Plantes' magnificent contribution to the field of academic botany in the past.

*The plant which florists and laymen call 'mimosa' is, of course, not a *Mimosa* but usually *Acacia dealbata* or some other *Acacia*.

AUSTRIA

BELVEDERE

Two botanic gardens in Austria are of special interest: they are not the only ones in the country—Graz, Innsbruck, Vienna University and St Veit have good small gardens. But the Bundesgärten in the glorious 400-acre park of the Schönbrunn Palace and the Alpengarten im Belvedere, both in Vienna, are outstanding, each in its kind. Both are under the general direction of Ingenieur-Diplom Franz Kargl and Dr Erhard Spranger. The Schönbrunn garden sets the example for those botanical gardens which find that their scientific purpose has vanished and which seek a means of survival in serving commercial horticulture by means of their command of horticultural botany. The Alpengarten im Belvedere, whose Curator, Herr Ferdinand Hodak, is a remarkable gardener and plant collector, is a model rock-garden of great beauty in a small space.

This Alpengarten im Belvedere is the oldest alpine garden collection in Europe, and since, despite the beautifully contrived ornamental planting of its specimens its collection of plants is made on scientific lines, it is a true botanic garden. Associated with it in the grounds of the seventeenth-century Baroque palace of Belvedere are a number of greenhouses and stoves which fall rather into the Schönbrunn than the Belvedere field of work. These, with heating, irrigation and ventilation automated almost as thoroughly as in the Schönbrunn garden, contain interesting collections of Proteaceae—*Banksia, Grevillea, Hakia, Callistemon, Leucadendron*, and some others—each in several species; species of tender *Ericae*, including many Cape Heaths; a collection of *Croton*; and a very good collection of *Philodendron* and related genera. The collection of Bromeliads is exceptionally rich in *Vriesia* and *Tillandsia*—in the course of raising these genera, some good new mutants have been selected for propagation. Sections of these Belvedere greenhouses are devoted to the plants required for botanical teaching. Experimental horticulture is confined to testing species and varieties for suitability as house plants.

The Alpengarten proper, alongside the greenhouse complex in the Belvedere park, derives from a botanic garden which was laid out and planted at Obere Belvedere for the Emperor at the instance of Thomas Host, his physician and botanist-in-ordinary, by Gerard van Swieten, and Herman Schuurmans in the last years of the eighteenth century. This garden must have been of considerable interest and even rather unusual in its day, because far from being a mere medicinal herb garden it was not only a true botanical garden but a deliberate attempt to display a complete *Flora Austriaca*; it is said to have included specimens of every plant native to the Austro-Hungarian monarchy in the early nineteenth century, a claim which must doubtless be accepted with reserve. But this garden was not the earliest botanical garden in Austria, for in 1803 the Archdukes Johann, Anton and Rainier von Hapsburg had had a botanic garden planted on the Gloriette slope of the Schönbrunn park. Why these three princes were interested in planting such a garden is not on record. The work was done for them by Heinrich Schott, curator of the Imperial Gardens. No trace of this royal garden is now to be found below the Gloriette, and when Schott died in 1865 the plants from the garden were transferred to Obere Belvedere, where Franz Maly was busy replanting the old garden there.

There is, in Franz Maly's work on gardens, a faint and rather curious echo of certain kinds of Chinese, later Japanese, gardening. In his rock work the intention is clear to represent symbolically certain kinds of alpine peaks and sheer rock faces, but the resemblance to Oriental work is probably a chance one. Maly was far in advance of the rock gardeners of, for example, the Edinburgh Botanic, at about the same time, and not until Sir Isaac

Below left No botanic gardens in the world have such settings of baroque magnificence as Vienna's gardens in the Belvedere and in Schönbrunn.

Below Set among baroque works of art and architecture the Alpengarten im Belvedere offers entertaining contrasts between Nature's works and man's.

Bayley Balfour redesigned the Edinburgh rock garden was any work in that field equal to Maly's. And this was so, even though Maly was influenced by 'Gothic' Baroque, so that by comparison with later, naturalistic, styles in rock gardening some of his work at Belvedere, still preserved, looks grotesque. Moreover some of it is very clumsy; for example, his practice of making little 'swallow's nests' of stone and mortar on his rock faces, to hold some of the plants.

Although the Schönbrunn *Flora Austriaca* garden, or rather its plants, had been shifted to Obere Belvedere, the only person really interested in botanic gardening was Maly himself. In botany and gardening, as in the arts and sciences generally, the Hapsburgs were but languid patrons, and until World War I Austria *was* the Hapsburgs, and what failed to interest them interested nobody who counted. Thus when Franz Maly died his garden was neglected, and by 1920 only 500 species of his large collection of alpines remained.

Then, in 1926, a man of parts with power to get things done took an interest in the Belvedere garden: Hofrat Franz Matschkal, director of the State Gardens, or Federal Gardens, as they had come to be called and which included Schönbrunn, decided to restore and enlarge the Obere Belvedere gardens and gave the job to Ingenieur Hubert Martin. He was to enlarge the garden, re-lay the rockwork, and plant it. Martin was responsible for the garden as we see it now. He himself went on plant-hunting expeditions and other plant collectors were employed. Seeds and plants were procured from other botanic gardens, and in due course a system of exchange of plants with many foreign institutions was set up and has been operated ever since. At the present time Herr Hodak has such exchange arrangements with 210 alpine garden societies and institutions all over the world.

By 1935 the Alpengarten had been brought to a high standard and had a great reputation among botanists and gardeners everywhere. But during World War II it was again neglected and, moreover, damaged; and by 1945 it had become a wilderness. Once again Hofrat Franz Matschkal came to the rescue and made Herr Adolf Munsch Superintendent of the Garden with the task of reconstructing its rockeries and restoring the plant collection. The work was well and quickly done and by 1947, fully restored, the Alpengarten was again opened to the public.

The garden is small, only 2,047 square metres, but in that exiguous space are grown approximately 4,500 species and varieties of alpine plants, representatives of all the world's alpine florae. The planting is superlatively good, and the maintenance meticulous, it is as once a valuable botanical collection and a model of good alpine gardening style. The genera which are particularly well represented are *Saxifraga* (350 species, varieties and cultivars); *Dianthus* (146); *Iris* (97); *Campanula* (80); *Anemone* and *Pulsatilla* (77); *Gentiana* (72); *Sempervivum* (70); *Sedum* (48); *Penstemon* (68); and *Arenaria* (40). The nature of the soil makes it easier to grow lime-loving than

General view of the Alpengarten im Belvedere. A number of aquatics have been cultivated here since the early nineteenth century.

Opposite: Top Gentiana cult. 'Belvedere'. A superb form of *Gentiana acaulis* selected and propagated in the Alpengarten. *Bottom Pleione limprichtii*, one of the alpine orchids superlatively grown in Vienna's Alpengarten.

calcifuge plants, but there are fine examples of the very dwarf alpine rhododendrons, some enkianthus and other lime-haters doing well in deep pockets of pure peat. But on the whole the Curator favours plants from high alpine, arid, low-rainfall regions because of the Viennese climate, which is hot and dry in summer, very cold in winter. So excellent is the quality of this garden that the following table of climatic statistics will be of interest to alpine garden enthusiasts everywhere:

ALTITUDE 552 FEET ABOVE SEA LEVEL

	Temperature (°C) (*max*)	(*min*)	*Humidity* (%)	*Sunshine* (*hours*)	*Precipitation* (*mm*)	*Wind* (*m/sec*)
January	16·7	−21·9	79	56	39	4·6
February	19·5	−25·8	76	81	39	4·7
March	24·0	−16·1	71	135	44	4·6
April	27·3	− 5·2	65	173	54	4·5
May	32·6	− 0·3	67	238	69	4·2
June	36·1	3·9	67	246	70	4·3
July	38·3	9·2	68	265	84	4·4
August	34·2	7·6	70	242	70	4·1
September	31·6	− 0·1	74	184	51	3·9
October	27·8	− 8·7	79	118	55	3·9
November	20·2	−10·9	81	58	51	4·4
December	17·1	−19·6	82	41	50	4·4

The garden has in its collections a number of plants which are very rare in cultivation, either because they are rare in the wild or because they are excessively difficult to grow. Among them are *Adonis amurensis, Ericaceae pungens, Rosa watsoniana, R. persica, Jankaea heldreichii* (which is found only on Mount Olympus) and *Genista holopetala*. Some of the oldest plants are truly magnificent specimens, a *Genista horrida* for example, which closely covers a number of large rocks. A big group of *Cypripedium calceolus* commonly bears fifty flowers at a time. The very large stock of *Pleione limprechtii*, a far more richly coloured species than the much commoner and hardier *P. barbata*, makes an astonishing show in early May. Perhaps the most striking flower in the garden, though to the gardener rather than to the botanist, is the Alpengarten's own cultivar of *Gentiana acaulis*, cv. 'Belvedere', a free-flowering plant with by far the largest and most richly blue trumpets, for the species, that I have ever seen: it was hoped that some plants for distribution would be available in a couple of years.

The Alpengarten im Belvedere specializes, as did its predecessor in the early nineteenth century, in the *Flora Austriaca*, or rather in the alpine section of that flora. But since it is a model for alpine gardeners—over three hundred amateurs whose admirable work we saw displayed collaborate with the garden in an association of its 'Friends'—as well as a botanical collection, the garden could not confine itself to Austrian plants. There are hundreds from all Europe's mountain ranges, from the Americas, from China, Japan, Manchuria, Persia and the USSR, and a few score from Africa and Australasia.

ALPENGARTEN
BELVEDERE
PLEIONE
LIMPRICHTII
47

SCHONBRUNN

Opposite and below When Austria wanted a national botanic garden her botanists found the nucleus of it ready-made in the Schönbrunn park, where generations of Hapsburgs had collected exotic trees and shrubs.

The great park of Schönbrunn was laid out and planted for the Hapsburgs by Adrian van Steckhoven, who began the work about 1700. It was completed later in the century by another Dutchman, Gerard van Swieten, who was assisted by Herman Schuurmans, van Steckhoven's son-in-law and business successor, who (see above) changed his name to Steckhoven and, in 1814, at the age of sixty-two, became Hortulanus of the Leyden *Hortus Academicus*. It was Herman Schuurmans Steckhoven who, under

van Swieten, planted the *Hortus Botanicus* mentioned above. As far as we can discover, there was no trace left by then of the physic garden which Clusius planted for the imperial family before the end of the sixteenth century. But it is at least possible that Schuurmans Steckhoven simply replanted that garden, for the pharmacy which sold drugs derived from the imperial physic garden appears to have been older than Schuurmans Steckhoven's garden.

However, it was doubtless due to Schuurmans Steckhoven that the garden in the Schönbrunn park was not a mere physic garden, which was probably what Thomas Host, the emperor's physician-in-ordinary, required, but an attempt, as we have already said, at a *Flora Austriaca* garden.

Although one part of the Schönbrunn park is planted and designated as a botanical garden, just as another part is laid out and stocked as a zoo (with, by the way, the best small, modern aquarium we have seen in Europe), it is of no particular interest in the context of this book; in any case, it is almost true to say that since the park's thirty-five gardeners are directed and superintended by the botanists of the Bundesgärten's *Verwaltung*, the whole park could be thought of as 'botanical'.

It is a pity that there is no tree list for the park; it would be interesting and valuable, as there are scores of magnificent trees; and since a certain number of new tree species were introduced to Europe by way of these imperial gardens, from Constantinople, some of the trees should be of historic interest. The fact remains that because the real work in economic and horticultural botany carried on by the *Verwaltung* is concentrated in the huge complex of glasshouses, we can here confine ourselves to some considerations of those glasshouses, their contents, and the work being done there.

The Austrian State has given the *Verwaltung der Bundesgärten* four assignments, four kinds of work to do in the enormous complex of palm houses and stoves and greenhouses inherited from the imperial past; it must provide some material for pure research, though it is not concerned with taxonomic or experimental botany; it must carry out experimental and plant-breeding work in decorative horticulture; it is engaged in the production of house plants on a very large scale, commercially, and it has to produce all the growing and cut flowers for state

Vienna's principal botanic garden is as it was when Clusius planted the first Physic Garden for the Hapsburgs, a part of the great Baroque garden of the Schönbrunn palace. The 'Roman ruins' are bogus, a relic of late eighteenth-century romantic gardening.

Below The great greenhouse range at Schönbrunn. Many of the tropical and sub-tropical trees, palms and shrubs originally collected for the pleasure of the imperial family still survive. The foreground is still recognizably 'Le Nôtrian'.

institutions, offices, occasions and ceremonies. If the Bundesgärten can be said to specialize in a particular branch of botanical gardening, it is that of display.

There are thirty-eight glasshouses in the complex and both display and service houses are worked by automation to an extent which is unusual. The Vienna Bundesgärten is very advanced in this respect. The big electrical plant, water-heating, purifying and mixing plant, ventilation and humidity-control plant and so forth are housed in extensive concrete galleries under the houses; our impression was that they are as sophisticated as anything of the kind to be found anywhere in the world. In the automation of ventilation, both by the control of fans and the opening of lights, and even the shifting of movable shading, all movable parts are controlled by both thermostats and light-sensitive devices. Everything possible has been done to keep the staff free to do the kind of work which no machinery has been devised to do.

Some of the experimental work being done is of great interest, for example, in the cold storage of flowers; in the experiments on growing flowers entirely by means of artificial light; in the effects of different light colours and spectra on plant growth and bud formation; in the control of light and darkness rations to bring various kinds of florists' flowers into bloom out of season. In the propagation department the meristem-cutting method was being used and experimented with not only on many genera of orchids but on other plant families. Meristem cuttings—microscopic cuttings taken from the few cells of the extreme growing tip of a plant—are not only free from any virus the plant may be infected with but make possible the raising of a very large number of plants from a very small amount of material.

Orchids are also raised from seed after cross-fertilization, in the search for good new cultivars. Mass propagation of anthuriums and other aroids, of cyclamen,

gerbera and some other ornamentals, in on a large scale.

Among the display houses the Mexican is the most impressive, maybe because a Hapsburg was once Emperor of Mexico. It is very rich in old and free-flowering specimens of *Selinicereus* and other night-flowering cacti. Many scores of the huge, scented flowers sometimes open on a single night. Among the most interesting plants in other collections are *Welwitschia mirabilis, Cissus juta* and *Dioscorea elephantipes*.

There are also some very remarkable *Fockea capensis* plants, the only ones in Europe, acquired for the imperial gardens in 1685, which means that the plants are nearly three hundred years old. Some other relics of the old Hapsburg stove and greenhouse plants are interesting, notably a small collection of *Passiflora* cultivars, some of which may, we believe, be lost elsewhere.

Separating the greenhouse complex from the old Orangery is a vast service yard, formerly the imperial flower garden, flanked by the covered, winter riding-school and quite given over to plant propagation for the supply of the Schönbrunn park itself and other State parks and gardens. There are about 800,000 plants in production all the time. The Orangery (built in 1767) is still, and very efficiently, heated by the old wood-burning apparatus of imperial times and it is crammed with enormous old evergreen shrubs, myrtles, citrus species, etc., in huge tubs and boxes which can be moved outside, or to another place when required to dress a hall or staircase. There are seventeen species of these plants; no cultivars; some of them are surprisingly old for container-grown plants, notably some of the oranges, which are over two hundred and fifty years old.

In terms of pure botany the Bundesgärten in Schönbrunn is not important, and although teaching and research are two of its activities, the academic work is mostly left to the Botanical Department of Vienna University

Left The hedges, originally planted and trained to satisfy the 'French' canon of horticultural taste, survive for the edification of modern horticulturists and botanists.

Below Yesterday's pleasure, today's lesson. Schönbrunn, like no other garden, teaches how science and art should go hand-in-hand.

Bottom This South African *Fockea*, acquired for Schönbrunn in the eighteenth century, is believed to be the oldest living specimen of its kind—a claim denied by the botanists at Stellenbosch University.

Below Inside the big Palm House. Today we can be in the lands of Palms in a few hours; when Schönbrunn was planted, Palms were a marvel.

DER SCHÖNE BRVNNEN
DER SAGE NACH ENTDECKTE AN DIESER STELLE KAISER MATTHIAS (1612-1619) ANLÄSSLICH EINER JAGD EINE QVELLE. WEGEN IHRES GVTEN WASSERS WVRDE DIE QVELLE VM 1640 GEFASST VND GAB ALS „SCHÖNER BRVNNEN" DEM FVR DIE KAISERINWITWE ELEONORA VON MANTVA ERBAVTEN SCHLOSS SEINEN NAMEN. DIESE ERSTE ANLAGE WVRDE 1683 DVRCH DIE TVRKEN ZERSTÖRT. 1758 ERRICHTETE DER HOFGÄRTNER ADRIAN VAN STEKHOVEN VBER DEM „SCHÖNEN BRVNNEN" EIN „GROTTA HAVS", DAS 1771 IM AVFTRAG DER KAISERIN MARIA THERESIA VOM HOFARCHITEKTEN ISIDOR CANEVALE AVF DIE HEVTIGE FORM ABGEÄNDERT WVRDE. 1780 SCHVF DER SCHWABE WILHELM BEYER DIE MARMORFIGVR DER NYMPHE EGERIA.

which has its own Botanical Garden across the wall from the Belvedere Palace grounds.

But the Bundesgärten is important in terms of horticultural botany and it has a lesson to teach certain botanical and other institutional gardens in many parts of the world which are beginning to wonder about their future, and even whether they have a future now that scientific botany has moved wholly into the laboratory, that systematic botany can still rely on herbaria, and that economic botany has moved into the field of agricultural and horticultural research institutes.

Where money is not hard to come by, a beautiful garden needs no justification but its existence, but in other conditions it may have to earn its living to survive: the Schönbrunn Bundesgärten shows that this can be done.

OXFORD

The *Hortus Botanicus* of Oxford University.

It is significant of the pre-eminence of England's Royal Botanic Gardens that the name of a Surrey village, now virtually a London suburb, has become world-famous, and the syllable Kew very frequently on the lips of Chinese and Peruvians, Alaskans and Indonesians, botanists and gardeners in every country in the world. Significant, likewise, that even when so great an institution as the Göteborgs Botaniska Tradgard* has to revise the nomenclature of its collections, as it did in 1962, it relies principally on the *Index Kewensis*. But before we come to an account of that great garden of plants which is at the same time a noble work of art, something must first be said about the much older and historically important *Hortus Botanicus* of Oxford University.

The commonly quoted date for the foundation of this garden is 1632. In fact, however, the garden was founded in 1621. It is true that R. T. Gunther, the historian of the garden, has a number of his foundation dates wrong; for Padua he is twelve years too early, for Edinburgh ten years too late, and his Salernitan (1309) physic garden (of Mathaeus Sylvaticus) is not recognized as a botanical garden by any good Italian authority. But for the Oxford foundation he bases himself on Daubeny,† who quotes his source. It seems that, probably inspired by John Tradescant's garden of exotic plants in South Lambeth, Lord Henry Danvers, Earl of Danby, a gentleman commoner of the House, being minded to become a benefactor of the University bought out the tenant of five acres of meadow for £250 and arranged that the University lease this ground from Magdalen College for 40s. per annum.

The reason why 1632 is the date commonly given for the foundation of this garden is clear: the land acquired by Danby had not been sufficiently raised above flood level—'by the addition of 4,000 load of mucke and dunge laid by H. Windiatt the Universities scavenger'—nor the wall and gate finished until 1632 and 1633 respectively. Lord Danby went to some lengths to provide for the Garden: at his death in 1644 'it was found that he had devised to the University the rectory of Kirkdale in Yorkshire for the use of the garden'. In the event, this endowment proved quite inadequate, and no Professor of Botany or Keeper of the Garden was appointed until, in 1669, Dr Robert Morison applied for the place.

Morison's is one of the great names in the history of botany. An Aberdonian and a graduate of the University of Aberdeen, he had fought on the royalist side in the Civil War and had been obliged to take refuge in Paris. There the Jardin des Plantes enabled him to use his exile in the study of botany, anatomy and zoology. He took a medical degree at Angers in 1648. His master in botany was the King's Botanist, Vespasian Robin, a great gardener, and through him Morison obtained the post of curator of the Duc d'Orléans' garden at Blois, the garden which (see above) inspired Sibbald to found the first Edinburgh Botanic garden. It was at the invitation of Charles II that Morison came to England in 1660, as King's Physician and Keeper of the Royal Garden, for which he got a house and £200 a year.

Oxford gave Morison £40 a year to deliver lectures as and when he liked. John Evelyn heard one of them as late as 1675. It also gave him the chance to make the Garden as he wanted it; and to put his botanical ideas and learning into writing. His published works include a *Praeludia Botanica* (1669), two, out of a projected three, parts of an *Historia Plantarum Oxoniensium* and several other works including a 'Socratic' dialogue about botany. He made a special study of variegated plants; he was a severe critic of Ray's system of classification and author of a better one sketched in his *Dialogus*.

Morison seems to have been an exponent, if that be the right word, of the ancient northern Celtic practice of *brag*, which is misunderstood by other peoples as mere conceit. Linnaeus, too, says‡ that Morison was, 'vain, yet he cannot be sufficiently praised for having revived a system which was half expiring'. He adds that Tournefort owed much to Morison but that Morison owed all that was best in his work to Cesalpino.

As for his influence on the Garden itself it must to some extent have demonstrated and reflected his theory; as we have seen, its arrangement was sufficiently impressive to lead Paul Hermann of the Leyden *Hortus Academicus* to reorganize the planting of that ancient garden. But some at least of the Garden's quality was due rather to the Brunswicker, Jacob Bobart, who had been appointed *Horti Praefectus* in 1632, kept that office until his death, and published, 1648, the first catalogue of its collections, a list which shows that by then the Garden had 1,600 kinds of plants—*stirpes* is the word used, it means, roughly, *races*—and many varieties; for example, there were nine kinds of *Primula* including a blue primrose (the seed of which must have come from Persia).

The art of grafting was advanced under Bobart but less by him than by an amateur, a Fellow of New College named Robert Sharrock, who was the first man to graft European grape-vines on to American wild species of *Vitis*. Moreover it was from observations made in the Oxford garden that Sir John Millington developed an idea which foreshadowed the definitive demonstration of plant sexuality§ by Linnaeus. Aesthetically, the garden as Bobart and his son and successor made it was criticized for an excess of clipped yew and box, but even so Celia Fiennes, who visited it in 1695, declared that 'the variety of flowers and plants would have entertained one a week'. Yet when Zacharias von Uffenbach, who was in a position to make comparisons, saw it a quarter of a century later he not only thought Professor and Keeper Bobart's appearance rebarbative—'Bobart had . . . an evil appearance. His nose was unusually long and pointed, his eyes small and deeply sunk, mouth awry with next to no upper lip, a great, deep scar furrowed his cheek, and his face and hands as black and coarse as those of the veriest labourer. His clothes, especially his hat, were in a bad state, and his wife

*See pp. 126–30.
†Daubeny's *Popular Guide to the Physic Garden at Oxford*.

‡In a letter to Haller, undated but *c.* 1737.
§Millington seems to have recognized that pollen was the male seed.

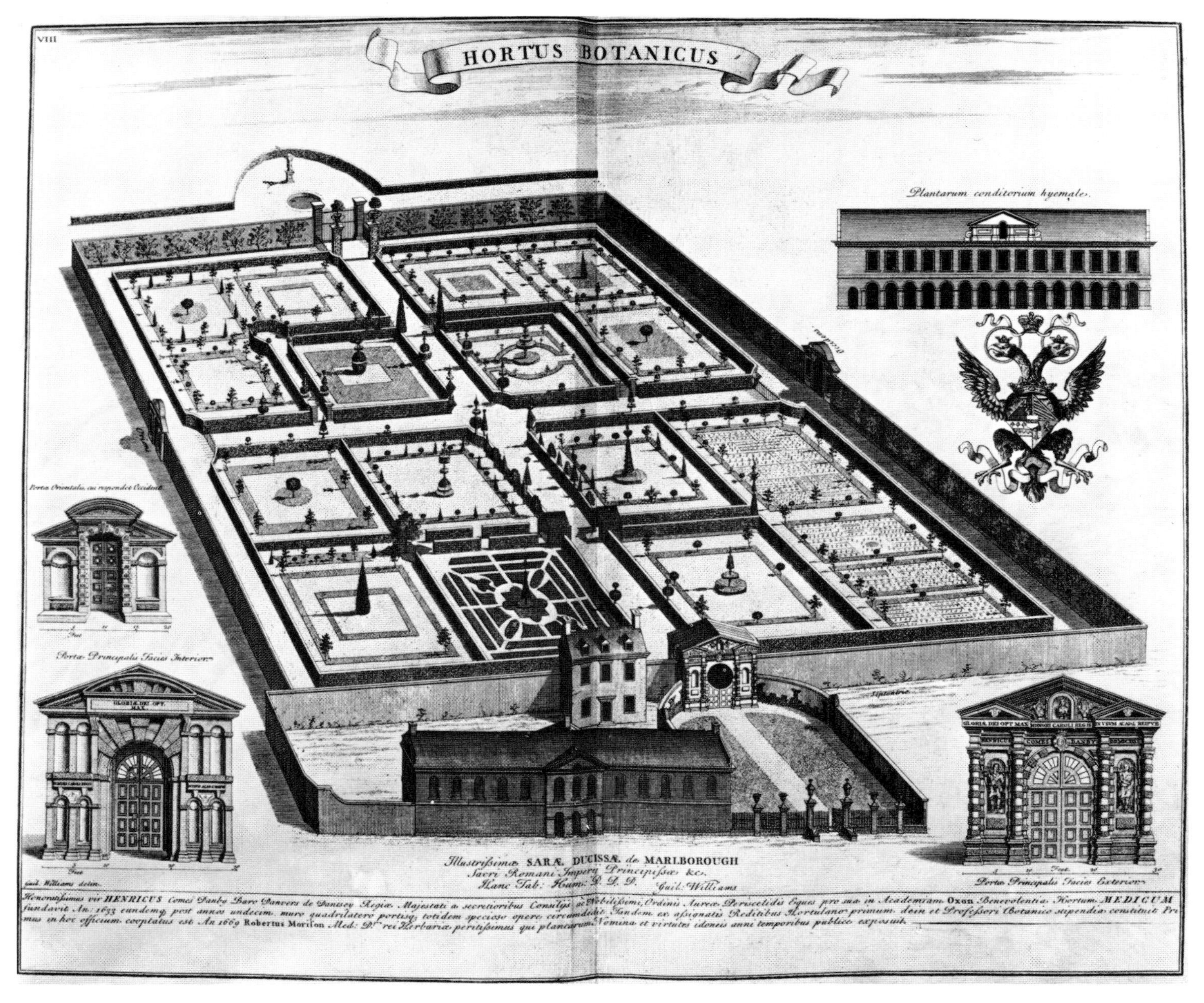

who accompanied him was old and dirty'—but says that

The collection was not equal to that of Leyden or Amsterdam. The arrangement was nothing remarkable but somewhat irregular. . . . Bobart complained that two years previously he had lost many plants through frost, to which Herr D. Büttner answered very neatly that that cold winter had done much injury to Botany for it had not only removed very many plants but three eminent Botanists. Tournefort in Paris, Hotton in Leyden, and Trionfetti in Rome.

After a rather bad spell, from 1726 onwards the Garden was immensely improved by the learning and generosity of an amateur botanist, Dr William Sherrard. He numbered not only Paul Hermann but Boerhaave and Tournefort among his friends and fellow workers; he had collected plants and made herbarium material in Greece and Anatolia; he had persuaded the great Doctor Dillenius to come to England and made him Superintendent of his own garden at Eltham in Kent, where the German made a *Hortus Elthamensis* for his patron. His benefactions to Oxford consisted in first presenting it with £500 to build a new conservatory, then presenting the Garden with many rare and curious plants; then presenting the Library with equally rare and curious botanical works; and finally making over his own herbarium whereby, according to Linnaeus, he made Oxford pre-eminent among all the universities of Europe for the study of botany. Nor was that all; when Sherrard died in 1728 he was found to have left the University £3,000 to endow the salary of a Professor of Botany, but on two conditions, that the University was to find £150 a year towards the proper upkeep of the Garden; and that Dillenius be the first Sherrardian Professor. These conditions were accepted by Convocation.

Dillenius did a very great deal to improve the Garden. In 1736 he was visited by Linnaeus, and although disagreeing with his system he conceived an almost passionate friendship for the great Swede,* finally offering him half his salary and half his house if only he would remain at Oxford, and weeping when Linnaeus refused.

When Dillenius died of a stroke at the age of sixty (he had become excessively fat), he was succeeded as Sherrardian Professor by one Dr Humphry Sibthorp of Magdalen, whose remarkable achievement it was to hold the Chair for forty years and in that time to deliver one 'not very successful' lecture; he did absolutely nothing else and yet he even contrived to 'bequeath' the Chair to his son John. That turned out to be a great service to his University and his only one to botany, for John Sibthorp was a notable botanist and a considerable plant collector.

Although the Oxford garden was to rise to greater heights in botanical horticulture, notably during the professorship of Dr C. G. B. Daubeny, the English contribution to the advancement of the art and science of plants was already being better made elsewhere—at Kew.

*Linnaeus named the genus *Dillenia* for his friend. For it was 'the most distinguished for its beauty of flower and fruit, like Dillenius among botanists'.

KEW

Opposite One of Kew's distinctions is the collection of over 800 plant paintings by Marianne North, in the Marianne North gallery bequeathed to the Gardens by the artist. An example of her work.

Below A plan of the gardens at Kew, 1763.

KEW: THE FIRST EPOCH

The history of Kew as a true botanic garden begins in 1759 with the appointment of William Aiton as Superintendent. But the gardens are older than that and something should be said of their origins before we return to a consideration of the Aitons, father and son; and of the prime maker of Kew, Sir Joseph Banks.

William Turner, the herbalist and 'father' of British botany, is supposed to have had a garden at Kew when he was physician to the Duke of Somerset at Syon House. But the first garden known to have been on the site of the Royal Botanic Gardens was the Kew House garden planted by Richard Bennet, who owned the house in the seventeenth century. Bennet's daughter, who inherited the house, had married Sir Henry (later Lord) Capel, who must have been an enthusiastic gardener, for there are several references to his garden in John Evelyn's *Diary*, where Evelyn refers to the Orangery, the Myrtetum and to some of the rare plants which he had seen at Kew. In a sense, then Kew Gardens were already 'botanical'. When Capel died in 1696 Kew House was inherited by his grand-niece, Lady Elizabeth Molyneux. When she, a widow, died in 1730 Frederick, Prince of Wales (the 'poor Fred' of the pamphleteers), who loved the house, leased it from the Capels and employed the artist, William Kent, first of the great series of talented garden architects who were to fill England with their lovely works during the next century, to lay out a new garden. So 1728 is the year when Kew Gardens became 'royal'.

In 1736 Frederick married the Princess Augusta of Saxe-Gotha—'long-nosed, long-necked and long-sighted'. The pair settled down at Kew, and although their life was notoriously stormy, it seems clear that they both adored their garden, the Princess in particular being a notable gardener and plantswoman. It so happened that they had a good friend who was not only a greater gardener than either of them but a very considerable botanist and in due course author of a curious botanical work probably based on the plants which his royal friends, under his guidance, cultivated at Kew.* This remarkable man was John Stuart, third Earl of Bute.

Plants were not the only things Bute raised at Kew; he had much to do with the education of the prince who was to be King George III. Later, as George III's Prime Minister he became one of the most detested men in England, but as a gardener and botanist he was admirable. His library included '. . . more than 300 folios strictly botanical and quartos and octavos in proportion . . .'.† He also (sure mark of a true botanist) had a very large private Herbarium.

Dr W. B. Turril, formerly Keeper of the Herbarium and Library at Kew Gardens, in his *The Royal Botanic Gardens, Kew*,‡ has no doubt that although the Gardens may have owed more to Frederick than has been acknowledged and certainly did owe much to Augusta, they owed far more to Bute, who continued to advise and help her with the Gardens after her husband's death.

The first catalogue of the plants in the Gardens was the *Hortus Kewensis*, made by Sir John Hill (1768); distinguished apothecary, he had some hand in managing the Dowager Princess of Wales's Garden.§ Another man of note who worked there was that singular dilettante and architect Sir William Chambers, who was employed by Augusta in about 1760.¶ Chambers had a mania, rapidly becoming fashionable, for buildings in alien, if not absolutely fanciful, styles, forerunning the 'romantic' revival, and he cluttered the Gardens with an Alhambra, a Mosque, a Gothic Cathedral, a Ruined Arch, a Pagoda (which set the fashion for chinoiseries) and an Orangery. By the merciful action of time and taste all but three (the

**Botanical Tables, containing the different families of British Plants distinguished by a few obvious parts of Fructification rang'd in a synoptic method.* Drawn and engraved by Johann Sebastian Muller, it is in nine quarto volumes and one of the rarest books in the world, for only sixteen copies were printed at a cost of £10,000.

†Smith, J. E. *Memoirs and Correspondence* (1832).

‡London, 1959.

§Turril, *op. cit.*

¶His folio of *Plans, Elevations, Sections & Prospective view of the Gardens and Buildings at Kew in Surrey* was published at Augusta's expense in 1763.

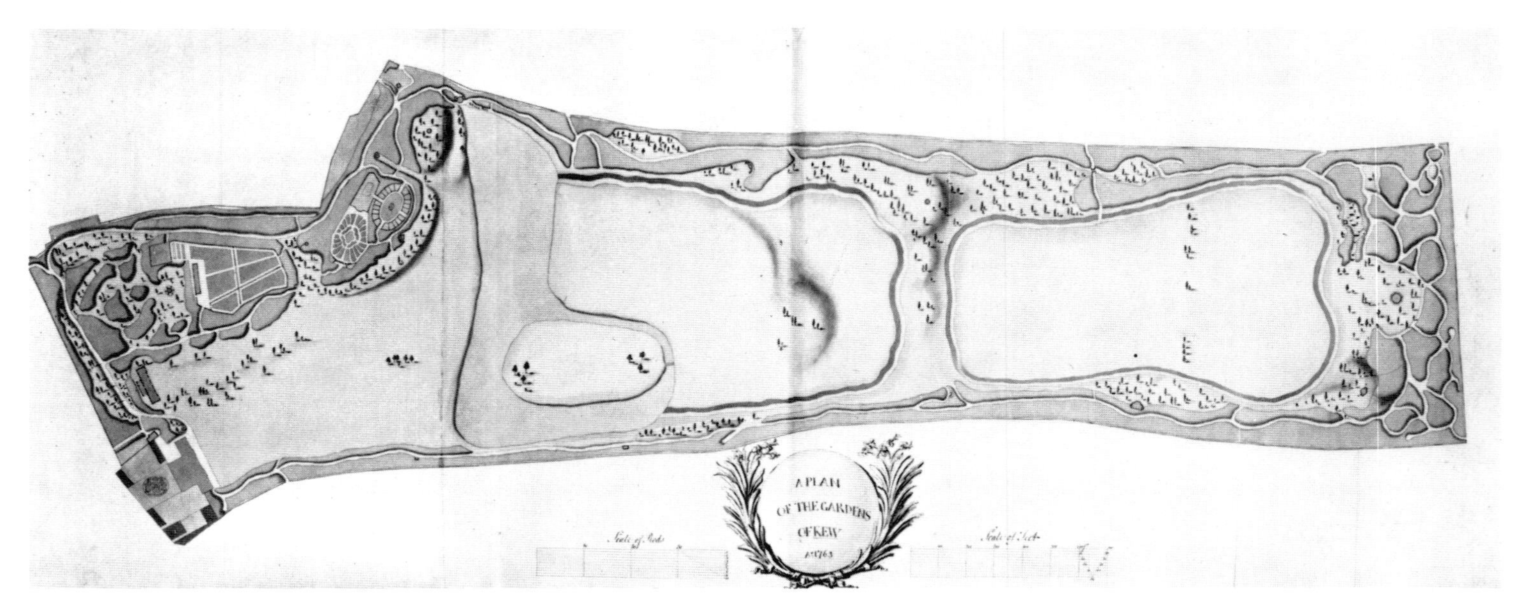

Camellia japonica, from the large collection of spring-flowering camellia cultivars at Kew Gardens.

Below The Ruined Arch, a result of the taste for artificial Roman ruins. Through the Arch can be seen the Temple of Victory (this is no longer there).

last three named above) of these and other of his follies have been demolished. His Orangery is admirable and the Pagoda at least amusing.

Now we come to William Aiton, only less important in Kew's history than his son and successor. Born in Lanarkshire in 1731, he emigrated, like so many ambitious Scots at that time,* to England when he was twenty-three and was given a job in the Chelsea Physic Garden by that remarkable popularizer of botany, exponent of gardening, and father of horticultural journalism, Philip Miller.

THE CHELSEA PHYSIC GARDEN

Philip Miller was a man whose recommendation in any matter of horticulture was the last word. A Scottish Quaker, he had made the Chelsea garden—today a pretty little enclave of peace, quiet and greenery between the base clamour of truck engines on the Embankment and the snarling scream of Kings Road sports cars—so famous that it would have rivalled Oxford, Edinburgh and even Kew had a man been found equal to him to carry on when he died.

The Chelsea garden was started in 1673 when the Apothecaries Company of London leased land for a Physic Garden from Charles Cheyne, choosing that particular site because they could easily reach it by wherry or barge from the city.* Not much was made of it until Sir Hans Sloane bought the whole manor of Chelsea from Cheyne and gave the land of their garden to the Apothecaries on condition that they took steps to manage the garden properly, a condition not surprising coming from the man to whom Linnaeus carried an introduction from Boerhaave when he came to England. Sloane got Miller, a nurseryman and florist appointed as Gardener at Chelsea

*'Sir, the noblest prospect ever Scotsman saw is the high-road that leads him to England'.—Samuel Johnson.

*Hadfield, Miles, *Gardening in Britain* (London, 1960).

The Pagoda, *left*, and the Orangery, *right*—examples of the *Chinoiserie* and neo-Classicism of eighteenth-century English garden architecture. Kew teaches not only botany and horticulture, but landscape gardening at its most exquisite.

Below left The Pergola Walk.

in 1722; and although Miller had to serve an undistinguished *Praefectus Horti* named Isaac Rand, it was Miller and not Rand who really made this garden. In the process, Miller made his own name so respected in the world of gardening and botany that Linnaeus came to see him during his English visit, and even persuaded him to alter the system of classification he had used in his best-selling *Dictionary of Gardening*, a not inconsiderable feat when one considers that Miller was notoriously obstinate and quarrelsome.

It was Miller at Chelsea who first raised from seed sent to London by the Jesuit d'Incarville, from China, *Ailanthus altissima*, Tree of Heaven;* it was the Chelsea Garden, thanks to Miller, which, according to a good authority,† 'excelled all gardens in Europe for excellence and variety'; it was at Chelsea that William Forsyth made the world's first real rock garden, using 'forty tons of old stone from the Tower of London, flints and chalk, and lava brought from Iceland by Sir Joseph Banks.'‡ But that was the Chelsea Physic Garden's last substantial achievement; and as Chelsea declined, so did Kew, thanks to the efforts of William Aiton and Sir Joseph Banks, grow in consequence.

KEW UNDER THE AITONS AND SIR JOSEPH BANKS

There must have been some initiative from Kew which made it known that a head man was wanted there. When Philip Miller recommended Aiton there could be no hesitation and Aiton was given the job as Superintendent, beginning his work there in 1759. 'At first,' says Turril (*op cit.*), 'Aiton had under his charge apparently only the relatively small botanic garden of the Princess Augusta. His responsibilities were probably gradually extended and in 1784 he succeeded to the entire management of both the Royal Gardens of Kew and Richmond.'

Five years later Aiton published the second, but first official *Hortus Kewensis*. A total of 5,500 species were described in three volumes arranged according to the Linnaean system, with a note of the habitat, date of introduction and introducer's name. By then Aiton had become a person of consequence with people like Sir Joseph Banks and Zoffany, the fashionable painter, among his friends. He worked on, and at, Kew for thirty-four years and died in 1793. By then his son, William Townsend Aiton, was the obvious man to succeed him, for the older Aiton had trained the younger for the job.

Helped by Sir Joseph Banks, W. T. Aiton set about revising his father's *Hortus Kewensis*, and between 1810 and 1813 he published a new, five-volume, edition of it, followed a year later by a catalogue which shows that there were then over 11,000 species of plants in the Royal Gardens. But at this time they were near the first great peak in their career, for it seems that between 1820 when both Sir Joseph Banks and King George III died, the

*Hadfield, *op. cit.*
†*Transactions of the Linnaean Society*, x.
‡*R.H.S. Journal*, quoted by Hadfield (*op. cit.*).

Gardens were declining. The old king, brought up at Kew House, had loved the gardens and lived much of his life in them.

During old Aiton's time the Gardens were about 9 acres, the Princess Augusta's original Botanic Garden; but both he and W. T. Aiton had charge of the neighbouring Gardens, that is, of parts of the old Kew House pleasure park and parts of the grounds of Richmond Lodge, adjoining where Queen Caroline had made great improvements. Her works, or most of them, were wiped out when George III, having inherited both properties, threw them together and employed 'Capability' Brown to landscape them. The final basic shape of the oldest part of today's Gardens is the shape Brown gave them. Meanwhile, he, the King, replaced Bute with Sir Joseph Banks as his botanical and horticultural adviser.

Born rich in London in 1743, Banks studied the natural sciences, went with Cook on his round-the-world voyage, and on his return became the scientific pundit of the age, favouring biology over the other sciences. The King appointed him to advise on the scientific side of the work at Kew and it was Banks who then initiated that policy of plant collecting abroad by planned expeditions which enriched England with 7,000 new species during his lifetime, and in the organization of which he used all his influence and a lot of his own money. A table of expeditions organized and financed by Kew under Bank's term there makes impressive reading:

Collector's name	*Countries collected in*			
Francis Masson	S. Africa	W. Indies	N. America	
David Nelson	S. Africa	Australia	Timor	
Peter Good	Australia	India		
George Caley	Australia			
William Ker	China			
Allan Cunningham	S. Africa	Australia	Brazil	New Zealand
James Bowie	S. Africa	Brazil		
David Lockhart	W. Indies	Congo		

Lockhart's Congo expedition was financed by Kew; but his West Indies (Trinidad) collection was sent there after he had become Director of the first West Indies Botanical Garden. Archibald Menzies was not employed by Kew but sent many valuable plants there when he went with Vancouver as botanist on that navigator's round-the-world voyage of discovery.

As we have said, the Royal Gardens began to decline at Sir Joseph Bank's death in 1720 and they continued to do so until in 1838 the Treasury appointed a committee to inquire into their management. Its members were two Head Gardeners, Joseph Paxton of Chatsworth fame; and Wilson, the Earl of Surrey's man; and their chairman was the famous botanist Dr John Lindley. As a result of their Report the Gardens were transferred from the Crown to Parliament; and the desirability of having a first-class, scientific, botanical garden was given parliamentary recognition.

The Great Palm House at Kew (*below* and *opposite*) was the joint work of Sir William Jackson Hooker, the Director, Decimus Burton, the architect and Turner, the Dublin construction engineer who greatly improved on Burton's designs. No longer the largest garden glasshouse in the world, it is still the most beautiful.

Overleaf From its beginning Kew has been particularly successful in combining botanical science and horticultural art. Planted as objects of study, trees become objects of beauty if the planter, despite a primarily scientific purpose, uses art in their placing.

THE SECOND EPOCH: THE HOOKERS

In 1821 Sir Joseph Banks had used his great influence to get his friend William Jackson Hooker made Professor of Botany at Glasgow University. Hooker made a great garden and a great name there and in 1836 was knighted for it. It was to him that the new management of Kew turned in 1841 and in April of that year he took up his appointment as Director of the Royal Botanic Gardens.

They then covered 15 acres. Hooker's first changes were to pull down the interior brick walls; to rebuild and modernize the greenhouses; to revive Banks's plant-collecting policy, sending Purdie to South America and Burke to North-west America. Next, he received and cultivated the new plants which his son, J. D. Hooker, off on the first of his travels, began sending home to Kew together with masses of herbarium material. He began the policy of distributing to other gardens and other countries the plants newly established in cultivation at Kew.

Within two years of starting work at Kew, Hooker had taken in 45 acres of the old Pleasure Grounds for the botanical gardens; shortly thereafter he took in another 200 acres. Three years only after becoming Director he had multiplied the size of the Gardens by twelve. Then, in 1844, he started building the great Palm House, still the most beautiful thing of its kind in Europe. The design of this is commonly attributed to Decimus Burton. But although Burton was called in and had a hand in it, it is probably not really his work. First the plans were discussed between Burton and Hooker over and over again. Then, the Dublin constructional engineer William Turner who actually built the Palm House eliminated ranks of supporting pillars which Burton, who knew nothing of stresses in curvilinear metal buildings, had included. Sir J. D. Hooker says that Burton and Sir William Hooker designed the Palm House; the Curator John Smith says that William Turner was the architect.* Hooker also employed the leading garden architect of the day, W. A. Nesfield, to lay out and landscape the newly acquired land. He made the lake and also built the Temperate House.

Thus Kew Gardens as we know them today are largely Hooker's work; he was, after Banks, their second real maker. But Sir William Hooker's work at Kew cannot be considered separately from his son's. Joseph Dalton Hooker became his father's Assistant Director in 1855, and his successor as Director on his father's death in 1865. A graduate (M.D.) of Glasgow, he had worked in his father's Herbarium even as a boy. As a Surgeon in the Royal Navy he had sailed the whole southern hemisphere, sending home a vast amount of new Herbarium material and seeds to Kew and, in due course, published his *Flora Antarctica*; and that was in 1847, after he had been for some time Assistant Professor of Botany at Edinburgh; he was elected a Fellow of the Royal Society in the same year. Some years later came his *Flora Novae-Zelandae*; and, after taking up his appointment at Kew, his *Flora Tas-*

*See *The Hookers of Kew*, by Mea Allen (London, 1967).

A typical Kew Garden prospect in the spring, with flowering *Prunus* in the foreground.

Daffodils and flowering cherry at Kew Gardens in the spring.

maniae. All this work was accomplished despite the fact that from 1847 to 1849 he was exploring and plant-collecting in India from the Gangetic Plain up to Darjeeling, and into Sikkim, Tibet and Nepal.

It is primarily to J. D. Hooker that Kew, and the rest of us, owe our garden rhododendrons, for by the time he had finished his Indian work he had sent home no less than 54 species until then unknown to science, most of them from the Sikkim Himalaya region. Hooker made botanical expeditions to three other parts of the world, to Syria and Palestine in 1860, to Morocco in 1871 and to North America in 1877. Meanwhile he continued to produce botanical works of the first importance at twice the speed any ordinary man would have taken to produce inferior work. He is, of course, the Hooker of 'Hooker and Bentham' the 'masterpiece of all botanical works dealing with the systematics of seed-bearing plants'.* His *Flora of British India* in seven volumes is a monument to his learning and industry. Despite the immense amount of work entailed by these and other publications, by his editorship of the *Botanical Magazine* and of *Icones Plantarum*, he had plenty of time and energy left for the Royal Botanic Gardens, and Turril sums up his work there as follows:

> He planted the Pinetum, brought the Berberis Dell into its present condition, created the Rock Garden, planted the Thorn Avenue, Atlas cedars, the Holly Walk, and the Sweet Chestnut Avenue . . .

He made large changes and renovations in buildings, walls and gates; he built the T-range of greenhouses and he built the Jodrell Laboratory, which was rebuilt in the mid-1960s. He added wings to both the Herbarium and the Museum, and he built the North Gallery. He was knighted in 1877, retired in 1885 and died in 1911 in his ninety-fourth year. The achievements of the two Hookers at Kew has made Kew the botanical centre for the whole world and, at the same time, one of the most beautiful gardens in horticultural history.

THE THIRD EPOCH: THE COMPLETED GARDEN

The next Director of Kew Gardens, William Turner Thistleton-Dyer, Sir Joseph Hooker's son-in-law, was, in a sense, the first Director to take over a completely 'made' garden with clearly established scientific, social and economic functions. A highly qualified teacher of botany, he had come into touch with the work at Kew when he started introducing the latest methods of teaching botany from Germany. His first work at Kew, that of helping Hooker with parts of his *Flora of British India*, led Hooker to make him his Assistant in 1875 and later Dyer married Miss Hooker. He was given a special charge, the Garden's work in economic botany for Colonial departments of agriculture and horticulture. It was as a consequence, and because of his keen interest in that work, that Thistleton-Dyer carried out two of the most important works ever undertaken by Kew in economic botany: the introduction from Brazil of *Hevea brasiliensis*, the rubber tree, to the Indian sub-continent by way of Kew and the Botanic Garden of Peradeniya in Ceylon (see below); and the introduction of *Theobroma cacao*, also to India and by way of Kew and Peradeniya, from Mexico.

Thistleton-Dyer was second only to Sir Joseph Hooker in the services he rendered Kew: he enlarged the Herbarium; he started the *Kew Bulletin*; he began, in 1895, the publication of the immense and world famous *Index Kewensis* which has been continued by Supplements ever since. To the Gardens he added the Alpine House, a sunken Rose Garden, the Bamboo Garden, and a Lily Pond by the Pinetum. Knighted in 1899, he retired in 1905 and died in 1928 aged eighty-eight.†

Colonel (later Sir) David Prain who succeeded Thistleton-Dyer had been a military surgeon who had served a considerable part of his life in India. A well-known taxonomic botanist, he had served as Keeper of the Herbarium and Librarian of the Royal Botanic Gardens, Calcutta. At Kew he had the difficulties of World War I to cope with; his principal achievement there was probably the advancement of pure and economic botanical research. Militant suffragette raids added to his difficulties. He was not in a position to make any considerable changes in the Gardens and it was left to his successor Sir A. W. Hill to get things on the move again by building a new wing to the Herbarium and by improving the landscaping of part of the Gardens. He might have done much more had he not been killed when the horse he was riding in the Old Deer Park threw him and broke his neck. His Economic Botanist at Kew, Sir Geoffrey Evans, filled his place until the appointment of Dr (later Sir) Edward Salisbury in 1943.

Salisbury's *forte* was plant ecology, and he worked in other fields, publishing a very successful botanical work

*Turril, *op. cit.* See also Allen, Mea, *The Hookers of Kew*.

†The age attained by great Kewites is remarkable. The average for the two Hookers, Bentham and Thistleton-Dyer was eighty-seven and a half and Sir Edward Salisbury is eighty-one at the time of writing.

for the ordinary reader (*The Living Garden*) and an important contribution to scientific botany in his *The Reproductive Capacity of Plants*. He supervised the post-war reconstruction work and built a new Australian House; during his term the Chalk Garden was added to Kew.

When Sir Edward Salisbury retired in 1956 he was succeeded by the Keeper of the Department of Botany at the British Museum, Dr (now Sir) George Taylor, among whose botanical activities had been plant collecting in south-east Tibet.

KEW NOW

Because Kew Gardens had such a very great nineteenth century there are people who believe, knowing nothing about the place, that it now lives only on its traditions and great name. We have met young and bright botanical scientists and gardeners in Germany and America who are apt to treat Kew as a dear old grandmother who was doubtless brilliant in the past but. . . . Nothing could be further from the truth. Not only does Kew remain paramount as a garden, but in its Herbarium with its five million specimens, in its Library and in the work of its laboratories, it is still the world's leading botanical institution.

Some idea of the life and energy which continue to be manifest at Kew can be derived from a glance at what the Director who retired in 1969, Sir George Taylor, had accomplished there since he started in 1956. Considering the scientific side first: the building of the new Jodrell laboratory completed in 1965 gave plant physiologists and cytologists, geneticists and others such working conditions and equipment as they had never had before and which are as good as any in the world. Emphasis now is on working not with dead material but with the living plant, so that the Kew scientists, with the Garden's incomparable collections to draw upon, are very well provided for. In taxonomy Kew's enormous Herbarium, to which a new wing was opened in 1966, makes it unchallengable; and the *Index Kewensis* remains the world standard for the names and descriptions of species. Kew's collection of botanical books remains the world's largest and was last year rehoused in a brand new and ultra-modern library designed for them. Meanwhile the Museums have been reconstructed and their collections properly related to the live collections in the Gardens.

Although Sir George Taylor is a botanist he has been remarkably successful as a gardener at Kew and there are half a dozen enduring new examples of this. In some ways the most interesting of these is the exquisite seventeenth-century garden, opened in 1969, which he designed and built and planted behind Kew Palace. Secondly, in the field of horticultural art rather than botanical science, there was his planting of the hitherto neglected peninsula in the lake to a colour-group of trees and shrubs in the manner of, but superior to, the colour-group planted by W. J. Bean when he was Curator in the late 1920s.

Then there is the more nearly 'botanical' gardening;

The new Jacobean Garden at Kew, one of the contributions made to the Garden by Sir George Taylor during his term as Director and opened by Her Majesty the Queen in the year of his retirement, 1969.
Far left Detail of one of the decorative figures in the new Jacobean Garden.
Bottom The Temple of Aeolus, one of Kew's eighteenth-century ornaments.

here Sir George's outstanding additions have been the Heath Garden, perhaps the best thing of its kind we have seen; the new Azalea Garden, planted with a few Japanese Cherries under some pines about a cedar and providing a lesson in how to use the difficult colours of azaleas; the new Pergola; the remaking of Hooker's Rock Garden so that Kew's alpine collections are among the best grown and displayed in the world, although conditions at Kew are far from good for such plants. There has been much culling of poor plants and replanting with good ones in many sectors of the Gardens and in, for example, the Orchid collection from which hybrids are being eliminated in favour of botanical species. The temples and follies have been opened perhaps for the first time since they were built. To Kew's statuary have been added the Apollo Fountain in the great pond before the Palm House; and the two magnificent Chinese lions topping one of the stairways. The old Kew statues have been restored to the Orangery. Finally, the new Filmy Fern House is unique.

Other, as it were negative, changes at Kew have not so much been initiated by the Director as they have been the results of social and other kinds of change in the world at large. For example, work in the field of economic botany is being run down, although even in that field Kew has still work to do; it is still a sort of clearing house for the exchange of economic plants between the old and new worlds, new strains of such plants as yield, for example, bananas, sugar and coffee are received from one quarter of the world, tested, selected, propagated and passed on to another.

A KEW SATELLITE

When Sir George Taylor took up his appointment at Kew he very soon realized that while improvement of the Gardens was always possible, expansion was not. But even supposing there had been room it was doubtful whether new plantings should be done there; bad air-pollution was killing some plants and damaging many more. That particular nuisance is now much abated and air-pollution at Kew is not as bad as it used to be. Still, it is serious even now; and the nuisance of aircraft noise owing to the nearby London airport is much worse than ever. There are other and more fundamental ways in which Kew falls short of being the ideal place for a botanical Garden. Poor soil has always been a problem; the Thames Valley is within the zone of lowest January mean minimum temperature for Britain; topographically, Kew is rather featureless, such features as it has are artificial, for example the lake, the Hollow Walk (later Rhododendron Dell) made by 'Capability' Brown, and the Temperate House terrace.

The problem of the future has therefore been solved by the acquisition of a Satellite to Kew from the National Trust, the great Wakehurst garden of over 400 acres; topographically beautiful, with a richly fertile soil, a mild climate and many fine trees, Wakehurst ensures the continuation of Kew traditions in conditions very superior to those which Kew has had to do with.

Magnolias at Kew.
Above Magnolia denudata.
Below Magnolia stellata.

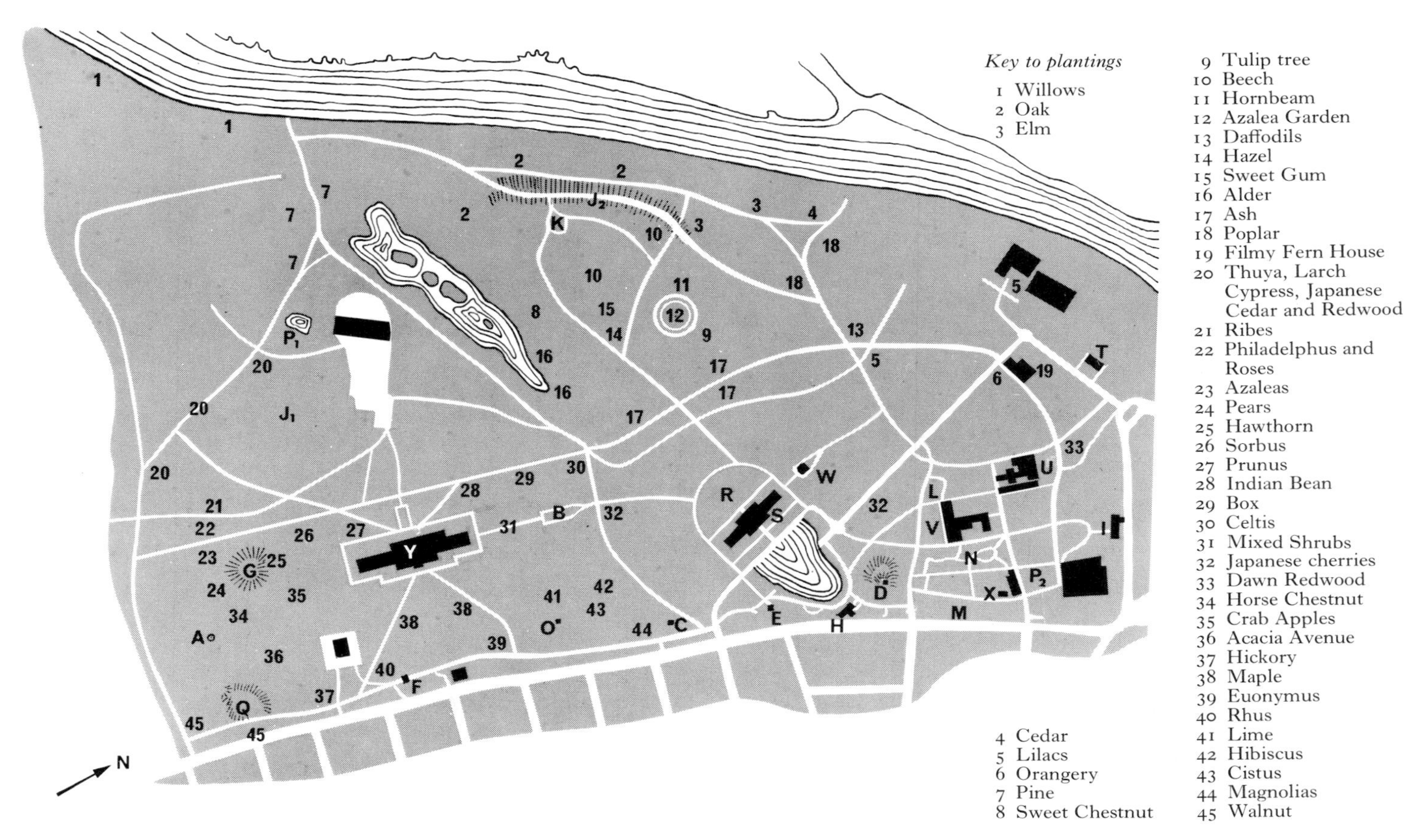

A KEY TO KEW

THE ORNAMENTAL BUILDINGS

A. The Pagoda. Built by Sir William Chambers in 1761–2, an example of the fashion for chinoiserie in gardens.

B. King William's Temple. Built by Wyatville for William IV in 1837 as an embellishment to the gardens. Opened for the first time to the public during the Directorship of Sir George Taylor.

C. Temple of Bellona. Another of Sir William Chambers' embellishments of the Royal Gardens, built in 1760.

D. Temple of Aeolus. Built by Chambers in 1761 and restored by Hooker in 1845, it is a dome on eight columns standing on a wooded mound and one of the prettiest ornaments in the Gardens.

E. Temple of Arethusa. Built by Chambers *c.* 1760, this classic portico or stoa is used as a memorial to Kewites killed in the two world wars.

F. The Ruined Arch, by Chambers 1759–60. The building of artificial Roman ruins became fashionable because English landscape gardening was inspired by French landscape painting (e.g. Claude, Poussin) of Romagna and Campagna scenes, idealized.

G. The Chokusi Mon, a replica of the ancient Gate of the Imperial Messenger at Kyoto, was presented to Kew in 1910 by the Japanese exhibitors at the Japanese–British Exhibition. It stands on the mound where Chambers built a mosque in 1761, since demolished.

THE MUSEUMS

H. The General Museum. Houses an incomparable collection of objects having economic, scientific or other connection with botany. Built by Decimus Burton in 1857. The Reference Museum is closed to the public.

I. The Wood Museum, formerly the cottage of the Duke of Cambridge houses a collection of economic timbers, especially of Commonwealth and British origin.

THE BEASTS

The Queen's Beasts are replicas in Portland stone of the ten royal heraldic animals which stood in front of the Abbey Annexe during the Coronation in 1953. They are by Mr James Woodford, O.B.E., R.A., and were the gift of an anonymous donor. They are *The White Greyhound of Richmond*, inherited from John of Gaunt through Henry IV; the purely mythical *Yale of Beaufort*, inherited from Margaret Beaufort through Henry VII; *The Red Dragon of Wales*, device of Owen Tudor, Henry VII's grandfather; *The White Horse of Hanover*, element in the arms of the House of Brunswick and coming to the Queen through George I; *The Lion of England*, inherited from Richard Lion-heart who used this beast as his device; *The White Lion of Mortimer*, inherited from the Mortimers, earls of March through Edward IV; *The Unicorn of Scotland*, which comes to the Crown through James VI of Scotland; *The Griffin of Edward III*; *The Black Bull of Clarence*, inherited from Lionel, Duke of Clarence whose devise it was, through Edward IV; *The Falcon of the Plantagenets*, badge of Edward III and his descendants.

Two Beasts relatively new to the Gardens are the Chinese Guardian Lions on the south-east side of the Pond. They are either of Ming date (1368–1644) or

VICTORIA
AMAZONICA
syn VICTORIA REGIA
GUIANA & BRAZIL

Victoria amazonica, the great Amazon water-lily, was first flowered in Europe in the big conservatory built by Joseph Paxton at Chatsworth. But it was at Kew that the general public saw this plant and wondered at its 6-foot leaves and huge flowers.
Below From a *Cymbidium* inflorescence in the cool orchid house.
Right One of the heraldic 'Queen's Beasts' which decorate the Palm House terrace at Kew.

perfect eighteenth-century copies of Ming originals. They were presented by Sir John Ramsden in 1958.

SPECIAL GARDENS

J. *Rhododendron Dell.* A winding artificial valley excavated by military labour in 1773 to a design for a Hollow Walk by 'Capability' Brown. Sir William Hooker first made it into a Rhododendron garden—the shade of gigantic cedars, beeches and a colossal plane tree make it excellent for the purpose. But the rhododendrons grown are all hardy hybrids and this garden is at its best in May–June. Botanical species rhododendrons are grown south-east of the water-lily pond (J1).

K. Bamboos. A collection of hardy bamboos chiefly from China, Japan and the Himalayas.

L. *Chalk Garden.* Collection of native and exotic plants thriving only on soils of high calcium content. First made and planted in 1944.

M. *Herbaceous* perennials, biennials and annuals collection planted with rhododendrons and other shrubby genera. About 6,000 species, varieties and cultivars arranged in parallel beds according to family.

N. *Rock Garden.* First made by Sir Joseph Hooker in 1882 with Mendip limestone and Yorkshire granite. Rebuilt and much improved by Sir George Taylor in the 1960s, using Sussex sandstone. One of the most attractive alpine gardens in the world, in which are cultivated a vast range of species from every alpine region in the world. It includes Stream and Bog gardens. It is at its best in early summer and again in autumn.

O. *Berberis.* A collection of barberries planted as a shrub garden in a hollow or dell excavated a century ago. Both evergreen and deciduous species from Asia, Europe and the two Americas are represented and the beauties of this large genus well displayed.

P. *Aquatics.* Apart from the aquatic plants grown in and around the lake and the pond, aquatic plants have this special garden to themselves, hedged about with *Berberis × stenophylla*. Both water-lilies and marsh and bog plants are included in this collection. At J1 is, moreover, a Water-Lily pond—formerly an old gravel pit—in which is grown a collection of hybrid water-lilies, primulus and irises.

Q. *Heaths.* The heath garden is one of Sir George Taylor's additions to the Gardens. It is one of the few special gardens which is as good in winter as in summer owing to the large collection of *Erica carnea* cultivars. But in summer and autumn the *Erica* and *Calluna* varieties make it very colourful.

R. *Roses.* There are, of course, species and other kinds of roses in many parts of the grounds. The formal rose garden near the Palm House, however, is devoted to the latest, or older, garden roses such as tea-roses, floribundas and so forth.

THE GREENHOUSES

There are twenty-four greenhouses in Kew Gardens.

S. Palm House. Built 1844–8. For long the largest, it remains the most beautiful glasshouse in the world. Attributed to Decimus Burton, it was probably chiefly designed by Sir William Hooker and the engineer, William Turner of Dublin, who built it. It houses a large collection of Palms, Cycads, tropical trees, such giant bamboos as *Gigantochloa verticillata*, Screw Pines, tropical shrubs, climbers and flowers of economic or ornamental value.

T. Aroid House. Oldest plant house in the Garden, originally built at Buckingham Palace and moved to Kew in 1836. Houses many genera of Aroids and should be of particular interest to house-plant growers since many of their favourites are here, including *Philodendrons, Anthuriums, Peperomias,* etc.

U. There are two Ferneries in this group. A third, the unique Filmy Fern House, is behind the Orangery. The combined collection of temperate, tropical and filmy ferns is a magnificent cross-section of this whole family of primitive plants. Also at U is the Conservatory built in 1890 (rebuilt 1962) with a permanent display of good decorative plants, both hardy and tender; and finally, in the same group, the Succulent House devoted to a fine collection of plants characteristic of the arid regions of all five continents.

V. The 'T' range of greenhouses. The hottest part of this range houses the famous giant water-lily *Victoria amazonica*. In other houses of this range are collections of South African plants, of tropical herbaceous plants, of the genus *Begonia*, of *Nepenthes*. Another houses the collection of economic plants; and four houses in this range are given over to the Orchid collection.

W. The Water-Lily House. Representatives of the genus *Nymphaea* in the tank, tropical shrubs and climbers.

X. Alpine House. The Alpines here collected are not tender but they are housed so that their flowers may be seen undamaged by rain.

Y. The Temperate House. Built by Decimus Burton during the Directorship of Sir William Hooker, this is planted with species not quite hardy at Kew.

UPPSALA

land of Linnaeus

Left Karl Linné, known as Linnaeus, whose system of classification of plants brought botanical order out of chaos.

Right Linnaeus's own garden in the University city of Uppsala is carefully preserved as his monument even to his method of labelling plants with painted slate labels.
Below right *Plantago media* selected as a potential garden plant at Uppsala.

Of Scandinavia's sixteen botanical gardens we have had to choose only three, each very different from the other two, to represent the rest: they are the *Botanisk Have,* in the heart of that very pleasant city, Copenhagen, founded in 1871; the *Botaniska Trädgard* in Göteborg, the planting of which began in 1916; and the University of Uppsala's botanical garden, which dates in its present form from 1787, although the University had a much older garden in which Linnaeus worked.

But first we should notice a little garden in Uppsala which is a monument to Linnaeus and which takes the form of a botanical garden reproducing that old one (founded 1655) which he found, albeit very neglected, when he moved from Lund to Uppsala in 1728. It was at Uppsala, despite the poverty of its resources in botanical teaching, that Linnaeus perfected the theory of plant sexuality which had been agitating European botanists for some time, and published his classic *Introduction to the Floral Nuptials*, dedicated to his patron Celsius. One of the results of that publication was that Professor Rudbeck, who was Keeper of the Botanical Garden, made Linnaeus his deputy, handing over to him the task of delivering the University course in botanical lectures. He was not yet twenty-three years old.

Linnaeus's connection with Uppsala continued one way and another throughout his life, and the country house he bought in its vicinity is now a Linnaean Museum. As for the monument garden in the town, it is a rectangular plot laid out strictly geometrically, some of the beds edged with clipped box. The plants in it are those which Linnaeus worked on; and the labels reproduce the ones he used, small, thin slabs of schist painted white on one side, and with the plant names exquisitely written in Indian ink. There is a little orangery (not now used for plants) and a prettily shaped pool for aquatic plants and water-loving ferns.

The University Botanical Garden took its present shape when King Gustavus III of Sweden donated a part of his royal park to the University for the enlargement of the Botanical Gardens in 1787; for by then Uppsala in particular and Sweden in general had been made very much aware of the importance of botany by the greatness and international fame of Linnaeus, whom the King had enobled before he died (so that Carl Linné became Carl von Linné). The piece of park handed over had been laid out and planted as the palace garden earlier in the century by the famous architect Harleman. That artist was under the influence of the Italian Baroque style in garden design and made the royal garden in that fashion. But for his evergreens he used the spruce *Picea abies*, clipping the trees and hedges as cypress or box or yew were clipped in the south. These trees and hedges were unique, and it was therefore fortunate that a condition of the donation was that they, and the garden layout in general, be preserved by the botanical gardeners. They were a great nuisance to the University botanists and gardeners for two centuries, but a nuisance well worth bearing, becoming so famous that English and German travellers went far out of their

Key to plan

1 Experimental plants
2 Benches
3 Rock garden
4 Road
5 New nursery
6 Systematic plant collections
7 Economic plants

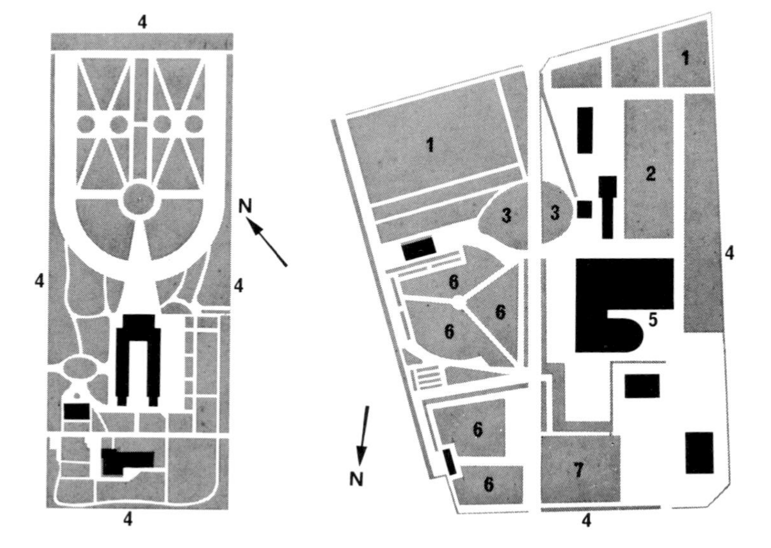

way just to see them; the big solitary clipped spruces were described as 'stack-shaped', a shape assumed with age as a result of the original pyramid clipping.*

The garden needs tall and fastigiate trees, for it has absolutely no topographical features, being a flat plane divided into areas. The principal plantings are as follows. There is a perfectly circular garden (one is reminded of Padua) devoted to a vast collection of herbaceous perennial species all raised from seed collected from wild plants and all of known good provenance.† Beds and paths are alternate in concentric circles; each circle of beds is broken into regular segments; each segment is devoted to a single plant family and the sub-divisional arrangement is geographical. The labelling is arranged for teaching: i.e., all labels face outwards and during lectures or tutorials in the garden the professor stands on the path facing the centre, the students on the path, across the bed from him, facing the circumference: he can read the names; they cannot.

Next in an adjoining section and arranged with military precision and impeccable neatness, is by far the largest collection of annual species—a speciality of the Curator, Dr Nils Hylander—which we have seen in any botanical garden, something just under a thousand species. Some

*For an account of these remarkable trees and hedges specialists are referred to N. Hylander and T. Arnborg, *En undersokning av granhacknarna och de klippa gransolitarerna i Uppsala botaniska tradgärd.* in 'Lustgarten' (1957). And see also Clarke, E. D., *Travels in Scandinavia* (London, 1799).

†By 'good provenance' botanists mean that the seed-bearer was quite certainly the true species.

The systematic planting in the Uppsala Botanic Garden makes the whole garden a living work of botanical reference.

work is done on producing new annual cultivars from these. There is a similar collection of biennial plants, although not, I think, so large. Most, but not all, of these plants are of northern provenance and many of arctic provenance. What struck us most forcibly in these collections was the number of wild plants which deserve to be, but are not, in garden use; for example, some fine species and varieties of *Dracocephalum*, some beautiful *Circium* species, notably a vividly ruby-flowered thistle from northern Japan; and an eastern Mediterranean plantain, *Plantago maxima*, which has beautiful silver-white flowers.

For the botanist and plant-breeder concerned with pasture improvement there is a fine collection of annual and perennial grasses. On a series of short, roughly piled, geometrically-ranged stone banks is a very complete and interesting collection of Scandinavian, including arctic, *Rubi*. For the gardener and seedsman, again, collections of the best cultivars of esculent vegetables are grown every year.

The old Orangery houses a few plants of interest, notably a night-flowering *Cereus*—we are not sure of the genus and species—which is described as pre-Linnaean. (Uppsala botanists and others tend to use the terms pre- and post-Linnaean as most of us use B.C. and A.D.) At two hundred and thirty years of age this cactus is a mighty one indeed, and very free-flowering.

Arranged, again, with perfect symmetry is the largest collection of economic plants in northern Europe, grouped according to their uses in commerce.

The alpine plants are not grown on the flat but in small, regular terraces of dressed stone; there, and in a large part of the plantings on the level, great use is made of sand as a mulch. It regulates the delivery of rainwater to the soil's surface; for some reason it tends to make up for poor drainage; and it greatly reduces water loss by evaporation. Particularly interesting in the alpine miniature terraces are the hardy cypripediums, especially *C. regina* from North America. The dwarf kinds of *Dodecatheon* are also notable.

Dr Nils Hylander, Curator of the Garden, is a distinguished botanist with a large body of important published work to his name. His major work, of which two of a projected five volumes have been published, is a complete *Flora* of North Europe. He is a botanist before a plantsman, a plantsman before a gardener. But his interest extends to cultivars so that an unusual feature of the Garden is a collection of good garden herbaceous perennials, planted and grown as systematically as are the species. This is complemented, in the Herbarium associated with the garden, by a unique cultivar herbarium which constitutes a quite remarkable record of old garden plants: there are something like 3,000 specimens. The two collections at Uppsala from which we derived most pleasure were those of *Paeonia*, which have a little garden to themselves, the plants growing in beds cut out of turf; and still more pleasing, the magnificent beds of *Hosta*. Dr Hylander's monograph on that genus, which he found in chaos and left in order, is now standard.

Left Statue of the young Linnaeus, Uppsala. His great work of systematization was completed in his early twenties.
Below Garden of herbaceous cultivars, Uppsala.
Bottom right From Uppsala's collection of *Paeonia* species.

GOTEBORG

The newest botanic garden in Sweden, at Göteborg, is laid out in part as a landscape garden, in part as a wild garden; but the design is stiffened with some formal elements.

The Gothenburg (Göteborg) Botanical Garden could hardly be more different from that of Uppsala. Lying just south-west of the city centre on latitude 57° 42′ and an acid soil which is for the most part weathered Göteborg gneiss, the top-soil is chiefly loamy clay with some bad patches of dense and infertile pure clay. By British standards the climate is extremely harsh: temperatures as low as −26°C (47 Fahrenheit degrees of frost) are recorded and severe night frost is usual until mid-April. Periods of very cold weather are accompanied by dry, bitter east winds; and as such conditions mean that there is no snow, the ground lies unprotected and damage to plants can be very considerable.

This garden owes its existence to a rich citizen of Göteborg, C. F. Lindberg, who, in 1908, bequeathed a sum of money to the town for the making of a botanical garden. Nothing seems to have been done about this until 1915, when the Town Council acquired the Stora Angarden manor with 150 hectares of land whose topography—hills, rocky gorges, small lakes and wooded heights—is enchanting. This was a very suitable site for a botanical garden in the modern manner, and in 1916, under the advice and direction of Professor Rütger Sernander of Uppsala, planting began. Yet perhaps 'planting' is hardly the word: Dr Sernander was doubtless too pure a scientist for this kind of project; his idea was to use the greater part if not the whole of the 375 acres as a sort of nature reserve, and to let it go back to the natural vegetation of the region which could then be properly studied. He had no patience with the public's demand for a 'proper' garden.

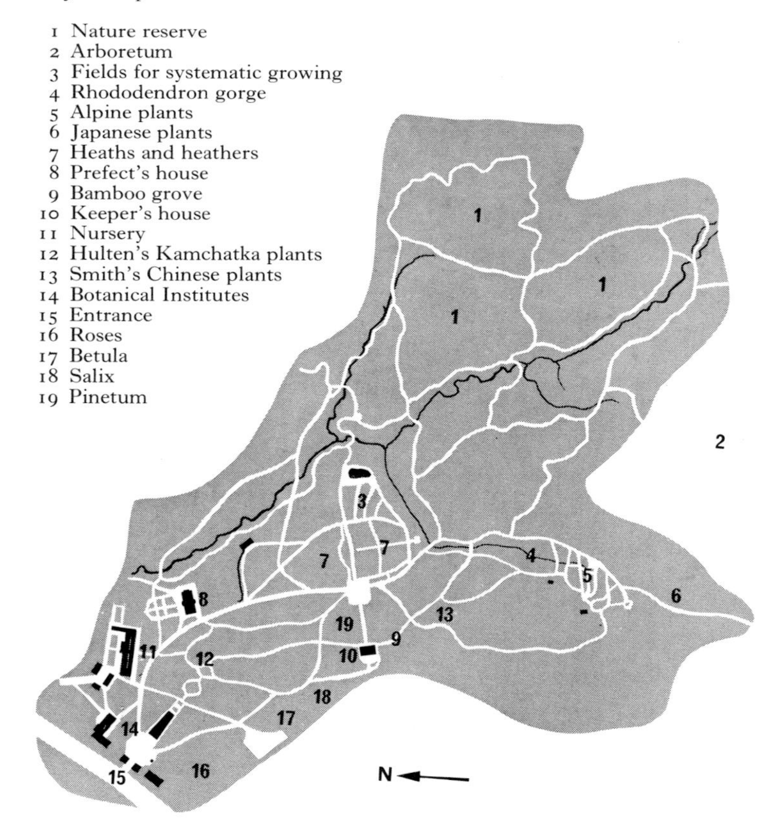

A rather more liberal policy was followed when the garden was taken over by Dr Skottsburg who, recently returned from a plant-collecting expedition to Hawaii with a collection which enriched the new greenhouses at Göteborg, became Director in 1919. Unhappily a good deal of the work which Skottsburg did in the gardens during the next decade was undone by the savage winters of 1939–41, which killed hundreds of normally hardy plants as well as all which were in the least tender, and so created a problem of plant replacement which could be solved only by a bold and enterprising policy of plant-collection. Such a policy requires a special kind of man, and the Garden did not get the right man until a new Director took over in 1949. Professor Dr Bertil Lindquist was a botanist, a dendrologist and a forester; he was downright, choleric and he made enemies, but he was brilliant, dedicated to his work and to the Garden; and although he is said to have put too much emphasis on the collecting and planting of trees, he served the place well. He had great force of character and, fortunately for him, so did the like-minded Chairman of the Society, *Sammans lut ningen Botaniska Trädgardens Vänner,* which, roughly translated, can be called 'The Friends of the Botanical Garden'. This was Dr Malte Jacobsen, erstwhile Governor of the province of Göteborg (*landhövding*), and a man of much consequence in the city. Between them he and Lindquist obtained such grants from the Royal Foundation (*Kungafonden*), from the City of Göteborg and from the said 'Friends' that it was possible to finance plant-hunting expeditions to Japan, Malaya, Ceylon, Alaska and the USA in general, Yugoslavia, Turkey, Poland, Burma, Thailand, India and Nepal. And these expeditions account in some measure for the great Garden's present riches in plant material. The Catalogue of the Garden's Plants which Lindquist published in 1962* contains about ten thousand names of wild and cultivated flowering plants, phanerogamae, pteridophyta, bryphyta, lichens, fungi and algae.† The Garden was and remains particularly rich in certain genera, as we noted during our own visit; there are about 43 species and 8 varieties of *Acer*; 63 species and 7 varieties of *Berberis* and *Mahonia*; 79 species of *Campanula*; probably about 50 species of *Clematis*. One head gardener at Göteborg had such a mania for *Clematis* that at one time there were about 80 species and very many varieties and cultivars, and half the trees in the place were overgrown. Other outstandingly good collections are of *Dianthus*; of *Gentiana*, with over 70 species; of *Hypericum,* with 29 species and 4 varieties; *Iris,* 117 species; *Lonicera,* 58 species and many varieties; *Meconopsis,* 20 species; *Opuntia* (under glass of course), no fewer than 31 species and varieties; *Penstemon,* 60 species and 6 varieties; *Potentilla,* 87 species and varieties.

It may surprise those who think of *Rhododendron* as on the whole a rather tender than hardy genus that no less than 253 species and nine varieties are named in Lind-

**Katalog över växterna i Göteborgs Botaniska Trädgard* (1962).

†Algae were first cultivated in a botanical garden by Jacob Bobart the younger at Oxford in the seventeenth century.

quist's catalogue—that, in short, more than a quarter of the whole genus was represented.* Since Lindquist's time, Dr Davidian, of Edinburgh, the greatest living authority on *Rhododendron*, has surveyed the Göteborg collection but a revised catalogue has not yet been published, for the present Director, Dr Per Wendelbo, is faced with the task of removing cultivars from collections which should be confined to species; and of remodelling the Garden's policy so as to reconcile the public's taste for (and right to) floral display, and the consequences of his predecessors' sometimes too austerely scientific attitudes.

We must on no account give the impression that the Göteborg garden is formal or regimented; the word 'collections' can be misleading; it should be made clear at once that this garden, planted 'after nature', is certainly one of the most beautiful and horticulturally exciting great gardens, in the style of the English 'paradise', garden, in all northern Europe. The richness of the plant material merely serves, from the strictly horticultural point of view, to give this lovely place a colour and texture above anything which could be expected of even a great 'picture' garden at anything short of, say, Bodnant standards.†

There is a very interesting and complete collection of *Salix*, because, for this genus only, Lindquist initiated a policy of planting cultivars as well as species (exception—a few of the rhododendrons obtained from bad provenances). There are 116 kinds, the trees being used ornamentally all over this enormous garden. Other genera represented by exceptionally numerous species are *Saxifraga* (96 and 38 vars.) and *Viola* (64 and four vars.). There are comparably good collections of very many other genera. The Orchid family, finally, is represented by one of the finest collections in the world, 2,000 species; Dr Wendelbo

*Only 28 under glass; the other 225 are hardy.

†Bodnant, at Tal-y-cafn in North Wales, is Lord Aberconway's spectacularly beautiful 'paradise' garden.

Left An aspect of the Japanese garden in the Göteborg Botanic Garden. It is 'Japanese' rather in its flora than in its design, as a result of two important plant-collecting expeditions made to Japan by scientific teams from the Garden's staff.

Right An illustration from the Göteborg Library of a *Pleione wallichiana*.

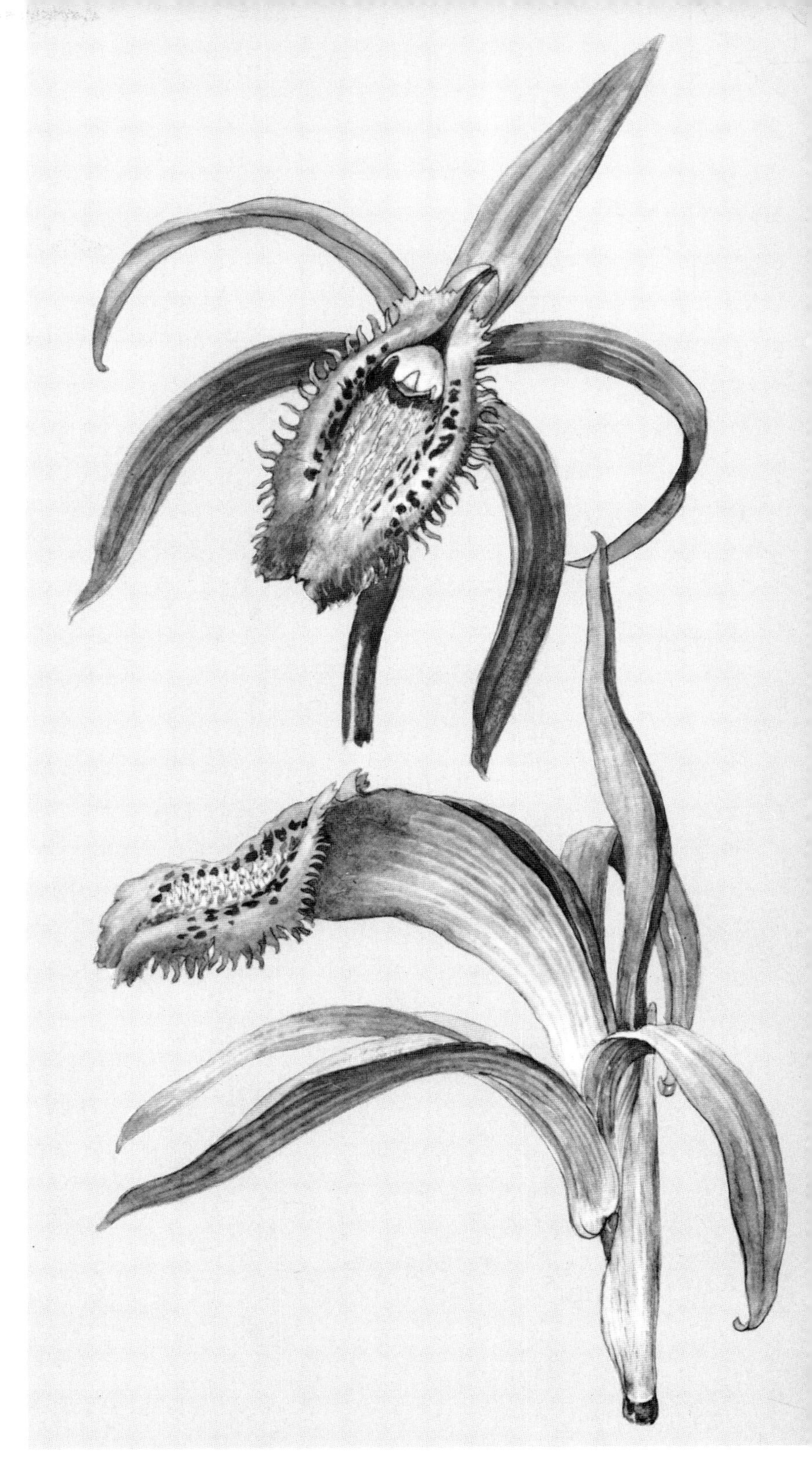

told us that the day is rare when there are not fifty species to be seen in flower.

In the collections we have mentioned, and in the three special collections yet to be described, it was the policy of Dr Lindquist, his collaborators and successors, to take a great deal of trouble to ensure good provenances; and by concentrating on high-altitude strains of exotics, material hardy in the climate of Göteborg has been obtained.

A special character is given to this Garden by three entire collections made for it in the Far East. These are Dr Harry Smith's Chinese collections from his 1922–3, 1924 and 1934 expeditions to Szechuan and Yunnan, the survivors of the 1939–41 cold spell; Dr Dick Hummel's collection made in the course of the Sven Hedin expedition to China; and Dr Bertil Lindquist's own collections made during his two plant-collecting expeditions in Japan. The Smith and Hummel collections made the Garden very rich in *Rhododendron* and in *Primula*, planted with an art as charming as that practised with the same genera in Scotland. Dr Lindquist's collection has given the garden a feature as delightful as any we have seen in any botanical garden in Europe, and a great distinction for Göteborg—the Japan-Dalen.

Centred on a pond decorated with one of those stone lanterns which are peculiar to Japanese gardens,* the Japan-Dalen is by no means a Japanese garden (certainly the most stylized and least 'botanical' horticultural style in the world), but an English-style woodland or wild garden, planted entirely and exclusively with Japanese native species—trees, shrubs, herbaceous perennials, bulbs and corms, annuals—everything is Japanese. All the plants in this part of the Garden were raised from seed of known, documented and impeccable provenance. It has very great charm: there are prostrate, ground-hugging, dwarf and semi-dwarf rhododendrons; fine young specimens of the lovely *Magnolia sieboldii*; numerous azaleas; maples; primulas, saxifrages; and many Japanese trees. The fine tree-climbing *Hydrangea petiolaris*, *Schizophragma* and *Acidinia* species are splendidly grown. There is the curious *Helvingia* flowering from the centre of each leaf; the Japanese *Deutzias*, groups of *Rosa acicularis* and *R. wichuriana*, of *Rhododendron albrechtii* and *R. brachycarpum*; and of Japanese *Syringa* species. The Japanese lilies are all there—their roots and bulbs protected by ground-hugging *Gaultheria* and other genera of like habit. Much of Lindquist's Japanese material was not collected until 1957, yet the Japan-Dalen looks surprisingly mature. the Garden now relieved of the burden of the University Chair in Botany—lay heavy stress on the importance of botanical gardens as conservatories of genetical potential—hence the emphasis on good provenances and complete documentation of each plant in the collections. In some of the older plantations, notably in the lovely rhododendron gorge with its fine cliffs of naked rock as background, this purity is far from evident, and it is that kind of imperfection which Dr Wendelbo has set himself to eliminate.

*It may be of interest that these lanterns are the only Christian element in the Japanese garden, which is so full of religious symbolism.

The very large Alpine Garden is well planted with pure species on a stony and hilly area of outcrops and screes. No doubt more purely 'botanical' than Edinburgh's, it is not quite such an exquisite work of art, yet it is without doubt one of the finest in Europe, and its streams and pools and little waterfalls are particularly well contrived and planted. The skirts of the rock garden are, incidentally, very rich in *Trillium*: we counted something like fifteen different species, some of them very rare.

The glasshouse collections are not, excepting for the Orchids, remarkable, but in the Orchid field the Göteborg garden has two remarkable distinctions: an unique collection of *Ophrys* species; and its extraordinary success in growing the notoriously difficult *Disa* species (*uniflora* and *tripetaloides*). To the best of our knowledge, no other garden has consistently been successful with this genus over a long period of years. From time to time an English specialist has had a good show of this spectacular orchid;

Part of the alpine garden at Göteborg, one of the finest collections of rock plants in the world.

for instance, in 1922 it is on record that the display of it at an R.H.S. Show was outstanding. But success is unusual; the method of growing *Disa* consistently well has only quite recently been mastered at Kirstenbosch in its homeland.* At Göteborg *Disa* is raised in such quantities that the Garden is always able to supply other botanical gardens with specimens.

Two thirds of the 375 acres of garden consists of Arboretum which Lindquist, in the face of strong opposition from the Town Council who thought he was planting far too many trees, planted in the beautiful area of hill and gorge and lake beyond the 50 hectares of garden properly so called, and the 'nature reserve' which is the surviving vestige of Sernander's influence. Lindquist planted 200 species of hardy exotic trees in 400 pure stands. The quality of the material is admirable, but pure-stand planting is forestry, not gardening, and there is a move to get a more pleasing irregularity into the Arboretum as thinning has to be done. There is a remarkable stand of *Cercidifolium* trees which, owing to close planting, have developed good trunks; the species is more commonly seen as a big shrub. A stand of large *Prunus sargentii* must be a magnificent spectacle in autumn. *Metasequoia glyptostroboides* is growing well and not experiencing that check to vertical growth now beginning to show in so many British specimens. Two interesting Japanese trees are *Acer tsonoski* and *Ligustrum tsonoski*. The prettiest of the conifers is labelled *Abies koreana*, but despite the cones and their violet-blue colour, it seemed too vigorous to be the true species.

The Garden is rich in *Magnolia* and grows *Stewartia* to perfection; likewise the woody species of *Paeonia*, and a great many kinds of shrub roses.

*It is a pity that Mr Conrad Lighton, in his admirable *Cape Floral Kingdom*, fails to mention Göteborg's success with this orchid in his account of its recalcitrance.

Copenhagen

There are at least two things which even the most careless and uninstructed visitor to Copenhagen's beautiful little Botanical Garden in the heart of the city are very unlikely to forget: the magnificent specimen of a 'London' plane tree just inside the main entrance; and the *Lilium martagon*, naturalized in the grass.

The Garden is affiliated to the Botanical Department of the University and is in some ways reminiscent of Edinburgh's, perhaps chiefly because it is so closely integrated with the city and a little because of its topography. It has a triple role: for the University it is a teaching garden; for the gardener a source of knowledge and inspiration; and for the general public the garden is the best of Copenhagen's numerous municipal parks.

The plantings are based on a pretty and beautifully landscaped artificial lake of considerable extent, with the 22 acres of garden all round it very pleasantly undulating. One's general impression is of being in an English 'paradise' garden, that is, of the kind in which eighteenth-century landscaping and nineteenth-century plantsmanship are combined. Thus the trees and shrubs, while they seem to be planted simply to please, are for the most part in systematic groups; one great shrubbery is entirely composed of *Berberis*, another, very beautiful in mid-June, of *Rosa* species, most of them huge specimens; yet another of various *Crataemespilus* chimaeras, and so forth. Much the same system is followed in the planting of herbaceous genera and of the special collection of native Danish plants.

The garden contains some very fine specimen trees: we have already mentioned the magnificent *Platanus acerifolia* (*P. occidentalis* × *P. orientalis*, the 'London Plane') near the main gates on the corner of the Gothersgade and the Voldgade. It looks like a multi-centenarian, but was planted in 1901. There is a remarkably well-grown, pyramidal *Taxodium distichum*, and fine *Ginkgo biloba* and *Sophora japonica*.

The System garden is on a semi-circular plan—that Paduan influence again. Its curved beds are pleasingly edged with dwarf box. The alpine or rock garden is one of the best we have seen, comparable for quality with Edinburgh's, though not, of course, in range and hardly in 'finish'. Fine specimens of *Salix retusa* are particularly impressive and, in the adjoining sand-garden the *Roscoea* species. The Copenhagen gardeners are skilful with moraine and scree plants; and it seemed to us that the aquatic plants, both in the circular pond in front of the greenhouse complex, and in the lake, are exceptionally good.

A special peat garden accommodates *Rhododendron* and some other ericaceous plants. There is a very pleasing little garden of *Rosa rubiginosa* cultivars.

Finally the greenhouses, with circular structures at the centre on the wings, are most attractive. The greenhouse complex includes a 'frozen' house for the cultivation of Greenland plants.

NORTH AMERICA

ARNOLD ARBORETUM

The Arnold Arboretum is the largest and best collection of woody plants in America, planted as a grand landscape garden near Boston.

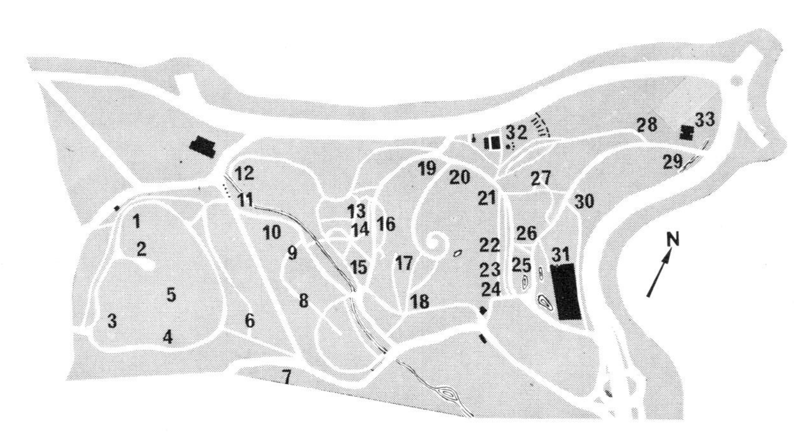

Key to principal groups of plants
1 English Hawthorns
2 Honey Locusts
3 Larches
4 *Crataegus*
5 *Sorbus*
6 *Pyrus*
7 Poplars
8 Fortune's *Rhododendrons*
9 *Sorbus* and *Kalmia*
10 Hemlocks
11 Yews
12 Larch and Golden Larch
13 Dwarf Conifers
14 Hazels
15 Hornbeams and Junipers
16 Daphnes
17 Brooms and Stewart
18 Asiatic Oaks
19 Walnuts and Viburnums
20 *Eunonymus*
21 Birches and Lilacs
22 Mulberries and Lilacs
23 Catalpas and Lilacs
24 *Forsythia*
25 *Philadelphus*
26 Tree *Leguminosae*
27 Horse chestnuts and deciduous *Ilex*
28 Tulip Trees
29 *Metasequoia*
30 Japanese *Acer*
31 Shrubs and Climber collection
32 Bonsai House
33 Administration

North America is rich in botanical gardens and arboreta. The United States has one hundred and three listed in the International Directory, Canada eight. The selection of gardens to represent this large number has been made with the help of a number of Directors of North American botanical gardens; and the order in which we describe them is that in which we visited them, and has no other significance.

THE ARNOLD ARBORETUM

In the third quarter of the nineteenth century Harvard University had a botanic garden principally planted with herbaceous material and which, since it could not draw upon the University's general endowment funds, was obliged to look for money elsewhere. Its Director, Asa Gray, naturally hoped for a share in the James Arnold Trust: Arnold had left about $60,000 to be used for an agricultural or horticultural purpose, but since no use had been made of it immediately it had, by 1872, increased to about $100,000 in the hands of the Trustees. But Asa Gray was unlucky; he got no money to extend his garden because the land which could be made available for some such purpose as Arnold had intended was remote from its site in Cambridge. This was land left to the University by another benefactor, Bussa by name, and its extent was 125 acres. To this the City of Boston added more, acquired by purchase, which it leased to the University at a peppercorn rent of $1 per annum.

The man chosen to be Director of the new Arnold garden was James Sprague Sargent; why this was so is still something of a mystery. But since the endowment of the new foundation was worth only $3,000 a year, it was desirable that the Director be a rich man who might contribute to the garden out of his own pocket and raise funds for it from his rich connections and friends. Obviously he ought also to be very interested in and to know something of botany. Sargent was rich; he had studied botany under Asa Gray; Sargent got the job. He found Boston ready to help by making and maintaining the roadways which would be necessary, through his botanical garden; but that he could not look to Harvard for any money.

Sargent's appointment ensured that the new Botanic Garden would take the form of an Arboretum: not only had he studied under Gray whom he did not get on with, but he was a director of Gray's Harvard Botanic Garden. If only because Gray was principally interested in herbs, Sargent decided to concentrate on trees and shrubs. In his first task, that of landscaping the site, he had the help of Frederick Law Olmsted, the landscape architect who was then busy laying out Boston's system of parks.

Not until 1898 was the Arnold Arboretum complete as we know it, 265 acres of soft, rolling hills, drives, walks, streams and ponds, beautifully wooded and richly planted with shrubs. But meanwhile the making of this had been going steadily forward; and that, of course, included the making of special collections of plants. Sargent's first great chance in that field came in 1877, when the Federal Government commissioned him to make a report on the nation's resources in forests. In the course of the travels entailed Sargent was able to send to the Arboretum a large number of American plants which were not then in cultivation. At the same time, by arrangements with European botanical gardens, he was also collecting exotic material.

Moreover, Sargent was able to buy the services of some outstanding plant collectors; thus in 1906, for example, he succeeded in tempting Henry Wilson away from England. Working for the famous nurseryman Veitch, Wilson had greatly enriched British horticulture with plants from China and had made a considerable name for himself. He became a member of the Arnold Arboretum staff in 1906 and remained one until 1930 when he died. He was Sargent's Assistant Director in the years before the latter's death, at the age of eighty-six, in 1927. Another very remarkable plant collector for the Arboretum was the Viennese Joseph Rock, whose work so greatly enriched the Edinburgh Botanic Garden's collection of Far Eastern alpine plants, and who, like Wilson, was tempted into American service by Sargent's high bidding. Rock collected in Tibet and North-west China. Purdom was yet another collector who worked for Sargent, but the Arnold Director's attempts to buy the greatest of them all, George Forest, failed. In its eight decades of effective working, the Arboretum has introduced into American cultivation no less than three thousand exotic species, a staggering record for one not very large and very under-endowed institution. The most recent introduction of popular note was that of *Metasequoia glyptostroboides*, the Dawn Redwood, in 1947.

When Sargent died there was an obvious successor—

Silver birch and weeping willow in the Arnold Arboretum, whose collection of hardy deciduous trees is the finest in the United States.

Henry Wilson. Perhaps because he was not an American, but more likely because he was poor and the Arboretum needed a rich man, he did not receive the appointment which he had anticipated even on his writing paper. A man of means, able to help the Arboretum to money as Sargent had done, was appointed; Wilson then adopted, without authority or precedent, the title of Keeper.* Perhaps he did so to cover his humiliation.

Sargent and his successors did not confine themselves to plant collecting, but also accumulated a good botanical library and an important herbarium. And they did not neglect the artistic side of horticulture, for they made considerable contributions to the advance and teaching of landscape gardening.

THE ARBORETUM TODAY

The Arnold Arboretum is situated in Jamaica Plains, just outside Boston, between US Route 1 and the Arborway, that is Massachusett Routes 3–28–138, fifteen minutes in a car and half an hour by trolley from the city centre. No signs point to its several gates, nor have Boston taxi-drivers or policemen ever heard of this world-famous institution, which is a tract of gently hilly country, topographically charming, richly wooded, and densely planted with collections of shrubs; in the planting the rules of aesthetics have been as much respected as those of science, with the result that the place is as pleasant for the casual visitor who knows and cares nothing about botany or dendrology, as for the specialist. For flower-colour it must be visited in the spring; but from mid-October until mid-November the flaming beauty of the New England autumn makes it almost unique among botanic gardens; for this colouring before leaf-fall is equalled nowhere except in Canada (including the Montreal Botanic Garden) and Japan.

The first flush of flower comes in April, in the *Forsythia* collection, which is probably the best in the world. Next, the Arboretum is very rich in the *Azalea* series of *Rhododendrons*, both native and exotic: they flower there from April to mid-June, thus overlapping and succeeding the *Forsythias*. Among the collections due to Sargent's special interests that of the Japanese and Chinese cherries is the finest. As early as 1878 the Arboretum was receiving cherry seeds from Sapporo in Japan; and it is fitting that one of the loveliest flowering trees in the world, *Prunus sargentiana*, was named for Sargent. These Far Eastern cherries, as fine in the Arboretum as in Japan itself, are at their zenith in late April, with a second season of glory when their leaves change colour in late October.

The *Magnolia* collection is a very fine one. Because the climate excludes the semi-tender evergreens, the collection is confined to the deciduous species; and the very early flowering ones, such as *Magnolia campbelli*, are missing, since their flowers would invariably be frozen. But for the rest the collection is complete, and the specimens magnificent. Their season is early May.

The Arboretum's collection of *Crataegus* and *Malus* used to be unequalled: this is still so for *Malus*, but the *Crataegus* collection has been reduced to thirteen species, the rest having, for some reason not yet determined, declined and failed. But the *Malus* (Crab-apples) are, whether in flower or fruit, the best we have seen in the world: flower is at its zenith in May, but it begins in April and continues into June.

The Arnold Arboretum is one of the few botanic gardens where both *Rhododendron*—its Rhododendron Glen is very beautiful—and other calcifuge (lime-hating) Ericaceous genera do well, side by side with *Chaenomeles*, which the English lime soil gardener regards as his consolation (together with the shrubby *Paeonias*) for *not* being able to grow *Rhododendrons*. The *Chaenomeles* collection—the Americans call them 'Japanese quinces'—is at its best for flower in May, the flowering being richer by virtue of Boston's hot and sunny summers than it usually is in Britain.

From April to June the *Loniceras* are in flower, and only in the Moscow and Kiev botanic gardens are the collections of this genus as good as in the Arnold Arboretum. The *Rhododendron* collection might, I believe, be better than it is: granted the harshness of the Boston winters, the Göteborg Botanic Garden in Sweden has shown what can be done with the evergreen species in cold climates; how-

*I am indebted for these facts to Miss Stephanne Sutton, BA, a member of the Arboretum's staff and author of a forthcoming *Life* of J. S. Sargent. Garbled versions of what happened have appeared in a number of biographies of botanists and plant collectors.

'Weeping Hemlock', one of America's most beautiful native trees, in the Arnold Arboretum.

ever, the Arnold seems to be the richer in deciduous species. The collection of *Syringa* (Lilacs) is comparable with and much more pleasingly planted than that of the Minsk *Botanichevski Sad* in Byelorussia; the shrubs line one of the principal drives and drift pleasingly up a hill-side, and are not drilled in rather boring lines as they are at Minsk, or grafted as standards as they are at Moscow. Moreover, the Arnold seems richer in species, although not in cultivars.

The Arboretum is rich in native roses, and in *Cornus*, not only the native dogwoods, but some of the Asiatic species. The tender evergreen species do not, however, survive the winters there. There are also fine specimens of the related *Davidia*, the Handkerchief Tree as we call it, Dove Tree in American: Wilson first collected this tree for Veitch, in China.

Two other collections are outstanding: the *Philadelphus*, Mock Orange, surpassing even the Earl of Rosse's fine collection of this genus at Birr Castle; and the *Kalmia*. This is a genus related to *Rhododendron* and confined to North America, and it is particularly effective when planted in masses. The Americans call it 'Mountain Laurel' and it flowers from about mid-June.

Not equal to the above but still remarkable are collections of *Wistaria, Euonymus, Hydrangea*, and *Caragana* among the flowering shrubs; and of numerous genera of trees, *Acer* being the most numerous and interesting.

Here, to give an idea of the pleasures which the Arboretum can offer to plantsmen, follows a brief account of some of the plants which I starred in my notes as particularly fine or especially interesting specimens: two Cork Trees, *Phellodendron lavalla* and *P. amurensis*; several species of the genus *Rhus* which make finer plants in North America than in Britain. Very fine specimens of various kinds of *Enkianthus*, almost incomparable for autumn colour and equalled, in my experience, outside America,

The Brooklyn Botanic Garden is not only an enchanting ornamental garden and one of New York's principal 'lungs', but one of the best teaching gardens in the world.

only at Mount Usher in Ireland; a very pretty collection of Birches; very graceful and well-grown specimens of *Albizzia julibrissin*, one of the most beautiful of its beautiful family (*Leguminosae*), with doubly pinnate leaves and pink flowers—its cultivar 'E. H. Wilson' is particularly fine. There are very good specimens of Sargent's Weeping Hemlock, splendid *Hammamelids*; *Kolkwitzia amabilis*, a beautiful and hardy flowering shrub introduced by the Arboretum from China in 1907; a fastigiate Sugar Maple; specimens of *Aesculus glabra leucodermis* (Whitebark Buckeye); an enormous *Acer rubrum* var. *schlesengeri* with leaves the tender colour of copper-beech leaves in spring; *Quercus borealis*, the Northern Red Oak, with its huge leaves; and a very large *Malus tarnisoides*, with masses of peach-pink, cherry-like fruits. As in most American botanic gardens, there are fine specimens of the Tulip Tree, *Liriodendron*; there are a very large number of *Cotoneasters*, in great variety of forms, and attractively planted; and a remarkable beech, *Fagus sylvatica* var. *asplenifolia*. The *Stuartias* are outstandingly good; and so, among the *Rhododendrons*, are the specimens of the species *schlippenbachii*. Had I to nominate one species in this large and diverse collection as being of very special interest, I should chose the specimens of that rare and difficult shrub *Franklinia altamaha*: at the Arnold Arboretum this species, always rare and now extinct in the wild, related to the *Gordonias*, and very hard to cultivate successfully, is magnificent, the white flowers like large single camellia flowers being at their best in October, just as the leaves are flowering into scarlet.

The Arboretum has a block of working greenhouses and, in the same enclosure, a display of dwarf and miniature conifers in great variety, very useful as a reference for gardeners planning a rock garden; and, in the same enclosure, the Larz Anderson collection of bonzai trees. This collection was presented to Sargent by Anderson as a gesture of gratitude for his work in creating the Arboretum
gesture of gratitude for his work in creating the Arbore-
tum, and it includes some of the finest specimens of the larger kind of bonsai trees (now called Cloud Trees), conifers and maples, we have seen anywhere. The oldest are about 175 years of age.

The Arboretum and its associated Herbarium, Library and Laboratories has a considerable programme of scientific work. The Director, Professor Howard, is emphatic that there is still an enormous amount of work to be done in classical taxomony as well as in the other fields; that we still have nothing like a complete survey of the world's flora, that we do not know exactly what grows where, and that it is high time we had an international index. A project of the kind originally mooted by Dr Avery, Director of the Brooklyn Botanic Garden, and espoused by Professor Howard, has been undertaken, or at all events a pilot scheme, at Longwood Gardens, to which we shall come below.

BROOKLYN

Our principal attention in New York was given to the Brooklyn Botanic Garden rather than to the one in the Bronx: while it is true that the Research staff of the Bronx Garden is engaged in the most advanced scientific work—notably in molecular biology—being done in any American botanic garden, the garden itself is very inferior to the Brooklyn one.

For a botanic garden of only 50 acres embedded in a city, the Brooklyn Botanic is extraordinarily diverse; it is beautifully planted and maintained; and it has an educational programme in horticulture, elementary botany and conservation unsurpassed and perhaps unequalled by any botanic garden in the world.

As an institution the Garden is part of a complex called the Brooklyn Institute of Arts and Sciences; it was founded and laid out in 1910, chiefly for educational purposes. Like many United States institutions it is a mixed public and private body: the land it stands on and its buildings belong to New York City, and the city finds about half its annual budget; but it has an independent board of governors and is also what is called a Membership Society with 4,000 subscription-paying members. So the other half of the Garden's annual budget of expense comes from members' subscriptions, from the income on a total accumulated endowment of $6,000,000 and from gifts.

Not all these gifts come from rich people: there is the example of the Danish immigrant seamstress, a lonely spinster all her life, whose one pleasure was the Brooklyn Botanic Garden, who left the Garden her life savings of $7,000.

The greater part of the Garden's 50 acres is planted as an arboretum of trees and shrubs which, standing as specimens or grouped in stands on lawns, are more important than the herbaceous collections. It is well and pleasingly landscaped, and considered simply as a park for recreation it is delightful, attracting about 35,000 visitors a year: two million people have visited the Garden since it was first opened, drawn from a much larger fraction of the United States' population than one might have supposed, for of the whole population of the United States one in twenty live within ten miles of the Brooklyn Botanic. To most of these people, a tree is a tree and a bush a bush; but what they are seeing and admiring is a collection of 10,000 species, varieties and cultivars carefully chosen to be both botanically representative and horticulturally attractive. The harsh New York winter confines the choice of species to the hardier kinds: as a measure, for the benefit of readers familiar with the plants in question, the tenderest evergreens which survive in the Brooklyn garden are *Magnolia grandiflora* and the *Skimmias*.

These just scrape through the average winter and receive some damage in exceptionally cold winters. They do not attain their optimum size.

Set into and distributed throughout the parklike body of the Garden are a number of small, special gardens and all of these are superlative of their kind. There are two, and will soon be three, Japanese Gardens, exquisitely made and planted under expert Japanese guidance, and

A distinction of Brooklyn Botanic Garden is its three Japanese gardens. This is a perfect replica of one of Japan's most famous Dry Landscape gardens, the Ryoan-ji.

about which I have more to say below; a series of Children's Gardens, connected with the educational programme, also discussed below; a Fragrance Garden originally planted for the blind, with labels in Braille as well as in ordinary letters; an exceptionally well-planted Boulder Garden, confined to such species as favour the scree and boulder fields left by the retreat of the glaciers at the end of the last (Würmian) Ice Age; and a number of pleasing little model gardens which amateur gardeners can study and learn from. From all this it will be clear that pure botanical science is not important at Brooklyn, as it is at the Bronx, although taxonomy still has a place, and plant morphology comes into the educational programme (see below). As in so many cases elsewhere, the emphasis has shifted from botany to horticulture. One other sector of this admirable little Botanic Garden to which I shall revert below, the collection of both ancient and new Bonsai trees, is in my opinion the finest in the United States and one of the finest collections of its kind in the world.

THE JAPANESE GARDENS

Because it is unique in the Western world and a perfect facsimile of the ancient original, academically the most interesting element of the Brooklyn Botanic Garden is its copy of the Ryoan-ji. The Japanese original dates from the fifteenth-century, and is a perfect example of a Dry Landscape garden of the kind made by Zen Buddhist monks, containing, in the disposition of its few rocks, the ancient elements of Horai and other symbols. It is an oblong of raked sand enclosed, on three sides, by beautifully coped walls and Japanese buildings, and on the fourth by a railed terrace above the level of the garden, from which the garden can be contemplated. Only a few people are admitted at one time and silence is enjoined on them. In the raked sand oblong are placed a number of single rocks and groups of rocks. It is in the disposition of those rocks, their

Below left The white painted trellis pavilion of Brooklyn Botanic Garden's Rose Garden where the All-America rose collection is displayed.

Below right Both tropical and temperate zone water-lilies are well grown at Brooklyn. The courtyard of the big greenhouse contain's the Garden's superb Bonzai collection.

sizes and shapes relative to each other, and to the dimensions of the sand oblong, that the beauty and the soothing effect of the garden upon the mind and spirit are to be found. Each rock has a meaning, of course; the Zen Buddhist monks incorporated into their Dry Landscape gardens very ancient pre-Buddhist symbols from earlier Horai gardens: Horai stands for a group of mountainous islands in the ocean which were inhabited by tortoises, cranes and immortal beings—all longevity symbols; and its symbol in the earliest Japanese gardens was a group of rocks of a certain kind and shape placed in the garden pond or lake—tortoise and crane islands. Actually a number of different symbols have become combined into a syncretic convention, and it is extremely difficult for a non-Japanese to be aware of all the implications of a given number, disposition and form of rocks. But the effect on the mind is undeniable; we were fortunate in having the Brooklyn Ryoan-ji to ourselves; it is the one perfectly peaceful spot in the whole New York conurbation.

The second Japanese Garden, actually older in construction than the Ryoan-ji, is a traditional hill-and-pond landscape garden of a kind older in Japanese history than the Dry Landscape garden. It was designed and laid out for Brooklyn in 1914–15 by Takeo Shioten, a Japanese garden artist working in New York; he had already, at that time, endowed the State with six or seven fine works, but the Brooklyn Botanic example is considered to be his masterpiece. Central to it is a small lake shaped like the Chinese character *shin*, difficult to render in a word of English (French, *esprit*), but *heart-and-mind* comes near. to it. This is fed by a cascade in the thirteenth-century style. The island in the lake is tortoise-shaped (the tortoise being one of the longevity symbols) as in the Japanese gardens of the eighth and ninth centuries, and is connected with the mainland by a Drum Bridge. The lake-shore has a Waiting Pavillion and a Tea House taken from two different periods in the art of Japanese gardening, but in a single tradition. The garden has two lanterns: these stone lanterns which we think of as typical were actually a rather late element in Japanese garden art (sixteenth century), and incidentally the only Christian element.

The principal plants of this garden are maples, cherries, azaleas and barberries, with some iris. The clipping of its evergreens is of course in the Japanese style, although the azaleas are, incorrectly, allowed to flower. This small Japanese landscape garden is an accomplished work of art.

The third Brooklyn Japanese garden, incomplete when we last saw it towards the end of 1968, is a Roji Garden. Essentially, the Roji garden is a winding walk, planted and gardened on both sides, connecting the dwelling house with the Tea House. The Brooklyn example consists of a wandering path through plantings; but the plantings are nearly all composed of dwarf conifers, a really exquisite collection, so that the garden is also a botanic collection.

The Boulder Garden consists of a level area of scree planted with a large range of scree species both native and exotic, and backed by boulders piled 'after nature' and planted. This garden has a long season in flower and in it, unlike most elements of the Brooklyn Botanic, botany is more important than horticulture, although most of the species cultivated in it are ornamental. Not far from this scree and boulder area is a two acre conservation enclosure left to grow wild and planted with natives of New York State.

Among the shrubs which, apart from some damage caused by air pollution, do well in the Brooklyn Botanic, are hardy hybrid (and a few species) *Rhododendrons*; *Wistaria* on pergolas or arbours decorating cross-ways; *Kalmia* and *Pieris* planted in close boskage; *Magnolia stellata*, of which there are many fine specimens and also a hedge, which is something we have not seen in any other garden; the evergreen *Magnolia verginiana*; an enormous *Magnolia stellata* × *M. kobus*; *Ilex crenata helleri*, a Japanese dwarf holly which makes a dense, hard, tough, small leaved evergreen dwarf edging hedge as attractive as dwarf box, as easy to manage, and much hardier. A very

Above One aspect of Brooklyn's Japanese landscape garden, designed and planted by a Japanese master-gardener. *Below* Both tropical water-lilies and tropical ferns are well grown in this garden in the heart of New York's most urban suburb.

fine specimen of Sargent's Weeping Hemlock. Among the genera which are well represented are *Syringa*, *Lonicera* and *Abelia*. Trees which are unusually interesting either because they are rare, or simply as exceptionally fine specimens, include a Sorrel Tree or Sourwood, *Oxydendron arboreum*, the only large Western Hemisphere tree belonging to the family *Ericaceae* and greatly valued as a source of honey; *Ginkgo biloba*, represented by some fine large specimens and also by a young avenue of the fastigiate form; *Cladastrus lutea*, the Yellow Wood Tree; a good collection of cherries, planted formally in rows to form a sort of quadruple avenue on fine lawns; and an avenue of Maples.

There are two rose-gardens, one large and one small. The large one is laid out formally, enclosed in white trellis fencing, and focused on a white trellis pavillion. It is devoted exclusively to cultivars, although some species, such as some *spinosissimas* and *rugosas*, are represented. The smaller Rose Garden has climbers trained on big hoops centred on a water-lily pool with a fountain. The roses chosen for both of these gardens are a guide to amateurs who want to know which kinds will stand up best to urban conditions and the New York climate.

In front of the conservatory complex there are two large water-lily pools, rectangular and formal, one devoted to temperate zone water-lilies, the other to tropicals. The conservatories have an adequate but unremarkable collection of tropicals; and a widely representative collection of tropical and sub-tropical orchids, as well as a succulent collection.

Bonsai is a special interest of the Brooklyn Botanic Garden. It began when, in 1925, Ernest F. Coe of New Haven, Conn., presented thirty-two fine specimens—a very valuable gift—to the Garden. Unfortunately it was neglected, there was at that time nobody on the staff interested in Bonsai. When Dr Avery became Director he found many of the little trees dead from neglect, but some capable of being saved and revived. With these he restarted the collection and began adding to it. Several friends of the Garden have presented their own collection to it, so that it is now not only large but rich in very fine specimens. Of these the best is a Japanese white pine growing in a flat dish and now over a century old; there are other pines, maples, some oaks. In addition to these ancient specimens there are many young ones; and a nursery in which Mr Frank Okamura, of the Botanic Garden staff, trains new ones. By the new techniques which have been worked out at the Garden, 'instant Bonsai' can be produced. As a part of its educational programme the Garden runs courses in making Bonsai for the public; and its important publications department has published both an elementary and an advanced handbook on Bonsai growing, by Japanese experts.* A kindred art which is studied and taught in the Garden is Bonseki, the making of miniature Dry Landscapes in trays, an art which originated in the Suiko epoch of Japanese culture (A.D. 593–627), but which was given its greatest impulse by a Japanese nobleman who was also an artist, Sansai Hosokawa in the late sixteenth century.

The Children's Gardens in the Brooklyn Botanic were started together with a series of courses in horticulture and botany, 'to introduce city children to the world of plants'.† These are vegetable gardens, and the children take home hundreds of pounds of produce every season. It is found that young children are much more readily interested in vegetables than in flowers.

'Basically, the programme consists of early spring classroom and greenhouse instruction and work, followed by the actual cultivation of individual outdoor gardens.'‡

Classes and garden work are on Saturdays and Sundays; there are two hundred plots each worked by two children in partnership, and very many more children apply for the courses than the 400 per season it is possible to accommodate.

I have referred to the educational courses for adults which are also an important part of the Botanic Garden's activities. Courses are diverse—Fruit growing, Bulb growing, Houseplants, Ikebana, Orchid growing, Gardening for Autumn, Evergreens, Flower Painting, Bonsai, Bonseki—these are only a very few of the short courses offered. The shortest courses last one day, the longest fifteen days, and all are popular with New York and out-of-town gardeners. About 5,000 people enroll for courses every year.

A still wider public is reached by the Botanic Garden's publications departments, which must, I think, be by far the most ambitious and successful thing of its kind in the world. The Handbooks are beautifully printed and illustrated—often in colour—reprints of the Brooklyn Botanical Garden's *Journal* for its Members (*Plants and Gardens*), for which the subscription is three dollars a year. These Handbooks are published at $1, and each is produced as follows. After a subject has been decided on, a guest editor is chosen, and he or she will be a world expert on the subject. This editor will then commission specialists to write on all the aspects of the subject and, having passed the contributions, the Garden's publication department takes over and produces the Handbook. About 10,000 of each handbook are sold to subscribers; there are fifty-six handbooks in print.

The special distinction of the Brooklyn Botanic Garden is beyond question its splendid record, and its continued and increasing activity, in educating an urban people in the fields of botany, horticulture and nature conservation.

**Handbook on Dwarf Potted Trees*. Ed. Kan Yashiroda. *Handbook on Bonsai: special techniques*. Ed. *Kan Jashiroda*.

†Miner, F. M., Curator of Education at the Brooklyn Botanic Garden, RHS *Journal*, June 1964.

‡Miner, *op. cit.*

LONGWOOD

Opposite The waterworks in one of Longwood's Italianate gardens are the most spectacular which have been made since the Renaissance; and no Renaissance garden had its fountains floodlit at night!

A visitor to Longwood Gardens near Kennet Square, Pennsylvania—and at a guess he was a New Englander from Vermont or Maine—was heard to remark as he left the Gardens, 'Just think what God could have done if He'd had money'. I quote it because it is apt: all the other botanical gardens in the United States, and most botanical gardens in the world, are short of money; Longwood is rich.

Longwood Gardens include one of the finest arboreta in the United States; they have a Herbarium; they employ, in the person of Dr Huttleston, a taxonomist, and a plant geneticist in that of Dr Robert Armstrong. They carry on experimental work; plant introduction is one of the Gardens' important functions; they acclimatize, propagate and distribute new plants. But Longwood is more concerned with cultivars than with botanical species, and with horticultural display than with pure science, though it is very much to the point that Dr Seibert, Longwood's Director, is an experimental botanist, a taxonomist, a plant collector, and a scientific worker in the field of economic botany.

In the year 1700 William Penn made a grant of land to one George Peirce, whose son Joshua built a house on it in 1730. In due course it was inherited by his twin grandsons Joshua and Samuel Peirce, and they were the men who started the arboretum which is now Longwood Gardens. If an account written by Mary Woodward and printed in *Westchester (Pa) News* in 1903 be true, no arboretum ever had an odder beginning. She says that the central 200-acre tract of the property was all peach orchard when the twins inherited it: and that Joshua and Samuel were so irritated by always having the best part of their peach crop stolen that they grubbed up the orchard and decided to indulge their—and particularly Joshua's—interest in natural trees and shrubs by planting an ornamental arboretum. It seems that their neighbours thought them at best eccentric and at worst insane. That was in the year 1800. Joshua Peirce had two sources for his collection: the woods around him, from which he collected young trees and shrubs in the course of a number of expeditions; and the famous Princes Nursery on Long Island, from which he procured exotics. Hedricks mentions the Peirce Arboretum in his *History of Horticulture in America*. But there were earlier tributes to the Peirce collection: Josiah Hoops refers to its double avenue of native conifers in his *Book of Evergreens* (1868) and adds 'This little select arboretum is probably . . . not surpassed in these United States'. And Darlington, in his *Memorials* . . . (1849), says that the collection on the Peirce estate had over 100 taxa:* he also says that the example followed by the Peirce brothers had been set by a neighbour named Jackson, who had started an ornamental arboretum of native woody plants in 1770.

A surprising number of the Peirce plants are still alive in the Longwood of today, among them a *Magnolia cordata*, whose history is quite well known; according to Sydney Stebbins' *Letters* this tree was one of the very first

*The present figure is about 12,000.

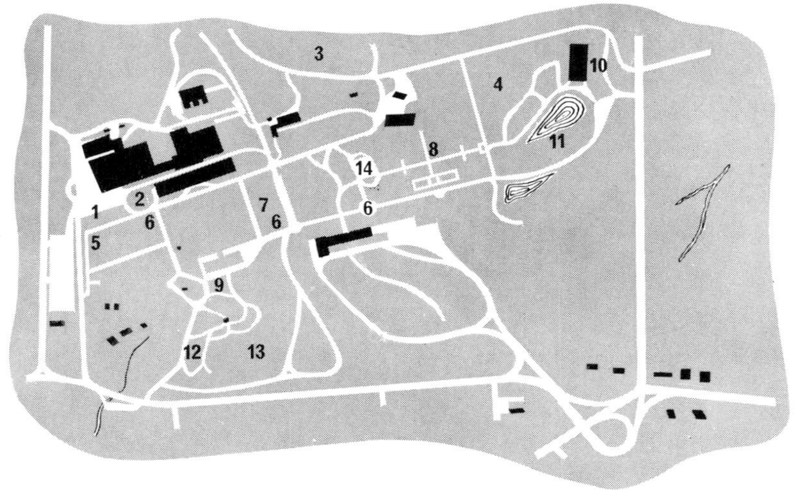

Key to principal divisions of plants

1 Herbaceous perennials
2 Annuals
3 Woodland
4 Apple orchard
5 Vegetables
6 Roses
7 Topiary
8 Flower garden
9 Rock garden
10 Water garden plants
11 Aquatics
12 Heaths and Heathers
13 Arboretum
14 Open air theatre

planting; Stebbins was Joshua Peirce's grandson-in-law and heir to G. W. Peirce. The first planting he refers to included *Taxodium distichum*, the Swamp Cypress or Bald Cypress; Firs, Yews, Beech, Holly, Box and Laurel, and also *Ginkgo biloba*, the Maidenhair Tree. In 1968 this immensely tall *Magnolia cordata* had a girth of 11 feet 6 inches. There are other remarkable old trees which I shall come to in their place. As early as 1853 Thomas Meehan was referring in his *Handbook of Ornamental Trees* to sixteen trees of 'notable import' including a *Magnolia acuminata* whose trunk was 4 feet in circumference; it was next measured in 1922 when it had attained 9 feet 6 inches; today the measure is 11 feet. Reverting briefly to the great *Magnolia cordata*, when Professor Sargent visited Longwood after Pierre S. Dupont had bought the place in 1906, he said that the tree was the largest of its kind he had ever seen, and E. H. Wilson also includes this remarkable specimen in his *Aristocrats of Trees*.

But the story of Longwood's greatness as a botanical garden really begins with the du Pont purchase. In 1906 P. S. du Pont was passing through Kennet Square (it is still a small and charming old Pennsylvanian town) and noticed a sawmill being set up near what is now Longwood. On enquiry he learnt that the numerous, quarrelsome and indigent heirs to the Peirces were about to fell the timber on their estate including the once famous arboretum, to pay their debts. Du Pont bought the estate to save the trees. He seems to have realized that wealth as well as nobility has its obligations. Happily, he was a much-travelled man and a trained engineer, as well as being as rich as a European government. I have heard persons who value themselves on their taste jeer at some of his works at Longwood; but how some of the Italian Renaissance new-rich, who created the great Italian gardens, would have loved the Longwood of today.

Longwood Gardens cover 1,000 acres. But the principal

Right Decorative detail at Longwood is either Italianate or French. For all its botanical importance, Longwood is one of the most spectacular ornamental gardens in the world.
Below Formal water-lily garden at Longwood. The central pool contains *Victoria* × 'Longwood' (*V. amazonica* × *V. cruciata*).

areas of interest to the million visitors who see them every year, are all within a sector of 300 acres. Nearest to the house which, enlarged, was lived in by Pierre du Pont, are the oldest plantings—magnificent trees including those already named, and also Lombardy Poplars, *Ailanthus*, and Norway Maples. (Each of these three species has, in its turn, been the most popular street tree in the United States. The last is being succeeded by Ginkgos, but now these, too, are showing signs of trouble due to air pollution.) The trees are densely underplanted with *Pieris* and *Rhododendron*, and with some other hardy evergreens. The whole of this park-like area is threaded by broad rides and narrow, winding walks. A typical ride, between a double avenue, is a noble vista ending in a circular lake with a fine Italian fountain. Among the most spectacular old trees are such natives of the surrounding woods as *Liriodendron*, White Oak, Hickory and Sugar Maple; the enormous Hemlocks are not locally native. Beyond the trees is one of the several lakes, with wonderfully contrived hill and tree vistas across water, rather in the Chinese landscape taste. All the lakes are spring-fed and feed, in their turn, the elaborate irrigation system. Near the water, and in the neighbourhood of other lakes or bodies of water, are some of the most magnificent stands in the world of *Taxodium distichum*, also individual 'kneed' specimens; and one big tree of the superlatively beautiful variety *ascendens*, with its jade-green foliage.

In several parts of the grounds are gardens of French-Italianate or Italian Renaissance inspiration. The influence of the Villa Gamberaia, of the Villa d'Este and of Versailles are all apparent. Notably, one very large garden of masonry, stairways, ballustrades, fountains, all embowered in clipped evergreens, seems to have been inspired by both the Villa d'Este and Versailles, but it still retains an American character, just as early Italian Renaissance gardens in France contrived to be French. The magnificent waterworks of this garden, with scores of fountains whose play composes an integral picture made of many jets, is flood-lit at night by a system of slowly changing and blending coloured lights. In white, yellow, amber and light green, this abstract painting in moving waters is beautiful; in the stronger colours, reds, purples, violet-blues, it is suddenly as vulgar as it remains spectacular; as vulgar, probably, as the musical fountains of the Italian Renaissance gardens, vulgar in the sense that the ingenuity of the contrivance was, and at Longwood is, more important than the taste of the effects.

In another, smaller, Italianate garden of masonry, sculpture, fountains and evergreens, the inspiration is the Gamberaia, and in this case the result is free from the flaw of excessive showiness. Here hemlock, clipped, is used instead of Italian cypress, and the flanking trees which contain the garden are *Tillia cordata*. Yet another Italianate garden is all topiary and lawns; for one of the distinctions of Longwood is that its 12,000 taxa of plants are used as material in the representation of every horticultural style ever developed in Europe or North America.

The flower gardens are of two kinds: there are a number of enclosed gardens, almost *giardini segreti*, since they are hedged by tall, clipped hemlock or arbor vitae used in the manner of cypress in Italy, and by masonry works pleasingly combined with these evergreens in contriving the enclosures. Each garden is devoted to a single genus: *Paeonia* and *Iris* are cases in point. In addition to these enclosed gardens, there are splendidly managed open, terraced gardens which are kept in vivid colour from April to November by bedding-out from nurseries. Here, as in the Vegetable Gardens and Herb Garden, and likewise in the formal Rose Garden, visiting amateurs can note the cultivars which would be most suitable for their own gardens.

In his Foreword to this book Sir George Taylor writes of botanical science as the handmaiden of horticulture; nowhere is it more so than in the great botanic gardens of the United States; and of them all, Longwood is in this respect superlative, whether in the practice of traditional methods or in the inventing of new ones.

Below The 'Water Eye' at Longwood, one of the most curious of the garden's remarkable collection of fountains and cascades.

Perhaps the only failure at Longwood has been the parterre garden of clipped box. *Buxus suffruticosa* does not seem quite hardy enough for Pennsylvania.

The open-air theatre at Longwood is remarkable for its size, for its extensive underground dressing rooms and control centres, and in half a dozen other ways. In the context of this book, however, the notable element is the entirely successful use of the ruggedly hardy American arbor vitae in the making of the clipped evergreen wings; in Europe box, yew or Italian cypress would have been used. The use of arbor vitae, of Hemlock, and of *Ilex crenata* in American botanic gardens, certainly enriches the materials available for Italianate gardening.

One of the most recent additions to Longwood is a large collection of heaths and heathers planted very pleasingly as a heath garden on an extensive hillside. Most of the *Ericas* are doing well and close coverage has been achieved; but in the hard winter of 1967–8 a large number of *Calluna* cultivars and some *Ericas* were lost. Longwood botanists

and horticulturists decided that this was for want of a snow covering. In Pennsylvania snow can usually—but by no means invariably—be expected to protect grounding plants from the severe frosts; when there is no snow, then damage is serious. Dr Seibert and his colleagues determined that Longwood collections must not be at nature's mercy. Consequently a snow-making machine has been added to the Garden's mechanical equipment, which is capable of providing snow cover over sufficiently broad areas to protect such collections as the heaths and heathers.

But however impressive all the other aspects of Longwood may be, one repeatedly reverts to its trees; and it is very satisfactory that a new arboretum of 260 acres is now being planted. As usual, I starred a certain number of trees in my notes, in addition to those already named as belonging to the old Peirce plantings. There is a *Ginko biloba* which is now about 85 feet tall; and at least one of the old *Magnolia cordata* trees bearing masses of yellow

An aspect of Longwood's great Victorian Conservatory which opens into the grand ballroom. This part of it is planted as a sub-tropical garden embellished by a year-round scheme of 'bedding out' tender herbaceous plants.
Below left Nelumbo, the oriental lotus, flower and fruit at Longwood.
Below right Victoria × 'Longwood'.

flowers, is about the same height. Several *Gymnocladus* are outstandingly splendid specimens. There are Hemlocks of great nobility and stature, one at least planted before the beginning of the nineteenth century. After a serious loss of tall old trees during a bad electric storm some decades ago, tall trees at 100-yard intervals were fitted with copper lightning conductors, since when there have been no more losses due to lightning. There are some magnificent specimens of *Betula lenta*, known to the timber trade as 'Cherry Birch', and formerly valued as a source of winter-green ointment. *Magnolia acuminata*, and *Liquidambar* are two genera represented by superb specimens. As well as specimens and stands of trees, there are several avenues and double avenues: one is composed of *Paulownia* and there are also good free-standing specimens of this species whose trunks, at Longwood have reached a circumference, in some cases, of 15 feet. A stand of *Libocedrus decurrens* is the finest thing of its kind I have seen anywhere. There are some gigantic limes; and some of the Sugar Maples are probably older than the Garden itself and must be among the largest in the country. Copper Beeches, early introduced from England, lose their copper colour quite early in the autumn. In the Pinetum, the Colorado Blue Spruces are the most beautiful of all the conifers.

The Conservatories at Longwood are very remarkable: in the large central greenhouse a small landscape garden of lawn, pool and trees, with shrubs and herbaceous plantings, is laid out. By the old Victorian method of bedding-out the harbaceous plantings are maintained in flower throughout the year. The trees are *Acacia*, *Dicksonia* (Tree Fern), and Norfolk Island Pine. Opening off this central greenhouse are a number of others: there are splendid horticultural Victoriana, for example, in the enormous and beautifully maintained espaliered nectarines. There are collections of *Acacia*, of *Citrus*, of *Camellia* and other genera tender or semi-tender in the climate of Pennsylvania. Under this vast expanse of glass there are formal gardens of tropical shrubs and herbaceous plants, including a collection of *Cycads* which is not, however, equal to Kew's. Many of the shrubs and small trees are Australian: these include some *Banksias*, and the Tree Ferns reach a height of 20 feet or more.

Longwood horticulturists are very skilfull in training plants for display: there is an enormous "chandelier" of ivy-leaved pelargonium; wonderful hanging baskets of 'Cascade' chrysanthemums trained into a shape I can describe only by saying they are like octopuses. By grafting *Hedera helix* in some of its prettiest variegated forms onto *Fatshedera*, the most enchanting little standards of coloured ivies have been produced, and also weeping standards, admirable for interior decoration. Chrysanthemums are also trained to form large, vertical, rectangular panels of flower.

The Garden has almost year-round displays of Orchids in the greenhouses, moved out as they come into flower, from the orchid collection. Every genus in cultivation can be seen here. And there is a fair collection of *Bromeliads*, well grown and displayed. The Palm House, the newest greenhouse built onto the conservatory complex is very beautifully planted to resemble a tropical landscape, and can be viewed from two levels, ground, and above; and also at night by electric light. Both the Fern and Cactus collections are set out 'after nature'. Altogether, the plant house complex at Longwood is one of the best in the world; although botanically less rich than those of some European gardens, it is more attractively planted than most.

In addition to the Garden's taxonomic work, based on both the living collections and on the Herbarium in charge of the taxonomist Dr Huttleston, Longwood Gardens undertake certain work in experimental botany, and in such related subjects as soil conservation and rehabilitation. In the latter field the Gardens are doing quite admirable work, for example in buying up mushroom farms on which the soil has been irresponsibly 'mined', and restoring them for arboriculture; this work is being done on a large scale. I have already mentioned the pilot project of putting the Garden's plant records into electronic terms. Longwood maintains laboratories and experimental greenhouses with controllable temperature, humidity, etc., and some work is being done by Dr Armstrong in phyto-genetics. The collection, acclimatization, propagation and distribution of exotics is undertaken. Two interesting plant-breeding programmes are in hand: a programme of *Camellia* breeding with the object of producing cultivars which are reliably hardy in the Pennsylvanian winter and like climates; and a programme of *Penstemon* breeding for new, hardy cultivars. Other trials in hand, and, in collaboration with the University of Delaware, are hardiness tests on a number of species; disease-control tests; chemical weed-control tests; and tests for the control of the European Red Spider mite which has become such a very dangerous pest in America. Two of the Garden's own projects are trials, in air-conditioned greenhouses, of the flowering of chrysanthemums by means of temperature manipulation instead of light and darkness manipulation; and propagation by cuttings without the use of any medium at all, that is, simply in moist air and darkness. Longwood is also very well-equipped for 'vernalizing' bulbs: 80,000 bulbs are annually prepared for spring flowering in automated cold and warm, light and dark rooms.

Longwood has a very fine collection of water-lilies; this was originally made for Pierre du Pont by Mr George Pring, whose original work in this, as in other fields, was done at the Shaw Garden in St Louis (see below). Among that Garden's achievements with these plants has been the crossing of *Victoria amazonica* with *V. cruciata* to produce a very vigorous F.1 hybrid, now known as *Victoria* × 'Longwood', which is now being grown in, among other places, the Edinburgh Botanical Garden.

Required to sum up Longwood as a garden, I don't think that I could do any better than repeat the witticism which I have already quoted: 'Think what God could have done if He had had money.'

MISSOURI

Opposite: above The Climatron in the Missouri Botanic Garden seen from across the large formal lily pool with *Victoria* and other water-lilies. *Below* Chrysanthemums in massed plantings are one of the features of the Missouri Garden in autumn.

That Garden which its maker called the Missouri Botanical Garden and gave to the people of St Louis, the people still call Shaw's Garden after their benefactor.

Henry Shaw was born at Sheffield, England, in 1800, educated at Mill Hill school, and then taken into the family business. On that business he accompanied his father to Canada in 1817, by which time he had picked up a taste for botany at Kew Gardens. In order to learn something about the cotton trade, he went to New Orleans. There, still in his teens, and using capital lent to him by an uncle, he bought sugar cheaply and, using the Mississipi as his means of transport and St Louis as his entrepot, sold it very dear into the then opening Middle West. With his profits as the basis of his fortune, he settled in St Louis in 1819 and began to import and sell English hardware and china. By the time he was forty he had made himself St Louis' richest citizen. The rest of his very long life was devoted first to a period of European travel, and then to the making of Shaw's Garden.

His determination to make the Garden was prompted by two other remarkable men: his friend and physician, the botanist Dr George Engelmann, a German immigrant who urged that the garden which Shaw was making at his country house should be a botanic garden and should become part of the State's heritage; and Joseph Hooker, Director of Kew Gardens, whom Shaw met while on a visit to England. After his travels, Shaw finally settled again in St Louis in 1851. Hooker, by then Sir Joseph, stayed with him at Tower Grove, his country house, while on his way to the Far West. That visit clinched Shaw's determination to make his garden a scientific one and to leave it to the people of Missouri when he died. Already an institution by reason of his wealth and worth to St Louis, he became even more so by reason of his soon famous garden.

Shaw offered the post of Director of the Garden to Dr Engelmann who was already well known as a botanist; Engelmann refused on the grounds that Tower Grove was too far from the city—it is now in the suburbs and about 20 minutes by bus from the downtown centre. But that did not prevent him from going to Europe to collect plants and buy Herbarium specimens towards the planting of the Garden and the enrichment of its Herbarium. Shaw then decided to take the direction of the Garden on himself; he kept it for thirty years, until his death at the age of eighty-nine. But Engelmann was his right hand. Dr Burch, the Botanic Garden's present Chief Horticulturist, says that Engelmann was

> . . . a remarkable taxonomist; his grasp of the relationships of plants was so strong that even some of the tiny fragments which came back by horseback, often soaked in fording rivers or battered by much handling, would be enough for him to be sure that this was a plant well known to science or something completely new. . . . Dr Engelmann became internationally known for his work with a number of very difficult plant families. . . .

The second Director, under the Trust and Board established by Shaw's will, was Dr William Trelease, an

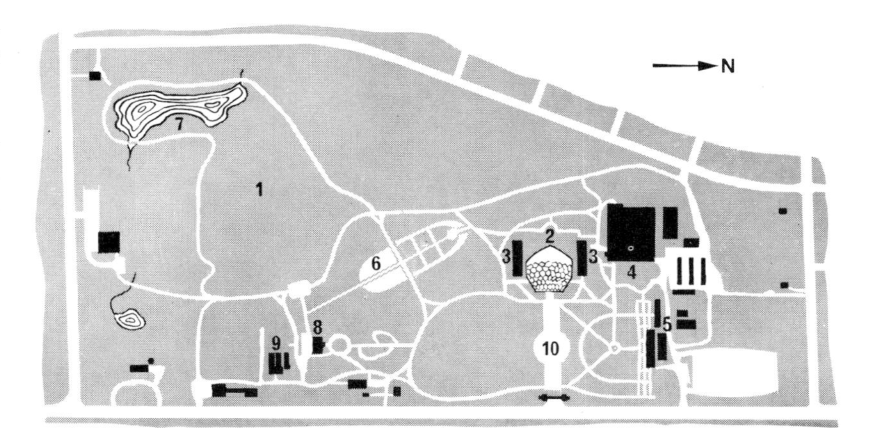

Key to principal divisions and areas

1 North American flora
2 Climatron area
3 Houses for xerophytes
4 Floral display houses
5 Linnaean house
6 Rose garden
7 Native flora and aquatics
8 Shaw's house
9 Experimental greenhouses
10 Water-lily area

able scientist during whose term the Garden's reputation was further increased. He brought more order into the layout of the Garden, his new designs being carried out partly by himself and partly by his successor, Dr George Moore, with the help of the first garden architects to take a hand at Shaw's, F. L. Olmsted and John Noyes.

The Herbarium was added to under each Director and it now amounts to 2,000,000 specimens or sheets; partly as a consequence, the Missouri Botanic Garden continues to be a taxonomic centre. The Library, too, was built up over the years and is an exceptionally good one. In experimental work of practical importance, the Garden has also made considerable contributions: its scientific staff, co-operating with its gardeners, developed the use of creosote as a wood preservative, and its results were subsequently used all over the world; Dr R. M. Duggar, its plant pathologist at the beginning of our century, was the first man to measure a plant-virus particle (tobacco virus) and so to contribute to the advancement of phytovirology. It was also in the Missouri Botanical Garden that the techniques of growing mushrooms commercially were first worked out in America. Introduction and acclimatization of exotics was and remains an important part of the Garden's work.

A striking example of the combination of botanical knowledge with horticultural skill to produce useful practical results is to be found in the work of G. H. Pring, sometime Superintendent of the Missouri Botanical Garden and associated with it for more than sixty years. Mr Pring had a particular interest in water-lilies. He started, in the Garden, a long-term breeding programme, working chiefly with the tropical species; he not only produced a whole series of cultivars now grown all over the world, but he gave the Garden such a 'bent' for these plants that he and his pupils have given us over forty new water-lilies. It was this work which left the Garden with its exceptional collection of water-lilies. In the early 1930s the Garden's staff brought the yellow-flowered

The Orangery (known as the Linnaean House) in the Missouri Botanic Garden. It is the oldest Orangery still in use in America.

Right Missouri Botanic Garden. This kiosk covers Shaw, the founder's, tomb. Shaw, an immigrant from England made his fortune as a young man and devoted the rest of his long life to the creation of one of America's greatest botanic gardens.

tropical *Nymphaea burtii* into their hybridization programme, and by so doing endowed gardeners with a greatly enriched range of colour in water-lilies. The cultivar 'St Louis' is the outstanding example.*

The Missouri Botanic Garden suffered a period of neglect and decline. Shaw's endowment was, of course, inadequate in modern, post-inflation, terms; and in the United States it is very difficult indeed to obtain municipal or other public funds for foundations such as botanical gardens or zoos, however valuable they may be to the citizens. The source of funds for such bodies are Trusts and Foundations, administered by Trustees or Boards which are apt to be swayed by current and ephemeral fashions in the arts and sciences. Botany and botanical gardening became unfashionable; the fact that they have contributed and do contribute to human well-being, perhaps more fundamentally and broadly than any kind of scientific institutions have ever done, counts for little with Trustees and Boards of Administration. In short, such institutions as the Missouri Botanical Garden are driven, in the United States, to fund-raising campaigns in competition with many others. Nearly all the botanic gardens in the United States are seriously short of money. The Shaw Garden is now recovering, rehabilitating and reorientating itself, to confront a world very different from Henry Shaw's and tasks of a new kind. But it is driven to appeal to the public for the means.

*Pring's work with water-lilies did not end at St Louis. I owe to Mrs G. H. Pring in a personal communication the story of how Pierre du Pont persuaded him to design and plant for Longwood the water-lily garden in which, with Pring's technique, *Victoria* × 'Longwood' was bred.

SHAW'S TODAY

Shaw's Garden now covers 75 acres in the city's suburbs; there are plans to extend it again if the funds can be raised. Its administrators are also responsible for the Grey Summit Missouri Arboretum, 1,600 acres, outside the city.

The garden's most spectacular feature, and its most original, is the Climatron, the first conservatory ever to be built on the Buckminster Fuller geodesic principle (in 1960). The begetter of this remarkable structure was a Director of the Garden, Dr Fritz Went, who had done much work at the California Institute of Technology on climate control and who made use of the geodesic principle in giving practical expression to some of his conclusions. The Climatron is a suspended dome, 175 feet in diameter and 80 feet high, the skeleton of aluminium, the glazing done with Plexiglass. It has no internal supports whatsoever. Under this great dome a number of different climates can be produced simply by the mechanical and automated control of warm and cold air, and of humidity. Consequently a number of different kinds of tropical flora can be accommodated, from the kind which require dry, hot conditions, to that which requires moist warmth. Dr Derek Burch explained that the Went method of air control, modified in practice by experience, gives a wide range of climates from one side of the building to the other, with a warm night temperature on the south and cooler nights on the north.

Taking advantage of these several different climates, the Climatron was planted with tropical species from a wide range of habitats; the plants were chosen for their botanical value, but planted with horticultural skill 'after nature' to create a tropical landscape. Growth was luxuriant and an appearance of satisfying maturity was accomplished within about four years. What they have now is a sort of romantic jungle on which a decent measure of order has been imposed, with representatives of florae from both arid and rain-forest habitats. 'The landscape of the Climatron,' says Dr Burch, 'includes a waterfall, an overlook from what was the portico of the old palm house, and an underwater tunnel which gives a good view of the structure of the *Victoria* leaves and of the underwater parts of other aquatic plants.' This tunnel is made of Plexiglass and passes under the water-lily pool. Incidentally, the water in that pool is kept clear of algae by the use of another aquatic plant, *Salvinia nutans*; and the pool is given a wonderful air of nature by the plantings of giant-leaved aroids, and bromeliads growing on rotting trees. The waterfall, too, is beautifully flanked by aroids, cycads and big tropical climbers. Like the rest of the Missouri garden, the Climatron was somewhat neglected during the time of decline and money troubles; it is now recovering and again being properly used and gardened.

One of the most spectacular plants in the Climatron is

the scarlet-flowered *Perovskia* with its orange fruits. Bromeliads planted naturally flower very well; there is an enormous specimen of one of the climbing palms which promises to span the entire building; one area is devoted to tropical bog plants. In all, the Climatron houses about one hundred taxa of tropical plants, and the type of structure is amply justified in the wide range of plants from different tropical climates, and in the wonderful vistas across the house, in every direction, unspoilt by the pillars or other supports which mar the interior of conventional conservatories.

The Climatron is flanked by two interesting houses of desert plants, American and South African respectively. Here, as in the Huntington Botanical Garden (see below), it is interesting to compare the different means taken by plants in the Old and New Worlds to solve the problem of survival in arid conditions, the New World cacti with the Old World succulents. A former Chief Horticulturist of the Garden, and still one of its officers, Mr Ladislaus Cutak, had a special interest in cacti and succulents and was largely responsible for these two houses. A by-product of this special interest has been the providing of leaves of *Aloe vera* for medicinal purposes: the pulp of these leaves has long been known to have healing properties; it is now being processed into ointments and even cosmetics, chiefly in Florida. The plant was first introduced into America by Spanish missionaries, specifically for its medicinal properties.

Contrasting sharply with the Climatron, the Missouri garden has a pretty Orangery dating from the mid-nineteenth century and certainly the oldest plant-house still in use in the United States. Known as the Linnaean House, it is not now devoted to systematics but to a collection of *Camellia* cultivars, Ferns and Mosses, its back wall being completely covered with *Ficus stipula*. In front of this Orangery are two more of the Garden's lily ponds, stocked with some of the smaller species which Mr Pring worked on; and a pretty garden of herbaceous perennials.

Another range of greenhouses is used for the propagation of plants for display: *Chrysanthemums* are important among these, and it was in Shaw's garden that some of the cascade training techniques, so brilliantly practised at Longwood, were developed. This range opens off the big display hall whose walls are planted with immense tropical evergreens like *Monstera deliciosa*, and which is used for massed flower displays. Other houses of this range contain the excellent Orchid collection—about 10,000 plants in a wide range of genera, and including many cultivars bred in the Garden; a good cactus and succulent collection; a Fern house; and the over-wintering and propagation tanks for water-lilies.

Of the formal outdoor features the two Rose Gardens at Shaw's are outstandingly excellent. Mr Alfred Saxdal, who is in charge of them, must surely be among the world's best rose growers; for it is unusual to see a Rose Garden which, in October, shows not a marred or diseased leaf, with every bush a model of healthy shapeliness, and covered with flowers which seem always to be at exactly their most beautiful point of growth. The principal Rose Garden is a circle within a semi-circle, prettily defined by white-painted post-and-rail fencing grown over by climbing roses. The beds are cut out of lawn, each an arc, arranged in concentric circles. To give the Rose Garden a little height, there are regularly spaced Arbor vitae trees, fastigiate and clipped to resemble Italian cypresses. Central, there is a pretty, very simple fountain and basin. The second Rose Garden is set in brick and grass paths, is on two levels, and is laid out in intercurved beds. The collection of roses numbers 5,000 plants representing over two hundred and fifty cultivars.

The wilder parts of the Missouri Botanical Garden have fine trees and shrubs. There are fine swamp Cypresses; *Maclura pomifera* in autumn sheds its fruits, like very large green oranges, but only the squirrels seem to enjoy them. There are many very splendid specimens of *Ginko biloba*, some of which have attained an enormous size; the female trees drop their stinking fruits very freely in autumn and self-sown seedlings are common. Incidentally, a study of the many old Ginkos in this Garden finally disposes of the idea that the male trees are always of the fastigiate form, the females spreading: both sexes occur in both forms. Hammamellids, Maples and Planes are all very fine plants; in one corner of a border devoted to dwarf conifers is an enormous old *Poncirius trifoliata* which looked very beautiful when covered, in October, with its vivid little oranges.

SCIENTIFIC BOTANY

The Missouri Botanical Garden's principal scientific activity is in the field of taxonomy: an encyclopedic work on the flora of Panama is in hand, and plant collecting in tropical America has been resumed. There is a very interesting collection of plant remains from about 2,000 archaeological sites in the Americas, which enables palaeobotanists to study and reconstruct the spread and development of economic plants in association with man. But the influence of Dr David M. Gates, the present Director, has taken the Garden into a new field of experimental botany—the measurement of the 'energy balance' of growing plants: what part of the energy in the form of daylight, falling on plants, is consumed, what part 'wasted'? That is a very crude and approximate statement of the question. Dr Gates is a physicist who came into botanic garden work by way of a special interest in such problems, and in ecology.

Shaw's Garden also has a considerable educational programme. In the past it ran a good gardening school, and it is the intention of Dr Derek Burch and the Director to reopen this in co-operation with one of the St Louis junior colleges. Courses for adults are run, in budding and grafting; in the raising of perennials and annuals; in Bonsai; in plant propagation; in the making and care of lawns, in greenhouse management, and so forth. There is a whole curriculum of courses for children, in botany, natural history, and horticulture.

STRYBING ARBORETUM

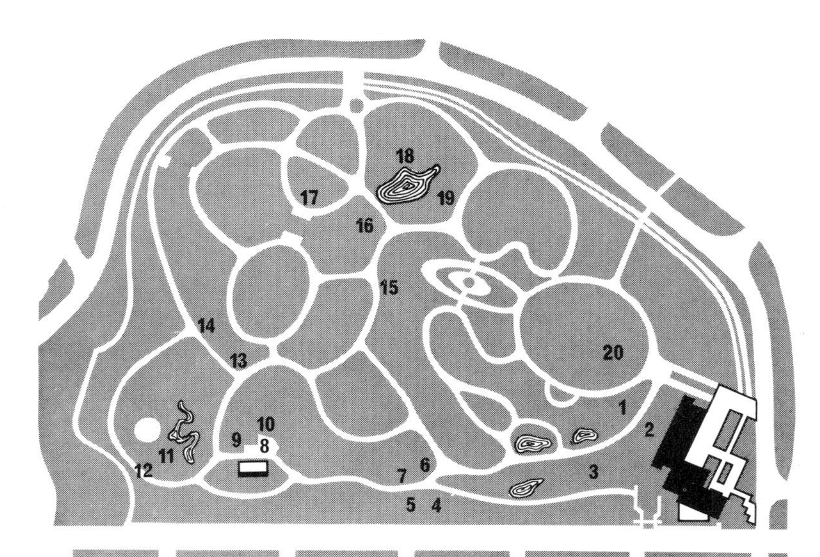

Key to principal stands of trees or specimen trees
1 *Pinus radiata*, Monterey Pine
2 *Chamaecyparis lawsoniana*, Lawson Cypress
3 *Thuya plicata*, American 'Giant Cedar'
4 *Metasequoia glyptostroboides*, Dawn Redwood
5 *Juniperus occidentalis*, Western Juniper
6 *Pseudotsuga menziesii*, Douglas Fir
7 *Taxus brevifolia*, Western Yew
8 *Pinus torreyana*, Torrey Pine
9 *Torreya californica*, California Nutmeg
10 *Libocedrus decurrens*, 'Incense Cedar'
11 *Tsuga heterophylla*, Western Hemlock
12 *Sequoia sempervirens*, Coastal Redwood
13 *Sequoidadendron giganteum*, Giant Redwood
14 *Picea sitchensis*, Sitka Spruce
15 *Pinus muricata*, Bishop Pine
16 *Abies grandis*, Grand Fir
17 *Abies bracteata*, Santa Lucia Fir
18 *Pinus contorta*, Beach Pine
19 *Pinus ponderosa*, Ponderosa Pine
20 *Cupressus macrocarpa*, Monterey Cypress

In 1868 the City and County of San Francisco bought 1,140 acres of barren, shifting sand-dunes beside the Pacific Ocean to make a park. It seemed an odd choice: the land was worthless but three years later a man of vision became Superintendent of this future Golden Gate Park. He was succeeded by another in 1887, and one who was, moreover, a man with exceptional horticultural skills, John McClaren. Using *Ammophila arenarica*, a grass imported from Europe, they stopped the sand from drifting and shifting. Having thus fixed it, they built a top-soil on the sand with trainloads of loam fetched from further inland, and of horse-manure from the city-sweepings which, in those days, were plentiful. Thus they made the 1,100 acres fertile. But they still had a terrible problem: high winds, often gales, off the sea. They planted screens against the wind with the native Monterey Cypress and Monterey Pine; and with Eucalypts, Olearias and other wind-resistant woody plants from Australasia. Out of the barren and shifting sand they made, by these means, one of the most beautiful parks in the world, in some ways *the* most beautiful. While the vigorous private enterprise of American farmers was busy destroying soils on a terrifyingly vast scale, these two municipal officials showed how a formerly sterile waste could be made to support thousands of species of useful and beautiful plants including enormous forest trees, for the Golden Gate Park is gloriously rich in magnificent trees, in fine sweetly rolling lawns, in hundreds of species of flowering shrubs including by far the most remarkable collection of *Fuchsias* we have seen.

In one corner of this man-made paradise is the Strybing Arboretum. Hall had dreamt of such a botanical garden; so had McClaren, and six years before his death in 1943 money for the realization of this project became available from the Helene Strybing Bequest. Thus in 1937 a start was made in an undeveloped but tree-planted area of 70 acres, providing San Francisco with a botanic garden.

The climate of San Francisco is exceptionally favourable for gardening, if the high winds be discounted. The United States Weather Bureau has described the town as 'an air-conditioned city'. There are none of the conditions which keep Los Angeles, further south, smothered in smog; on the contrary, the air is kept constantly on the move. Summers are cool and pleasant, winters are mild, mists and fine rain favour lush growth. The sun shines for sixty-six out of every hundred hours of daylight; the mean maximum temperature is 62·7° F, the mean minimum 50·6° F. The record low temperature is only 2° of frost, the record high 97° F, lower than London's. The 21·75-inch mean annual rainfall is supplemented by atmospheric humidity, and, fairly easily, by irrigation, much of which is done economically with purified sewage water.

The Arboretum's first Director, Eric Walther, started by planting according to a geographical plan; but this had, for want of space, to be abandoned in favour of a plan whereby species of similar requirements and diverse origin are shown together. Since the Garden was, and still is, primarily educational (this was insisted on in the Helene Strybing Bequest), labelling is excellent, and very well maintained, with a specially devised immovable label to forestall the usual botanical garden problem of vicious or thoughtless label-stealing. Since Walther's time over 5,000 taxa have been added to the Garden. In 1960 a landscape architect, Robert Tetlow, was employed, and between then and 1965 the hitherto flat garden was given a much more interesting topography by digging hollows and raising small hills. Today, apart from the motor-roads designed to give access to every part, the Strybing Arboretum has the general look of a very good English landscape garden of the modern, romantically planted rather than the eighteenth-century classical type.

THE PRINCIPAL COLLECTIONS

The Strybing Arboretum is chiefly remarkable for its collection of *Magnolias*, and related genera. About half the known species are represented, most of them by fine, well-grown specimens; and this includes a whole range of forms of the lovely *M. campbelli*, with gigantic flowers varying in shade from pale pink to almost crimson. There are very good specimens of *M. grandiflora*, and such rarer, also evergreen, species as *M. nitida* and *M. delavayi*. Magnolia relatives include *Talauma*, *Michelia*, *Illicium*, *Liriodendron* of course—that fine native is in almost every American garden—*Kadsura*, and *Drimys*.

Central to the whole Garden and a useful point of reference when wandering in it, is a fountain and basin, and about this has been planted a collection of *Malus*—105 trees in thirty-five varieties. These were planted partly as a trial of *Malus* cultivars or varieties likely to be

A courtyard garden protected by windbreaks planted to climbing plants in the Strybing Arboretum. The Arboretum is only one corner of the immense Golden Gate Park in San Francisco, all of which is on artificially made top-soil.

of most use in the soils and climate of San Francisco, and partly as an ornamental feature. Differences in performance are already apparent; but so difficult an enemy is wind in this botanic garden that the Strybing's chief botanist Mr Arthur Menzies believes that these differences are due rather to siting, degree of shelter, than to variety.

There is an interesting collection of *Rhododendron*. The Strybing gardeners have concentrated on tender species which cannot be grown in climates less favourable than San Francisco's. So the *R. maddenii* series is well represented and also some of their choice hybrids. The collection includes *R. crassum*, *R. burmanicum*, *R. javanicum* and others as rare in cultivation. There are numerous species of the *azalea* series, both spring- and autumn-flowering, and other deciduous rhododendrons. Many of these exquisite flowering shrubs are planted, with *Camellia* cultivars in a Rhododendron and Camellia walk under tall shade trees both coniferous and broad-leaved, and which is very charmingly planted. Shade for both these genera is important in this garden because one day's midsummer burning sunshine can result in very bad scorching. In all about 250 species of *Rhododendron* are being propagated and planted out, and these include some of giant-leaved primitives such as *R. macabeanum* and *R. sinogrande*.

The Strybing collection of dwarf conifers is a good one. One hundred and seventy-two varieties are planted most pleasingly about a pool, on hillocks and rocky bluffs. Most of these came from the famous Noble collection, the rest of which are in the National Arboretum (Washington, DC). This collection includes thirty forms of the Lawson cypress, and many dwarf, prostrate or colour forms of *Chamaecyparis, Pinus, Abies, Picea, Cryptomeria, Sequoia* and others.

During our own visits to the Strybing Arboretum the following species were starred in my notes, either for their size, quality, special interest or simply for their beauty. The stand of *Cedrus atlantica* var. *glauca*; the enormous Cape Heaths, larger than any we saw at Kirstenbosch; the fine specimens of such tree ferns as *Dicksonia* and *Cyathea* and some much rarer genera; the Australasian collection, including *Leptospermum, Pittosporum, Callistemon, Grevillea, Olearia, Hebe, Metrosideros, Eucalyptus, Acacia* et alia; ivies, bergenias and periwinkles are used for ground-cover displays. There are myrtles from both the Old and New Worlds. I have stars against the sixty-year-old Monterey cypresses; the big *Eucalyptus globulus* trees, their white branches very beautifully displayed against the clear blue sky; the very pretty specimens of *Acacia baileyana*; the puce-fruited *Acmenia smithii*; the South African shrubs, including *Proteas* and the tree *Leucadendron*; the enormous *Geranium palmatum* from Madeira presented to the Arboretum by that great plantsman Lord Talbot de Malahide; *Mantenus boaria*, a very lovely tree from Chile, like an evergreen willow; the tree-climbing near-hydrangea, 60 feet tall, *Pileostegia viburnoides*; the lovely golden-flowered *Freemontias* and notably the new

cultivar with beautiful 4-inch flowers, called 'California Glory'.

There is a very pretty Garden of Fragrance with the labels in both letters and Braille. Here are all the principal aromatics, great mats of thyme, lavenders and sages and many rarer fragrant plants. There is a curious story connected with the beautiful stone-work—drystone walls and rockeries—of this area. William Randolf Hearst bought an abbey in Spain and had it demolished and shipped it to California with a view to re-erecting it there, all at a cost of over a million dollars. But he lost interest in this, as in most of the enormous number of things which he bought, and presented it to the City of San Francisco; the city had no use for it and gave the stone to the Strybing Arboretum: the consecrated stones now provide a setting for plants of particular value to those who are blind.

A small area of the Arboretum is devoted to little demonstration gardens designed for the guidance of house-owners. Other educational work includes very thorough courses organised for botany teachers. Four hundred San Francisco school-teachers have passed through such courses and it is estimated that through these teachers the Garden's educational staff have reached something like 50,000 children. Many of these children have been to the Garden to follow the Redwood Trail, a track among Redwoods underplanted with native woodland flowering plants and ferns; or to take the conifer walk through the Strybing Pinetum, rich in native and exotic conifers.

One of the Strybing's most valuable activities has been and still is the introduction and trial of exotics which may be suitable for the climate and conditions of San Francisco, like the forty-three species and cultivars of the *javanicum* series of *Rhododendron*, already mentioned. The list is too long to give here, but some of the most interesting introductions should be named, the Australian *Acmenes* and *Eugenias* for example, including the Lilly-pilly trees; the fragrant and graceful *Boronias* from Queensland; *Callunas* from Scotland; new *Camellia* cultivars from several countries; *Clethra arborea* from Madeira; *Echium*, the gigantic borage from the Canary Islands; many epiphytic *Fasicularia* and other *Bromeliads*; *Hebes* and *Hoherias* from New Zealand; Chilean *Lapagerias*; *Lobelia gibberoa* from near the source of the Blue Nile in Ethiopia; *Luculia gratissima* from the Himalaya; and many more.

In pure taxonomic botany, the Strybing Arboretum serves the San Francisco Academy of Sciences, where Dr Elizabeth McClintock makes constant use of its material.

Favoured by the social and horticultural importance of the Golden Gate Park, and by its growing value as a centre for teaching elementary botany, natural history, ecology and nature conservation, this very young botanic garden has a better chance than most of growing up into the sort of tasks which modern institutions of the kind can justify themselves by performing.

HUNTINGTON

The Huntington garden, surrounding the Huntington Foundation collèction of books, MSS and paintings, is not only one of America's grandest ornamental gardens but a major botanical centre.

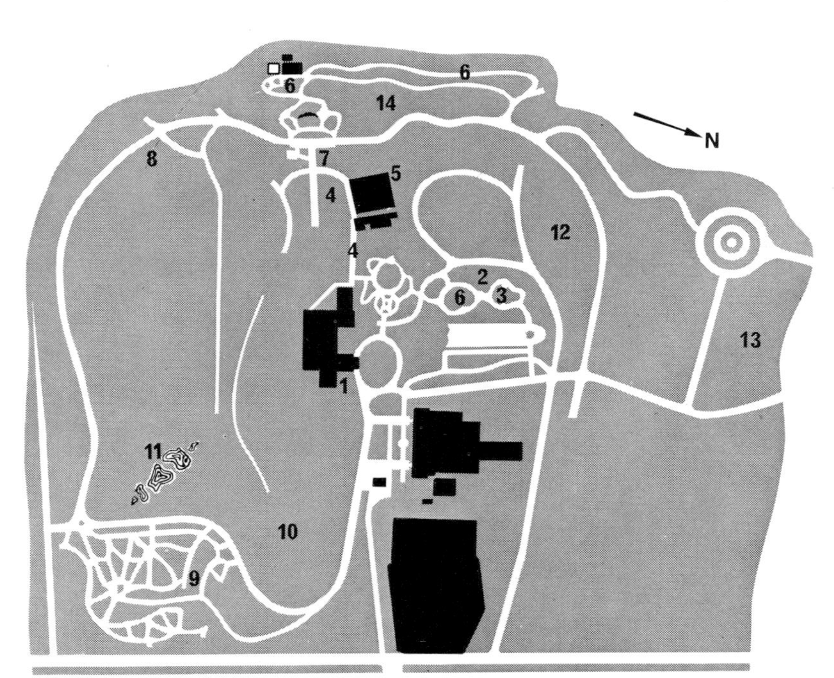

Key to principal botanical and cultivar collections

1 Cycads
2 *Rhododendron*
3 Azaleas
4 Roses
5 Culinary and medicinal herbs
6 *Camellia*
7 *Wisteria*
8 Australian plants
9 *Cactus* and *Euphorbia*
10 *Palmae*
11 Aquatics
12 Deodars
13 Oranges
14 Japanese garden

Henry Edwards Huntington was the son of Solon Huntington, one of those adventurous New Englanders who opened up the West in the 1840s and 1850s and who, together with his brother Collis, both helped and exploited the new territory. Collis Huntington, the abler of the pair, became one of the great railway tycoons; Solon became a hardware baron. As a result of his own ambition and his uncle's help, Henry Huntington went into railroads after showing his mettle by first managing, then owning and developing, a sawmill business which cut the sleepers for his uncle's railways. He sold out this business profitably and had a spell managing the paternal hardware business. He became Superintendent of Construction of Collis Huntington's Chesapeake, Ohio and South-west Railroad. Later, he reorganized the failing Kentucky Central Railroad; still later, at Collis Huntington's suggestion, he moved to San Francisco to manage the Southern Pacific Railroad and to represent the Huntington interests on the West Coast. It was while he was on his way to take up this new work that, passing through San Francisco, he stayed a night at the San Marino estate of J. de Barth Shorb; the place made a very strong impression on him.

In August 1900 Collis Huntington died and left his nephew a legacy which made him a very rich man. Meanwhile H.E., as he was known, had made himself rich in any case by developing a system of electric tramways in and all around Los Angeles, providing a system of public transport all over that region. Had the city of Los Angeles, now one of the most unpleasant conurbations in the world, retained that system, it might conceivably have saved it from its present fate of being choked to death by the automobile.

It would serve no purpose to go here into E. H. Huntington's other business activities. In 1906, as an immensely rich sexagenarian and a widower, he retired, or half-retired, and thereafter devoted himself to collecting books, manuscripts, and later paintings, with all the energy he had displayed in business, and with virtually unlimited means. He also bought the San Marino estate which had so taken his fancy when he first went to California, and in due course built the large and beautiful house in which he lived and kept his collections. Huntington was far too active and clever a man simply to become a dilettante farmer—the estate was primarily a big citrus farm—and art collector. He decided to use his estate—500 acres but later reduced to 200 acres—to discover by trial and experiment which economic and ornamental plants would grow and flourish in that part of California to which he had become attached. To this end he appointed the botanist William Hertrich superintendent of the projected botanical garden, and sent him all over the United States and Mexico to collect plants. The combination of Huntington's money and drive with Hertrich's skill and knowledge was the origin of one of the most remarkable botanical gardens in the Western hemisphere, and in some respects in the world. It is outstanding for its *Cycads*, and for Cacti and Succulents grown out of doors it is superlative, not surpassed by any botanical garden in the world.

The origin of this collection is curious. In the spring of 1907 Huntington asked Hertrich what they could do with a certain piece of barren, sun-baked and over-drained hillside which was beside the main drive leading to the house—now the Library and Art Gallery. Hertrich suggested that only desert plants would flourish there, and proposed an open-air collection of such plants. After some hesitation Huntington, who had no taste for Cacti, agreed to try this, setting aside half an acre for the trial. Hertrich went off on a tour of Arizona, Texas and New Mexico, returning with three railway trucks full of Cacti, including nine fully-grown specimens of the gigantic *Carnegiea gigantea*, the 'Giant Sahuaro'. That was in 1908. In 1912 Hertrich went to Mexico, returning with so much material that the Desert Garden had to be enlarged to 5 acres. For two more decades the collection was added to: we shall return to it in its place.

I am not here concerned with Huntington's activities as a collector of books and works of art; but it is at least possible that to some extent his books and pictures and statues helped to shape his taste as a gardener. Another influence was that of his second wife, Arabella Huntington, whose European education and sophistication had to be allowed for. True, he did not have the taste or judgement, or rather confidence in his judgement, to emulate those earlier tycoons, the Italian Renaissance princes, and commission original works. His landscaping, in the spirit of his collecting, was a series of clichés: but the best

An aspect of the exquisitely designed and botanically interesting Japanese garden in the Huntington Botanic Garden. Many of the plants in this Japanese garden are botanical rarities, to match the quality of its works of art and architecture.

Below One of the Italian statues which decorate the Huntington's many small gardens.

clichés, just as the statues which he bought in Italy for such parts of his Garden as the grand North Vista, were very good of their kind. But in his attitude to plants Huntington had a genuinely enquiring mind and a scientific approach, and he had the very great merit of giving Hertrich a fairly free hand.

THE HUNTINGTON NOW

The Botanical Garden consists of 207 acres, most of it planted, but a part more or less wild. It is, like the great Library and Art Gallery which it surrounds, administered by the Huntington Trust.

The climate of Los Angeles is virtually sub-tropical: San Marino is an outer suburb adjoining the residential suburb of Pasadena. As the mean annual rainfall is only 15 inches, the whole Garden is under more or less continuous irrigation. The buildings which house the Library and the Art collection, with their noble terraces and porticos and fine statuary, are among the most beautiful ornaments of the grounds; and at various points, focusing a vista, providing a proper end to a walk, or central to a parterre, there are other statues, fountains, miniature temples and finally the great Huntington Mausoleum, designed by John Pope, prototype of the Jefferson Memorial in Washington, DC, and flanked by splendid specimens of *Eucalyptus citradora*. These artefacts, most of which are Italian or Italianate, help to make the Huntington Garden one of the most beautiful in America.

In other ways, too, the aesthetic horticultural aspect has been as much respected as the botanical. For example, a great many of the Huntington's fine collection of Palms have been disposed in avenues rather than being planted at random or in groups according to some systematic or geographical plan. There is a particularly good avenue of Guadalupe Palms. Some other plant families are treated likewise—there is, for example, the avenue of *Cedrus deodara* known as Deodar Lane. There is a Rose Garden of a thousand bushes, constituting a permanent trial of new cultivars, planted formally in beds cut out of lawn, and enclosed, as seems the usual practice in American botanic gardens, in white-painted wooden fences and pergolas for climbers.

The Huntington Garden is, in fact, a complex of many gardens; too big to be an integer like the Strybing Arboretum, it also differs from that other Californian gardens in other respects. Thus it is ideal for arid-soil plants but cannot, without great difficulty, manage such high-rainfall and cool summer shrubs as *Eucryphia* or *Fuchsia*. As one might expect, its Australian garden is a good one, not only in the many fine specimens of *Eucalyptus* in variety, but in the numerous representatives of *Callistemon*, *Melaleuca*, *Leptospermum* and many other beautiful flowering shrubs and trees.

The *Cycad* garden, among the half-dozen best of its kind in the world, falls short only of the Fairchild, Florida (see below) example and, of course, of the superlative collection at Kirstenbosch. Among the other, smaller gardens are a Herb Garden, prettily laid out in a formal but not rigidly rectangular design, and in which very good use is made of great mats of silver-leaved varieties of, for example, thyme, the sages, and the curry plant. Here, as in San Francisco, flower and leaf scents are remarkably strong. Culinary and aromatic herbs predominate in this garden, but it also includes some rare shrubs and trees: the Huntington is very rich in the genus *Myrtus*, and particularly in forms of *M. communis*. The small-leaved *M. communis* var. *tarentina* is much used for close clipping and for simple topiary. The Herb garden also contains a small collection of Crab apples; and it is centred on a very fine well-head of eighteenth-century rococo German workmanship. The Herb Garden is separated from the Rose Garden by a big pergola or arbour, as it is called locally, carrying old *Wistarias*, *Clematis*, and a good collection of Jasmines.

The uprights of this and other pergolas and arbours at the Huntington are interesting: they appear to be old, rough-barked, forked, grey tree trunks: in fact they are cast concrete simulacra, but so admirably made even to the small details of scars and saw-cuts, as to be most decorative: it appears that this kind of work was a feature of nineteenth-century American gardens.

The Camellia Garden is a large one of 5 acres. It was started as a testing ground or trial in 1942, was opened to the public in 1951, and is now mature. It is planted in a canyon in the shade of Californian evergreen oaks, and pines—heavy shade is essential during the heat of summer

Left and right The Huntington 10-acre garden of *Cactaceae, Euphorbiaceae* and other succulents is unique. No other garden in the world has so many species of so many genera, from both the Old World and the New World, growing together in the open.

—and it includes 1,500 varieties and cultivars of half a dozen species. The principal species are *C. reticulata*, a Chinese species known chiefly in its ancient cultivars, with flowers 6 or 7 inches in diameter; *C. japonica*, the hardy spring-flowering *Camellia* which has many hundreds, and perhaps as many as two thousand cultivars; and the autumn- and winter-flowering *Camellia sasanqua*.

The collection of *Magnolias* is not an outstanding one but the specimens of the native *M. grandiflora* in several parts of the grounds are magnificent, one in particular, handsomely set off by a marble statue of Hebe in front of it, must be among the half-dozen largest in the world.

In many parts of this Garden there are boskages, chiefly of two kinds; near to the house, formal, grouped on a statue or fountain, and composed of large-leaved evergreens. Further from the house, much use is made of both evergreen and deciduous *Rhododendrons*, principally of the *azalea* series, and which can only flourish in this garden by virtue of the copious irrigation. In these boskages much use is made of ferns. Small, winding paths, terminating or joining at a fine marble seat, a statue or a miniature temple, thread these plantings of shrubs, ferns and some herbaceous perennials, with many spring-flowering bulbs.

Not far from the house is the pretty little Shakespeare Garden. This is a flat, formal garden of flower-beds and winding paths, in which a majority of the plants are flowers, shrubs and trees appearing in one or more of Shakespeare's plays. The plants are labelled not only with their names but with the source quotation.

Most charming of all the many small gardens which compose the Huntington is the Japanese Garden: Californian horticulture has been very greatly influenced by Japanese gardening—indeed, United States' gardening in general has come under that influence. The Huntington Japanese garden was made in a small canyon or gorge, and its history is curious: Huntington decided that he wanted this rather dismal and unsightly corner of his estate transformed very quickly; a well-known San Francisco nursery had a good Japanese garden by way of showpiece. Huntington bought it as it stood, and while his architects and contractors changed the topography of the little canyon, and made a lake in it, the rocks and plants, together with a number of very valuable Japanese works of art from a great dealer in such things, were transported to their new sites. The Garden contains not only a whole series of rare dwarf conifers including specimens of the Dragon's Eye Pine, maples, and a most handsomely sited and grouped stand of Japanese cycads, but several superlative Japanese artefacts, including a drum bridge, a bronze Buddha and stone lanterns. Above this Japanese landscape garden, is first a new and well-executed Zen Dry Landscape garden, and a Tea House or rather a Japanese dwelling house, used for flower shows and lectures; and adjoining it a courtyard designed as a setting for the Bonsai collection.

Further along the canyon in this quarter of the Garden there is a wild garden on both sides of a stream, planted with suitable natives and exotics. Associated with the Japanese complex there is a Ginkgo grove and the beginnings of a moss lawn.

THE DESERT GARDEN

What distinguishes the Huntington Botanic garden is its unrivalled Desert garden which now occupies 10 acres, and which, as I have said above, was made by Hertrich over a period of many years. It has about 25,000 plants representing hundreds of species of xerophytes, plants adapted to arid conditions. Whether this collection surpasses or equals that of Monsieur Marnier-Lapostolle in range of species, I do not know: but it is not simply a collection, it is a garden. It is interesting—in fact it is fascinating—on several different levels: in the almost infinite diversity of form in its plants, the grace of some, the grotesqueness of many, the symmetry of a few species; in the beauty and diversity of its flowers; and in the extraordinary diversity of the means taken to solve identical problems of survival.

For most laymen the *Cactaceae* family of the New World xerophytes is the most interesting element of this great collection; it is represented by very numerous genera: the *Pereskias* from Mexico, the West Indies and tropical America have thin stems and real leaves, some species being shrubby, others climbing or scandent; the *Opuntias* from the South-west United States, Mexico and Central America are here in several species of each sub-genus, with vivid satiny flowers springing from the edges of thick oval, eliptical, spherical or almost shapeless lobes of tissue, and their often edible fruits—Prickly Pear is the most familiar species—but the two sub-genera have together over a thousand species varying in stature from 4- or 5-inch cushion plants to branching shrubs 20 feet tall. Then there are the *Mammillaria,* spiny cushions sprouting lesser cushions and covered in season with flowers which may be white, yellow, pink or crimson; the many-stemmed *Echinocereus* species with their large lavender, purple, red or yellow flowers; the *Echinocactus* and *Ferocactus* species, including the Barrel Cacti, burly cylinders of spine-covered tissue up 8 feet tall. The *Echinopsis*, of which about fifty species have from time to time been grown at the Huntington, bear some of the loveliest flowers in the family, especially *E. robinsoniae*, named after a former chairman of the Huntington Trust. *Cereus* is fully represented by species which are mostly like small, leafless, stylized trees, very spiny, with numerous white or pink flowers and big, colourful fruits: the collection includes night-flowering species of the genera *Selenicereus*, *Hylocereus, Monvillea* and *Harrisia*, for the most part climbing or scandent, the flowers often 12 inches or more in diameter, white, pink or yellow. The Garden has splendid specimens of *Carnegiea gigantea*, although none has, I believe, attained to the height of 30 or more feet found in wild specimens. Doubtless they will do so, but it takes about 200 years for a *Carnegiea* to reach its full stature.

Some aspects of the great Huntington garden of succulents. Every shape known to geometry, sizes varying from an inch tall to 30 feet, flowers of exquisite beauty issuing from grotesque masses of armed tissue—the diversity of these plants is infinite.

Carnegiea is not the only genus in the Huntington with species capable of reaching that size, and some may even surpass it. Thus, there are representatives of *Pseudo-domitrocereus*; some *Pachycereus pringlei* and some *Lemaireocereus weberi* from Mexico, and also of the potentially gigantic *Trichocereus pasacana* from Argentina. One fine *Cereus, C. huntingtonianus*, commemorates the founder of the botanical garden.

Those are just a very few of the most curious or beautiful Cacti in the collection, but Cacti are not by any means the only desert plants in the Huntington. In the Old World, plants solved the problem of surviving in arid conditions rather differently and in more than one manner: the *Aloes*, of the family *Liliaceae*, by thickening their sword-shaped leaves so as to store water in their tissue; the *Euphorbias* still otherwise; and the so-called Living Stones more in the manner of tiny cacti, but using camouflage instead of an armament of spines against grazing animals. The two hundred *Aloe* species from Africa, Madagascar and the Canary Islands are not all to be found in the Huntington collection, but very many of them are, including, of course, *Aloe vera* which has become naturalized in Mexico since its introduction by Spanish missionaries for its medicinal value.

Euphorbia is well represented: the fine flowering shrubs of *E. milii* are most attractive. Very interesting is the way in which some *Euphorbias*—for instance *E. echinus*—have arrived, in the struggle for a rainless existence, at a cactus-like solution which is yet not identical. And very pleasing are the noble stands of the tree-shaped but cactus-like *Euphorbia ingens*. All Euphorbias seem to do well in the Huntington conditions, whereas the *Mesembryanthemums* and their allies, all from South Africa, are not all happy there; some, indeed, flourish, but others suffer from the relatively cold winter rainfall. But the *Crassulas* do well, many of them flowering freely in winter in California: there are splendid 5-foot specimens, in groups, of *C. portulacea*. In the same family, the *Cotyledons,* with their stiff leaf-rosettes and tall flower spikes, are attractive, as are the similar Mexican *Echevarias* and the Californian *Dudleyas*. Also in that family are the *Sedums*; most of the Huntington species are from Mexico.

The familiar *Compositae*, the daisy family, is represented by xerophytic species of *Klemia* and *Senecio*: some grow erect with stems of bright blue more colourful than the flowers; others are mat-forming plants often covering several square yards of ground.

The South African *Stapelias* are difficult in California. Only a few of the five hundred species can stand cool, winter rainfall however slight it may be; but a few species, some with flowers like blotched, liver-coloured starfish, do survive.

There are many interesting genera of *Agavaceae*, including of course the *Agaves* themselves. The smaller *Beschornias*, beautifully glaucous and with much more attractive flowers, are very ornamental; and the colossal *Furcraea* species, with 30–40-foot flower stems, are most impressive. Some of the *Yuccas* in the Huntington collection are tall trees. The allied genera—*Dasylirion, Nolina* and *Beaucarnea* are all present.

There are several genera of xerophytic *Bromeliaceae*, and of these by far the most spectacular are the *Puyas* from the western watershed of the Andes. The huge spikes composed of hundreds of flowers are often blue-green, aquamarine or some other unusual shade, and they have an extraordinary metallic sheen; the allied *Pitcairnias* are almost as curious in form and colour.

Planted in 10 acres of stony soil and red porous rock, these and thousands of other desert plants compose a fantastic picture of imposing shapes, monstrous forms, formidable armour, weird camouflage; colours shrill and tender, textures varying from fine satin to the most warty, spiny and rebarbative. Nothing we have seen in botanical gardens anywhere in the world more effectively and impressively demonstrates the infinite variety of solutions which plant evolution has produced, to a very small number of identical problems.*

Although the Desert Garden dominates one's impression of the Huntington Botanical Garden, there are other things well worth mentioning. In the first place, some of the trees are exceptionally interesting: the tall, spiny-trunked *Choisya insignis*, a great cloud of white flowers; the splendid stands of *Libocedrus decurrens*; the scarlet-flowered *Erythrina crista-galli*, reminding us of the Carlos Thays Garden in Buenos Aires; the *Dombeyas* and *Pittosporums*; the *Eugenias* handsomely clipped into tall, burly cylinders yet still bearing their bright puce fruits; the enormous *Magnolia grandiflora* specimens already mentioned; the live oaks, *Quercus engelmanni*, including one specimen believed to be the largest in the world; the 60-foot tall tree aloe, *A. baynesii*; a very fine *Agathis robusta*; and the evergreen, giant-leaved *Magnolia delavayii* in flower. Among the conifers there are good specimens of *Pinus canariensis*, and the very lovely weeping *Taxodium mucronatum* from Mexico, and *Metasequoia glyptostroboides*, Dawn Redwood, here taller than in any other botanical garden we have seen.

Among the trees of economic importance there is a collection of *Persea* (Avocado Pear), another of *Citrus*.

It was heartening to know that Mr Boutine, the Huntington's taxonomist, was spending part of the autumn of 1968 collecting plants in Mexico, for the Garden. Plant collecting is in the Huntington tradition; it was part sponsor of Frank Kindon-Ward's last (Manipur) expedition, and continued activity in that field is very much a sign of life and of looking forward to a very long future.

*My acknowledgements are due to William Hertrich's own *Guide* to the Huntington Collection of xerophytic plants in helping me to interpret what I saw there.

FAIRCHILD

Opposite The fruit of one of the *Cycads* in the Fairchild Tropical Garden's (Florida) collection of Palms and *Cycads*. The Fairchild *Cycad* collection is second only to that of Kirstenbosch in South Africa (which has every known species).

The Fairchild Tropical Garden was founded in 1935 on the outskirts of Miami in Florida by Colonel Robert A. Montgomery. His first Florida garden, about a mile from the present foundation, was started early in the 1930s, partly in order to accommodate the plants which it was his hobby to collect. At that time Colonel Montgomery already had one of the world's largest pinetums at his summer estate in Connecticut; the best part of that collection of conifers is now in the New York (Bronx) Botanical Garden. When Montgomery settled in Florida his interest shifted from conifers to Palms and Cycads, for which he had plenty of room in his 80 acres of grounds. The importance of the collection and the pleasure which he took in conveying to others his own interest in and knowledge of certain families of plants led him to decide to found a public tropical botanical garden. It is probable that this decision was inspired by his reading of *Exploring for Plants* by David Fairchild, sometime head of the Plant Introduction Bureau of the United States Department of Agriculture. Dr Fairchild's introductions of tropical plants to Florida were nowhere collected in one place: Montgomery thought that it would be of use to botanical science and to horticulture, and a proper tribute to the great plant collector, to found a Fairchild Tropical Garden.

The climate of Florida is peculiar. Although it is subtropical and can, as a rule, be considered to be virtually tropical, in the matter of temperatures and humidity, about once in every twenty-five years freezing air from the Arctic is drawn south as far as the tip of the peninsula, and there is a frost capable of wiping out a tropical garden in one night. More frequently, a degree of cold may be experienced in winter quite capable of doing serious damage to tender plants. Dr Fairchild, consulted about a suitable site for the garden which was to bear his name, pointed out that the warmest part of the Florida mainland is the limestone ridge between Miami and Cutler and as near as possible to Biscayne Bay. A suitable 83-acre site was found south of Matheson Hammock (a 'hammock' in Florida is a native wood). Montgomery bought it; founded the Fairchild Tropical Garden Association to administer it; and in 1935 landscaping and planting was started. Meanwhile the Colonel approached the local (Dade County) government which agreed to take title to, and accept responsibility for, 58 acres of the new Garden, leaving 25 acres to the Association, a body of subscription-paying members.

The Association's 25 acres were to accommodate the great Palmetum which was to be planted and which they undertook to maintain. The County agreed to vote funds for the maintenance of the rest. Responsibility for plant records, overall supervision, plant nurseries, research and educational programmes rested with the Association. An interesting result of this arrangement is that the Association members do, nowadays, garden the entire area, the county's part as well as their own; and that as funds are short excepting for special scientific projects, many Association members contribute their labour free towards

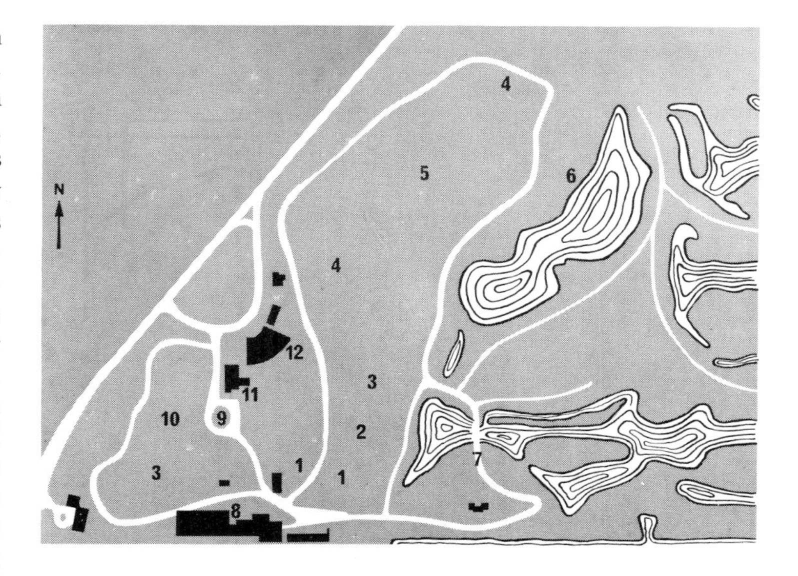

Key to principal plantations and divisions

1 *Philodendron* collection
2 Palm glade
3 Palms collection
4 Flowering trees
5 Banyans
6 *Hibiscus* collection
7 Amphitheatre
8 Display house
9 Cycads collection
10 Sunken garden
11 Museum
12 Auditorium

the maintenance of a Garden which they regard, with some justification, as their own.

THE PRINCIPAL COLLECTIONS

The great distinction of the Fairchild Tropical Garden is its Palmetum. There is no doubt that this is the best and largest collection of Palms in the United States and it is probably the best and largest in the world. It contains about 500 species and it is steadily being added to and may well, in due course, reach the avowed goal of a thousand species, for both known, and new, unnamed species for study, reach this Garden from all over the world. Among the most interesting Palms in this pleasingly planted collection—for the rules of landscape-gardening were respected in the planting—I made a special note of the enormous and incidentally monocarpic *Corypha* palms; the *Nephrospermum*, very rare in the Western hemisphere although there are a number at the Hope Garden in Jamaica; valuable oil palms from both Central and South America—*Atalea* and other genera; the curious Bamboo Palm, *Ptychosperma*; the Triangle Palm, *Neodryposus decaryi*, with its highly decorative triangular bunching of the leaf stalks· the strangely Victorian-looking Petticoat Palms from Cuba; the big collection of *Phoenix* (Date Palm) species; *Saboa palmetto*, which is the State tree of Florida; the bulb-shaped Bottle Palms of the genus *Mascarena*; the very ornamental Fish-tail Palms; the spiny-trunked *Acromias* and fan-leaved *Latanias*—and a great many more. There are palms which yield sugar, like the *Arengas* with their fragrant flowers and bright red fruits, hedging palms such as *Rhapis*, wind-resistant Hurricane palms (*Dictyosperma*), ornamental palms for horticulture such as the scarlet-fruited *Oxyan-*

dra, palms like the Royal Palm which make magnificent avenues, palms, indeed, for every purpose under the sun; they are, at the Fairchild, the principal object of scientific study and experiment.

Equal in interest and second in importance is the *Cycad* collection. The plants are concentrated in two areas, because the original collection did not leave enough room for enlargement. The genera represented are *Ceratozamia, Cycas, Dioon, Strangeria, Zamia, Encephalartos* and perhaps some others from both the Old and New Worlds. The beautiful symmetry and diverse forms of the leaves and the magnificence of the cones make these plants particularly attractive. A few of these Cycads have some economic importance, for example the Sago Palm (not a palm at all, of course) *Cycas revoluta*. Others are very ornamental in tropical gardening, notably the 40-foot *C. circinnalis*.

At one time Montgomery took a particular interest in *Ficus*, and the Fairchild has a good collection of these, including huge specimens of *F. religiosa*, *F. sycamorus*, and the curious *F. roxburghii*, with its dense clusters of figs growing from all over the naked wood. *F. fairchildii* was first studied and described in this Garden; in all there are about a score of species, grouped for the most part in two plots, although there are big specimen trees here and there in the Garden and also on the Montgomery estate which, although closed to the public, is regarded, for scientific purposes, as part of the botanical collection.

Dr Fairchild had a particular love of climbing and scandent plants and collected them from all over the tropics. A massively strong pergola about 400 feet long, spanning one of the drives, was built to accommodate scores of these plants while others are grown as specimens on poles. Some of these climbers are very spectacular, among them the scarlet and crimson flowered *Passiflora* species (Passion Flowers); the Malay Jewel Vine, *Derris scandens* which makes an enormous plant whose stem may be 15 inches in diameter at the base; the magnificent *Combretum grandiflorum* from Africa, with its large, scarlet flowers; the equally showy *Congea tomentosa*, the so-called 'Shower of Orchids' plant from Burma; and *Heteropterus beecheyana*, valued in tropical ornamental gardening for its bright red, winged fruits.

Bromeliads are, of course, grown outside in the Fairchild Garden, both as epiphytes and as ground plants according to their kind. The collection is often brilliantly colourful when in flower: it includes nine species of the genus *Aechmea*, familiar as a house plant; several kinds of Pineapple, some being very decorative; *Areococcus* species from Brazil; half a dozen *Bilbergia* species from Brazil and Venezuela; the lovely red and purple *Bromelia balansae*; some native *Catopsis* species; *Glomeropitcairneas* from Dominica; West Indian and Brazilian *Guzmanias*; the tricolour *Neoregelia carolinae*. There are *Tillandsias*, those strange plants like bundles of string which subsist on what they get from the air around them; *Witrokias, Pitcairnias,*

The Fairchild Tropical Garden has the largest collection of Palms in the world—over 500 species. It is constantly being added to and the Garden's scientific staff are chiefly concerned with Palms.

Below Tillandsia, 'Spanish Moss'. A *Bromeliad*, it grows as an epiphyte on hundreds of trees in the Fairchild Tropical Garden.

A special plant house built on two levels, the higher covered with glass, the lower with slatting which leaves it open to the air but shaded and protected, was completed in 1967. It houses a very attractive display of rare palms, species which are tender even in the climate of Florida, the rarer *Cycads, Ferns* and *Bromeliads, Helicornias*, etc., and a beautiful display of orchids. These rarities are planted to simulate a natural plantscape.

Notable are the *Pileas*, some giant Maidenhair Ferns, giant mosses, the yellow-flowered *Ludovia lancaefolia*, a great rarity from New Guinea. Some of the *Vandas* are particularly fine ones.

So much for a glimpse of the Fairchild plants. As to the gardening, one of the most admirable features of this botanical garden is its design and landscaping. An absolutely flat site is the despair of landscape architects, who must rely on topography of a site, on slopes, hills, differences of level, to provide the basis of their design. But so skilfully is the Fairchild Tropical Garden landscaped that its level flatness is masked, and a considerable number of fine vistas, some across water, have been achieved, vistas which give a remarkable illusion of looking down from a height and into great distances. The artist who accomplished this, William Lyman Phillips, who was a pupil of the famous Frederick Law Olmsted and a graduate of the Harvard School of Landscape Architecture (Arnold Arboretum), succeeded in satisfying the requirements of botanical science as well as of art: thus, each principal plant family has its own plot; in the case of the very large collections, then a plot is allotted to a genus or to one or two genera. Plantings are nowhere crowded together, and broad sweeps of lawn give perspective. According to one account of his landscaping theories Phillips '. . . compared the use of open areas in the garden to open spaces in an art gallery, wherein one has room to walk and to view the pictures from close range or from a distance, according to inclination'.

and still others—a representative collection of fascinating plant family.

Dr John Popenoe, Director of the Fairchild, has a special interest in the flora of the Bahamas. Some years ago he realized that the pace of development and urbanization in the Islands threatened the extinction of many unique species. He obtained a special grant of money from the Michaux Fund of the American Philosophical Society to carry out a rescue operation. The majority of the native woody plants of the Bahamas are now under cultivation in the Fairchild Tropical Garden, and under study in the laboratories. This special collection, planted in the extreme eastern section of the garden's extensive lake area is still being added to. The operation has certainly saved a number of species from extinction.

The Garden has one of the world's best collections of *Philodendron* and other aroids, growing in the open. Planted as objects of study by the Garden's scientists, they also serve an ornamental purpose. Of the genus *Philodendron* alone, nine species are grown, some of them with gigantic leaves which have a surface area, for the largest leaves, of up to 20 square feet. Most species flower and set and ripen fruits. Other aroids include the genera *Acorus, Aglaonema, Alocasia, Amorphophallus*; half a dozen *Anthuriums* with spathes like brilliantly coloured patent leather; species of *Monstera*—but there is not much point in continuing to make lists: the fact is that the aroids are as well represented as the *Bromeliads*.

SCIENTIFIC WORK

The Fairchild Tropical Garden has received grants for its work in scientific botany which have made possible the building of new laboratories. The scientific staff consists of first, the Director, Dr John Popenoe, who is a plant physiologist. I have already referred to his special interest in the Bahamas flora. On the side of economic botany, he maintains and studies and experiments with a collection of tropical fruits, notably the *Annonas* (a genus which includes *Cherimoya*, Sour Sop, Custard Apple and others). Then, Dr P. Barry Tomlinson is engaged in fundamental work on the anatomy and morphology of the Monocotyledons, especially the Palms. Dr William Gillis is the Garden's taxonomist, and Dr Berg its phytophysiologist.

Dr Tomlinson's work is carried on in association with the Universities of Cornell, Harvard and Miami, and the Royal Botanic Gardens, Kew; it is concentrated on the Fairchild collection of Palms and Cycads. A distinction of the work done in this department, of a kind which the layman can appreciate at a glance, is the exquisite botanical drawing of Miss Priscilla Fawcett: Dr Tomlinson places photography a poor second when it comes to illustrating morphological point.

Another distinction of the Fairchild is Dr Tomlinson's respect for traditional methods, which by no means excludes the use of sophisticated techniques and apparatus. The new Research Library is making as much effort to acquire the classics of nineteenth-century botany, as the latest works: 'One of the pleasures available to the modern research worker,' Dr Tomlinson insists, 'is the reading of older scientific literature and the appreciation of the way in which ideas and concepts have developed. One thing this reading teaches is that a fact correctly observed is inviolate. Scientists who made these correct observations were those who looked carefully, who did not turn aside from their dissecting trays and microscopes to reach for their pens, too soon.' And he continues: 'Continuing to examine plants in whatever way he pleases, the observer

Below Patterns in Palms, *Pandanus* and *Cycad* in the Fairchild Tropical Garden. These families of plants are studied there not only in their morphology and cytology (and taxonomically), but for their great 'architectural' value in garden design.

will discover new facts. After all, what is research? The good eye, without the textbook.'

Dr Gillis is building up the Fairchild Herbarium. He points out that voucher specimens are still essential as the only reliable references for checking descriptions. His laboratory, like Dr Tomlinson's, is new and well equipped. His new additions of Palm specimens to the Herbarium are already considerable.

One research project involving all the scientific staff, is of practical importance to Florida, other States, and a number of tropical countries. The Montgomery estate and part of the Fairchild site were originally mangrove swamp, and some of this still remains on the periphery. Advantage has been taken of this not only to make a collection of *Rhizophora* species (and other genera of the *Rhizophoraceae*), but to make a special study of them as soil-builders.

Mangrove forests, their trees very tolerant of salt water, are important in certain parts of the world because they help to hold back the encroaching sea, and, in time, to form new land. Some species, moreover, also have a shorter term economic importance: the massively heavy wood does not float and can be useful in marine construction; the bark of some species is rich in tanning substances; and the small wood makes a useful form of fuel.

The Fairchild Garden has, finally, an educational programme managed by the Curator of Education, Mr Scott Donachie. Under the National Science Foundation there is, first, an advanced seminar in tropical botany for post-graduate students from the north who have had no chance to work with tropical plants. There are courses for Members of the Garden, for school-children there are Garden tours with accompanying lectures, not only the Garden itself, but in the Rain Forest sector which is a wild part of it, and in which many epiphytes, including numerous magnificent orchid species, are planted as they grow in nature.

MONTREAL

Opposite The Expo '67 'climatron' in Montreal, seen from part of the city's great botanic garden which pioneered the extension of botanic garden teaching of botany and horticulture to schoolchildren.

In 1931 Canada's greatest living botanist, Brother Marie-Victorin, EC, who died in 1944 and whose monument is as much the great Garden he created as the actual memorial which stands in that Garden, persuaded the Executive Committee of Montreal to reserve 180 acres of the Maisonneuve Park in the eastern quarter of the city for the laying out and planting of a botanical garden. It is actually a complex of many gardens, the layout having been designed by the first Curator, Dr Henry Teuscher. There are some thirty of these specialized, component gardens surrounding a fine building which houses the Herbarium and the Botanical Library, an Information Centre and the Administration Offices. The easiest way to give some impression of this Montreal Garden will be to list the principal component gardens in more or less the order taken by the usual guided tour of them.

The first garden is of purely horticultural interest, a collection of about a thousand cultivars of herbaceous perennial plants suitable for the gardens of the Montreal region and Eastern Canada in general. This is focused on a formal pool and three smaller pools, and it is given a third dimension by a pergola of climbers. From it one moves into a more interesting garden, botanically speaking, a collection of economic plants which, from the educational point of view, is better contrived than any we have seen. Here the chief esculent vegetables, cereals, oil-bearing plants, dye plants, etc., are displayed; as are also their probable primitive ancestors, so that one can see how man has remade his economic plants. Labels in this garden not only give the vernacular and scientific names, but the use of the plant and something about its history in cultivation. Grouping is not botanical, but according to the part of the plant—leaf, root, stem, fruit—which is useful. In the case of the vegetables which amateurs commonly cultivate in their own gardens, the best new kinds, thoroughly tested in the Montreal Garden's own trials, are grown for the guidance of private and market gardeners. A collection of fruit trees is planted as an extension of this Economic Garden.

Next comes a Physic Garden, and this, again, is an interesting one. It has five collections: first there is an historical collection, of the medicinal herbs which were grown in the monastery gardens of the Middle Ages, gardens which were ancestral to the modern botanic garden; then a collection of plants used medicinally by the American Indians, growing after nature among native trees prettily grouped about a pool; the third collection is one of plants still in the modern pharmacopaea, and hardy in the latitude of Montreal; finally, as a means of educating both children and adults in what plants to avoid, there is a collection of poisonous native plants including those provoking allergies such as skin-rashes, asthma, hay-fever and so forth.

This garden is followed by a plantation of hardy garden shrubs, and then by a rock garden in which species and cultivars are mingled so that it is of interest both botanically and horticulturally. Next comes a useful trial and demonstration of possible hedging plants: it is better than most of its kind, firstly because it displays about fifty species, varieties and cultivars which can be used to make hedges and screens; secondly because various ways of training and clipping are demonstrated in practice.

The Water Garden, charmingly enclosed in stands of birch and larch, is one of the best in the world. For anything like a representative collection of aquatic plants and near-aquatics, it is not really sufficient to have one or two lakes and ponds: there are very large differences in the requirements, for example in depth of water, of aquatic plants. The Montreal Garden has 109 pools, some deep and some shallow, some clear and some muddy: and in them are cultivated a very diverse collection of aquatics, marsh plants and bog plants native to Canada, together with some of their Asian and European allies.

The Trial Gardens are horticulturally admirable. The first, confined to two genera, *Iris* and *Hemerocallis*, both of which do well in the region of Montreal, has about 200 cultivars of each genus. Other trials are devoted to garden annuals; the Montreal garden accepts new commercial kinds from nurseries and seedsmen, for trial; and those emerging successfully are then planted for display in the Garden of Annuals.

One section of the Montreal garden is given over to educating children in gardening: here, as in the Brooklyn Botanical Gardens, city children are able to try, in practice, the theory they learn in the classroom. Adjoining this Children's Gardens area is one devoted to a number of model gardens demonstrating what householders can do in the space available to them.

GREENHOUSES IN MONTREAL

The range of greenhouses in the Conservatory complex has fourteen connected houses of which ten were com-

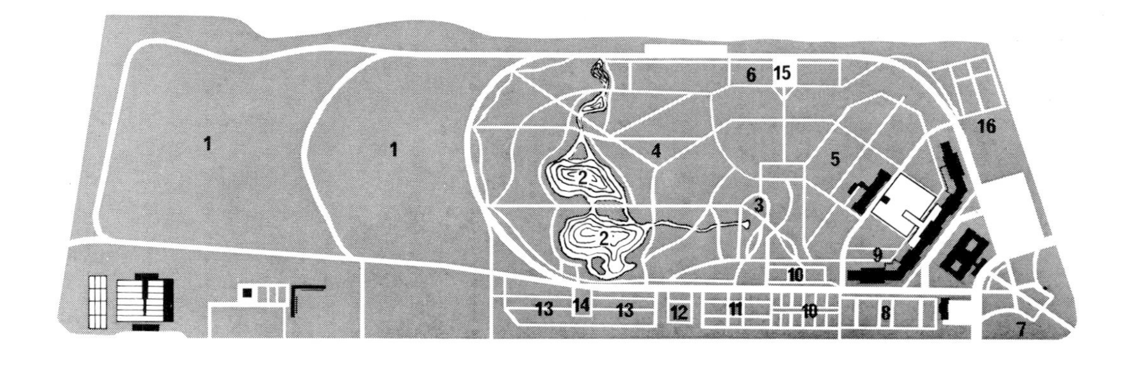

Key to Map

1 Arboretum
2 Water garden: aquatic plants
3 Alpine garden
4 Ecological groups
5 Nurseries
6 Demonstration of hedging plants and methods
7 Annuals
8 Display garden
9 Shade plants
10 Economic plants
11 Fruit garden
12 Physic garden
13 Shrub garden
14 Rock garden
15 Water and bog garden
16 Children's garden

Flamingoes and other wader birds with sub-tropical plants inside Montreal's version of the 'climatron', an unexpected legacy of Expo '67.

plete and planted at the time of our visit. The big central house is used for seasonal displays, each display having a special theme and being stage-set to express it—for example, spring flowers of ancient Greece. Opening off it to right and left are houses devoted one to the genus *Begonia*, and the other to a simulated tropical rain-forest with typical flora—aroids, ferns, etc., in the usual way, and epiphytic plants cultivated after nature, including orchids. One house is confined to tropical economic plants, one to aroids in many genera, another to tropical ferns. A representative collection of Cacti and Succulents is housed in two greenhouses.

A fine alpine garden has been completed relatively recently: rocky outcrops embellished with a waterfall have been planted in geographical order: there is an area given over to East Canadian alpine plants, another to plants of the Rockies, others to plants of the Alps, the Caucasus, the Pyrenees and the great Asian mountain ranges. Among other purely botanical sections reaching completion are one devoted to the ecology of Eastern Canada; and an Arboretum and a garden of botanical systematics. In all some 20,000 species are cultivated at Montreal and used as a living laboratory by the scientific staff, a vast and diverse object lesson by the teaching staff. The scientific papers and bulletins which issue from this Canadian garden have a high reputation among botanists all over the world.

Formal garden layout in Montreal, and part of the collection of hardy conifers.

USSR

MOSCOW

Krantzia capitata. An example of botanical painting from the Herbarium of the Moscow Botanic Garden, headquarters of the network of botanic gardens in the USSR.

The USSR has many more botanic gardens than any other country in the world. This is not simply a function of its enormous territorial size, for there are more such gardens *per capita* of the population or per square kilometre—about 150 botanic gardens in all. In each of the fifteen republics there is one central botanic garden, and a group, more or less large, of satellite gardens, the density being greatest in the Ukrainian SSR and least in Central and East Asia. This complex of botanic gardens is presided over by the Council of Botanic Gardens of the USSR whose Chairman is Academician Tsitsin, famous for, among other things, his production of perennial wheat (of which more anon). The master-garden of this complex is the Main Botanic Garden of the USSR Academy of Sciences in Moscow, whose Director is Professor Lapine.

The Garden is in the suburbs of Moscow, in the district Botanieskaya Street, and it covers about 900 acres (361 hectares), most of which is planted woodland. All of its principal divisions are extensive; and a programme of development is in train, a programme which includes the making of a string of new lakes, and the building of a vast new greenhouse on a crescent plan. The USSR national Herbarium is in Leningrad, where the botanic garden is relatively unimportant, but the botanical laboratories and scientific workers are concentrated in Moscow. In any case, Soviet botanists are less concerned with taxonomy than our own, and more concerned with experimental and economic botany. The Moscow Botanic Garden is also the centre of a vast botanical survey of the USSR, and for a major work of plant collecting all over the Union. Members of its scientific staff lead or take part in plant-hunting expeditions to various parts of their enormous and very varied country. The principal object is the discovery of new economic plants, but potential new garden ornamentals are not neglected.

The tasks entrusted to the Garden by the USSR Academy of Sciences include the investigation of the plant resources of foreign as well as domestic florae in the service of the national economy; research into the problems of urban horticulture; and the dissemination of botanical and horticultural knowledge. And since, of course, the raw material for the scientific and economic work are plants, the number of species has been built up steadily since a big reorganization in 1945, until there are now more than 3,000 herbaceous and 1,800 tree species, as well as about 6,000 decorative cultivars, and 2,000 crop cultivars.

A large area of the Moscow garden is devoted to a nature reserve of trees, shrubs and undergrowth native to the Great Russian geographical zone. This is allowed to grow naturally and its self-regeneration, ecology and changes are studied by students and post-graduate workers.

An area of 75 hectares (180 acres) is devoted to trees and shrubs hardy in the Muscovite climate but which are introductions, the under-storey being self-generating. The plants come from the far North-east of Asia, from North Europe and from North America, and there are about 1,500 woody species, planted systematically but also in the correct geographical order. You can either walk or drive on good roads through this part of the botanic garden, and stop at such plantings as are of particular interest as, in my own case, the genus *Sorbus*, with 58 species; the plantations of *Larix*, of *Eleagnus*, of *Abies* varying from small conical and prostrate forms to forest giants and including the lovely *Abies sibirica*; the *Picea*, the *Taxus*, and the *Euonymus* plantations. One of the discoveries made by means of this great Arboretum is that forty-eight species of conifers can be grown in the rather harsh Moscow climate. But perhaps one of the most impressive collections, at least to the visiting foreigner, is that of *Lonicera*; I counted up to fifty species of these honeysuckles, many of them native, with *Lonicera involucrata* as the star. Almost as pleasing are the great *Viburnums*.

Sixty-two acres of the wild garden of trees and shrubs are devoted to plants native to the USSR, planted after nature and according to the Union's principal geographic regions. Here, among specimen trees planted park-wise, to allow plenty of light, are groups of the most important kinds of shrubbery and herbaceous plants, very beautifully tended, and separated by winding paths so that no single group is inaccessible. As well as the 'natural' kinds, the collections of *Populus*, *Berberis* and *Caragana* being very good, there are a great many hybrid trees and shrubs, planted to compete with and be compared with the species; most of these have been bred by the man who is regarded as the greatest creator of new plants in the USSR, Academician Sukachov. It is in the spirit of Marxist–Leninism to look upon man as, above all, *Homo faber*; and he must be a maker even in the world of plants. The work of Tsitsin and Lapine in the field they call 'remote hybridization' leads Soviet botanists to talk of new, man-made species (not simply, as we should, of hybrids); it leads also to a greater interest in the breeding of wild plants with ancient cultigens than in our own case; and it leads also to the large-scale propagation and planting into the wild of such new plants. It is probably true to say that Soviet botanists are most active in trying to make use of hitherto unused genetical material than botanists elsewhere.

Typical of the manner in which the herbaceous plants of particular regions are grown for display and teaching in this Moscow garden are the sections devoted to the small plants of the Altai, and the Caucasus. In both cases, artificial hills of rock and earth have been constructed in great clearings in the woodland, and planted with representatives of the regional florae. Wild tulips whose leaves were everywhere apparent were over, but I starred in my notes species of *Paeonia* including the lovely golden-cupped *Mlosklosewitchii*. There were sheets of yellow and orange *Chamaenerion* 'poppies' broken by patches of bright blue *Aquilegias*; *Ranunculus*; *Viola*; many *Alliums*, in which the USSR is rich, and very many quite unfamiliar plants including a rock-climbing, purple flowered daisy which I failed to identify. So well are these wild gardens made and planted that the excitement and

Moscow's television mast serves to identify the city's great botanic garden. The Moscow garden is used for recreation, as well as for botanical science and garden craftsmanship.

pleasure of seeing them was like the excitement and pleasure of finding the flowers really in the wild. Clumps of *Eremurus* were coming into flower, and among the shrubs were several *Ribes* including gooseberry and black-currant. These Altai and Caucasus gardens are properly tended and impeccably labelled, and crowded with plants of the greatest interest to the gardener as well as the botanist.

A third large section of the Moscow Botanical Garden is planted on both sides of a long, winding drive, backed by tall trees—the 'Garden of Permanent Blooming': it is not, in the climate of Moscow, literally possible to have plants in flower all the year round; but the planting here has been done with skill and taste in such a way, and with such species, as to have flower from the first possible moment in spring until the last possible moment in autumn. I do not know how many kinds of plants are used, but they are planted in very large numbers of each kind, and although cultivars are used as well as species and varieties, the planting is done after nature rather than as a garden.

There are two Rose Gardens, planted formally, and laid out geometrically and in rectangles. Two thousand five hundred species, hybrids and cultivars are grown, including most of the West European and American favourites and novelties. The roses are well grown and cared for. In one of the Rose Gardens an attempt is being made at an Italianate style, with rising terraces, a formal pool, and a water staircase. It is less happy than the rest of the Garden, partly owing to an injudicious use of concrete. Nevertheless this section, with its evidence of expansion, and of new features being constantly added to the Garden, is heartening in a world where big gardens are more apt to shrink than to expand.

A large section of the Moscow Botanical Garden is devoted to horticulture rather than botany. This is rectangular in plan, with beds contained by clipped hedges. It is not beautiful but it is certainly useful: there are large blocks of such perennials as paeonies, irises, etc., planted in variety for purposes of comparison. A feature, partly within and partly outside this garden of decoratives, are the lilacs, certainly one of the largest collections of cultivars in the world, although not equal to the wonderful lilac collection at Minsk (see below). An avenue of these are all standards on stout trunks; elsewhere they are grown in blocks, in mown grass, separated by narrow walks, and either as half-standards or as bushes. Lilacs are of particular interest to Muscovite gardeners for they are ruggedly winter hardy, yet stand the blazing heats of summer, and they flower with magnificent freedom in May. One even sees great bunches of lilac offered for sale in the streets of Moscow, along with lilies-of-the-valley gathered in the woods and of which there are an enormous number also in the Botanic Garden.

I have already mentioned that a vast new, crescent-shaped greenhouse is projected as part of the Garden development plan. There is an older greenhouse, a big one, already in use, with a considerable collection of tropical plants. It plays an important part in the botanical education of Moscow's schoolchildren and is quite often crowded with parties of boys and girls receiving object lessons and lectures. The collection is much what it is in all such greenhouses and in no way remarkable, excepting, perhaps, for the pleasing way in which orchids are grown after nature, among ferns, and not simply set out on benches or hung from the rafters. This greenhouse is also a source of material for the laboratories.

An important section of the Garden and of its associated experimental farm on the Black Sea is that devoted to the perennial and other new wheats associated with the practice of 'remote hybridization' and the name of Academician Tsitsin, former director of the Moscow Botanic Garden where much of the original work was done. The first reliable account of this work to appear in English, a work which could transform the economy of several countries outside the USSR as well as much of the USSR itself, came in 1959, when N. I. Tsitsin and I. F. Lubinova published in *The American Naturalist* (Vol. xciii, No. 870, May–June, 1959) some results of their experiments—long since published in Russian of course—in crossing wheats with *Agropyron* ('Couch Grass') species. These crosses were made over a very wide range, both geographical and varietal, of both parent kinds; and some hybrids were obtained which the two botanists considered as new 'species' of wheat (*Triticum agropyrotricum perenne* et al.). Among the characters of some the most promising new ones was the perennial habit; another was that of producing a second growth of stalk and leaf, useful for fodder, after harvesting. Tsitsin and Lubinova reported of Perennial Wheat:

> These perennial and forage wheats make possible a new use of the oldest cultivated crop, wheat, not only for grain with a high protein content, but also for green fodder or hay, with a protein content equal to that of ordinary soft wheat.

Other qualities claimed for the new wheats were that they did not 'lodge' and were immune to fungus diseases. By 1965 thousands of hectares of the new wheats were being planted, but in their most important new character they were failing, the perennial habit. It was not that they failed precisely as perennials; but that the winter loss of plants to cold was in excess of 70 per cent in the first winter, and a further 30 per cent of the remainder in the second winter. In short, selection for hardiness had been insufficiently thorough. Work on this problem is still going forward in the Botanical Garden and elsewhere and it seems likely that satisfactory perennial and grain-plus-fodder wheats will be produced. This is one part of a big plan to combine useful, but not yet used, characters in wild plants with the useful characters of old cultivars, in new cultivars. It is perhaps because Soviet botanists are enthusiastic over the prospect of the valuable economic qualities which still remain to be found in the world of plants that botany is more active and more honoured in the USSR today than it is elsewhere.

Another and similar example of the kind of work under-

taken in the Moscow *Botanichevsky Sad* is that concerned with the winter-hardiness of timber trees, fruit trees, and shrubs, in the climate of North Asia and North America. Of the 2,330 kinds of trees and shrubs representing 247 genera and 70 families of plants cultivated in the Moscow Botanical Garden, only about half have proved perfectly winter-hardy, a fact which has given rise to experimental work in determining—and finding ways of increasing—winter-hardiness. The problem has been found more complex than one would have expected. Professor P. Lapine, now director of the Moscow Botanical Garden, writes:

Winter and frost-resistance of woody plants is a complicated phenomenon. It depends on the ecological character of the species, variety or biotype; it varies within populations and it changes considerably during the development of the individual plant with age and during the yearly seasonal development cycle. Professor I. Tumanov has shown that plants of *Betula pubescens* are damaged by temperatures lower than 23°F in active growth, but after a complete winter hardening can stand—319°F in test chambers for long periods of time.

The point being to emphasize once again that the chief function of the Moscow Botanical Garden is to carry out work which has economic and social implications.

Were I required, however, to 'identify' this garden by a single visual impression, it would be taken from the wild part: it would, perhaps because of literary associations, be the strikingly beautiful impression of sunlight shining through the tender green May foliage onto the lovely silver-white trunks of innumerable birch-trees rising out of vast green, clean and beautifully fragrant pools of lily-of-the-valley.

MINSK

The Botanical Garden at Minsk is the Central Botanic Garden of the Byelorussian Academy of Sciences and like all 'Central' gardens it has its satellites, in this case two, away from the capital of the Republic.

Minsk was totally destroyed during World War II, and is now a new city; although without architectural grace or boldness, it is a pleasantly spacious city, its building style chiefly marred by the excessive conservatism which is such a blight in the USSR (one is often reminded of Australia in this respect). It is a great centre of science, light industry and technological education.

The Botanical Garden was first laid out and planted in 1932, is about 250 acres in extent, has about 8,000 kinds of plants (*taxa*), and is far more interesting in its wild or natural parts—the greater part—than it is in the formal and strictly horticultural parts. 'We Russians,' I was told by Professor Nicolai Smolsky, Director of the Botanical Garden, 'prefer the English, the natural style, when we make a botanic garden. In fact they carry that taste much further than we do.' Like all the other 'Central' botanic gardens of the USSR, this one is as much concerned with economic botany as with pure science. As in Moscow, the wild plantations are arranged on geographical lines; but with an added refinement since an attempt is made to reproduce the correct plant associations, the plant ecology, of each region. From the aesthetic point of view this garden suffers from the serious handicap of being absolutely flat. In Western eyes it lacks topographical

Below One of the beauties of the Minsk, as of the Moscow Botanic Garden, is its birchwoods, with their exquisite patterns of sunlight and their birdsong. It is in birchwoods that the lilies-of-the-valley, beloved of the Muscovites, are gathered, but in the botanic gardens the wild flowers are respected by the citizens.

charm: but then Russians love the steppe and hate the mountains, so per haps this flatness may even be a merit in their eyes. And the flatness is somewhat mitigated by the lavish and loving use of trees.

Apart from the strictly systematic and scientific sections, the horticultural part of the Minsk garden consists of a large formal Rose Garden; and the geometrically laid out, rectangularly planted trials of ornamentals (in very large numbers), the principal genera being *Iris*, *Paeonia*, *Tulipa*, *Narcissus*, *Lilium* and *Gladiolus*. 'Our garden flora,' I was told, 'is relatively poor. By 1944 the garden had been totally destroyed by the Germans and we had only 200 kinds of plants left, horticultural plants; now we have 1,500.' The trials of ornamentals include

Nymphaea and *Lotus* in suitable ponds. Nothing much need be said about this horticultural section of the Garden excepting that it does valuable work in selecting and testing ornamentals suitable for the light and dusty soil of the region, and for the light rainfall (total annual mean precipitation is under 600 mm), not to mention the bitter winter cold. For whom is this work carried on? Chiefly for the Municipal Gardens and Parks authority, and for the State-maintained gardens of the Byelorussian SSR. But to some extent, also, for amateur, private gardeners. The Botanic Garden provides the State and Municipal Gardens with advice, supervision, seeds and plants. It helps amateur gardeners not as individuals but through their clubs or societies: of the Republic's nine million population, one million belong to such a society, either concerned with horticulture, or with nature conservation. Some of these own *dachas*, country cottages, where they cultivate small gardens. Their most beloved plant is the rose, followed by the lilac. Tulips and gladioli are favourites. About 4,000 cultivars of the relatively small number of species are kept on permanent trial. Amateurs, acting through and with their societies, are encouraged to make their own contribution to the advancement of gardening.

For educational purposes a large and, alas, absolutely hideous range of greenhouses is maintained with subtropical and tropical plants. There is a good collection of tender succulents. It is significant of the climate of Byelorussia that even such hardy evergreen as *Camellia japonica* and box are grown under glass, and most of the evergreen azaleas too.

As we noted in Moscow, a genus with which the Northern peoples of the Soviet Union excel is *Syringa*, and at Minsk the trials of lilac cultivars are even more impressive, and much more extensive, than in Moscow. Trial grounds and nurseries are vast and 150 kinds of lilac are cultivated. In no garden in the world have I seen more, and more beautiful lilac flowers. The Garden has bred, propagated and distributed four new cultivars of its own. Just why lilacs flourish so in the poor soil and harsh climate of Minsk is not clear, but there is no question about the fact. Grafting as standards seems to be preferred; but whether as standards or bushes the plants have stouter trunks and a much more rugged bark than our own lilacs ever have: is this a reaction to the cold winters?

In economic botany the Minsk garden concentrates, for practical as well as social reasons, on two kinds of plants: new plants capable of being made into fodder silage; and plants yielding tanning elements. An extensive area is given over to finding out how to cultivate plants which members of the Garden's scientific staff have chosen and collected during plant-hunting expeditions, studied in the wild and in the laboratory, and decided might be valuable. Plant-collecting has been concentrated chiefly in the Altai, Caucasus, Pamirs and Crimea. The principal genera under trial as fodder plants at Minsk are *Polygonum*, *Herecleum*, and *Lymphytum*,* all in more than one species

*The merits of this 'Russian Comfrey' are already recognized in Britain by some market gardeners, poultry keepers, and the Soil Association.

Large areas of the USSR botanic gardens are given over to the trial cultivation of plants which promise to be of economic value and whose seeds must be saved. Not even scientific communism seems to have found a new way of protecting seeds in the field.

and varieties. But these are not the only wild plants under trial on a large scale. Professor Smolsky never makes the old-fashioned mistake of dismissing folk tales about the merits of plants as eyewash; instead, he tries them, to find out. Australian *Acacias*, with the highest yield of tannin for the leather industry, cannot be grown in Russia excepting perhaps in the extreme south by the sea; very numerous plants were tried for tannin yield and in the Tenshi mountains a *Polygonum* was found which, ruggedly hardy, gives a percentage weight yield of tannin, from its roots, not much less than the best *Acacias*. Another example: *Rhaponticum carthamoides* has long been known to the Siberian peasantry as 'Plant of Life'. It has some of the properties of the Chinese 'Ginseng'. Eaten by wild deer, it is supposed to be responsible for medicinal qualities in the powdered antler-horn. It was tried as a fodder-silage plant by some of the Byelorussian Academy of Sciences workers in the Minsk Botanical Garden: results of feeding cows on it in spring have been curious and are being studied, for these cows had a higher fertility than the 'control' cows which received none. Smolsky believes that it may turn out to be important.

The Minsk Botanical Garden is also used for the introduction and acclimatization of cultivars from overseas, and for hardiness trials of trees and shrubs. Another of its tasks is the teaching, by suitable propaganda, of 'conservation': a part of the Minsk garden exactly reproduces a typical Byelorussian countryside, with all the native plants; it is made round a large ornamental pond, almost a lake; here children are taught to recognize and respect their native flowers and trees. Another section confined to Byelorussian plants is part of the Arboretum which also includes a Central Asian section, a Siberian section and a North American section. These sections are separated by drives which are avenues of trees some of which are already well-grown, one of Black Poplar, another of the Birch, *Betula vericosa*. Other avenues, which are as yet half-grown, are made of a maple, of a *Cerasus* species, and of *Picea excelsa*.

Apart from the lilac cultivars, the following were the plants, all growing 'after nature', which I starred during my visits: *Iris Sibirica*, massed in grass and making a charming effect; species of *Crataegus*; *Berberis amurensis*, a dense, rounded bush with myriads of pendent golden flower racemes; a local dwarf, yellow-flowered *Cytisus*; and some of the fine shrubby *Loniceras* with which all the Russias seem to abound.

When the Minsk Botanical Garden is mature, in about twenty years time (and before then it will have received the new laboratory and administration buildings under construction and its new Research Institute), it will be as beautiful as it already is useful. Really hard problems of poor soil, low rainfall and cold have had to be overcome; in fact there seems to be only one physical problem left to solve—how to prevent the innumerable magpies which haunt the garden from pulling all the rubbish out of the litter bins which Soviet citizens are meticulous in using.

KIEV

Below Ginkgo biloba, the ancient temple tree of China, is represented by fine specimens in several Russian botanic gardens.

The Central Botanical Garden of the Ukrainian SSR is in Kiev on the Dnieper. This ancient and very beautiful city is at the centre of one of the world's oldest agricultural and horticultural regions: men were living there by the cultivation of plants when all round them tribes still at the hunting stage had yet to learn how to till, plant and harvest.

The *Botanichevsky Sad* here, fifteen minutes in a trolleybus from the city centre, has, unlike those of Moscow and Minsk, an interesting topography: the site is hilly and it commands the most splendid views over the Dnieper. Like botanical gardens everywhere, it has ranges of greenhouses for the display of subtropical and tropical plants: they are inferior to those of Britain and Germany and are of no particular interest to the foreign visitor, although, of course, valuable for the education of students of botany and as a source of material for the Garden's scientists. The distinction of this Garden is to carry the Soviet botanical policy, that of planting 'after nature', a stage further than either Moscow or Minsk, and on the way towards the perfecting of this method which one finally sees in the Botanic Garden at Tashkent. The florae of the different geographical parts of the USSR are collected and planted into reproductions of natural landscape; but with this much which is unnatural, that species are crowded together and can be seen within an hour or two which, in nature, are much more widely scattered. This means that the Minsk scheme of 'ecological' planting is not adhered to in Kiev.

There is almost no formal garden: only near the entrance gates is a small area of straight formal walks, dividing groups of trees including a magnificent stand of *Abies concolor*, and small 'mazes' of neatly clipped beech providing, inside themselves, sheltered places for seats where mothers sit supervising children at play, and old men in retirement read the newspaper aloud, in quiet voices, to their old wives. But closely surrounding this small tamed area on all sides is a simulacrum of wild country, and this is intersected by walks and drives.

Soviet botanists are only mildly interested in exotics and then, as a rule, for 'economic' reasons. They have enough work coping with the immense flora of their huge country. What 'botanical garden' means to the new generation of Soviet botanists is well demonstrated by the special Kievan Ukrainian section of the Botanic Garden: an area of gently undulating meadow-land, with clumps of trees and pleasing views of the Dnieper and its busy traffic of steamers, is planted with what seems to be, and ultimately aims to be, a complete collection of the Republic's wild flowers. They are planted at random, in large groups or clumps, and the only unnatural aspect is the labelling, which is excellent, and, perhaps, the close juxtaposition of species which would not be found so close together in the wild. The Ukraine, like Siberia, has a very rich flora of herbaceous plants, perennial and annual. You wander among these wild flowers and can study at your ease and within a few-score acres the flowering herbs of a country the size of France. Some of the flowers were familiar as West European garden subjects; others were not, and by them I was forced to admit the truth of a point so often made by Soviet botanists (though

A belated fall of snow in Ukraine's principal botanic garden at Kiev. As is fitting in a city rich in magnificent trees, most of the garden's chief rides are avenues of handsome trees.

Left Works of modern Soviet sculpture in Kiev's Botanic Garden. The garden is almost as much concerned with garden design and horticulture, as with botany. *Right* The rose named 'Gagarin' for the pioneer cosmonaut, was actually bred in the Tashkent Garden.

in a sense British gardeners and plant-collectors were a century ahead of them in this), that we have by no means finished the long, long task, which began in Neolithic times, of domesticating, and often remaking plants, whether for economic or ornamental purposes.

I have mentioned labelling: only one or two clumps of each kind, though it may be represented by many more, are labelled, presumably to reduce the element of unnatural intrusion to a minimum; but by searching you can find the identity of every plant, in Latin and Russian. There are, among the grasses which are likewise part of the collection, great wide pools of blue composed of *Linums*, *Adonis*, *Campanula* in many species. Acres of this section were, at the time of my visits, strongly scented with the spicey fragrance of the huge, branching inflorescences of *Hedysarum grandiflorum* which were alive with bees. There are silvery and lacey *Artemisias*; several species of *Onobrychis* with their erect racemes of butterfly flowers in shades of purple; beautiful red-flowered spurges; many roses, including some floriferous *spinosissimas*, for here and there are groups of shrubs and trees.

In some of the other (geographical) divisions of this big garden there are splendid boskages of the native *Philadelphus* species, tall borages, *Iris* in charming variety; a great many different barberries, some with red leaves or plum-red fruits in hanging clusters. There are many handsome umbels; and even wild rhubarb. The genera of these sections are numbered in hundreds, the species in thousands. They constitute some of the most attractive and interesting features I have seen in any garden, and this 'wild' planting gives one the pleasure of discovery as a garden in which the plants are set out for display in beds can never do.

That, then, is an impression of the botanical side; it is very extensive, but to write more about it would simply to print lists of names, nothing can convey the look and feeling of it, not even photography. There is also an important horticultural side; and, of course, both pure science and economic botany. The horticultural sections include a large-scale trial of lawn grasses; a clematis trial on a complex of pergolaed terraces: this genus does very well in the conditions of Kiev, and the garden has produced some cultivars of its own. In the scientific and economic field there is a massively equipped open-air experiment in the use of powerful growth lamps to vary day-length, and to study the effects of extra hours of 'daylight' on crops.

Trials of ornamental plants are on a large scale and are concentrated on herbaceous perennials and on *Iris*. The collection of paeony cultivars is a very complete one, the plants are very well grown, and in this section there is much hybridization in progress, so that new cultivars are raised, some of which would seem good enough to win Royal Horticultural Society Awards. One long hedge in this part of the garden was composed of a small, compact and very fastigiate oak; it was produced in the Tashkent Botanical Garden and I shall have more to say about it. The flag iris trial is on the same large scale as the paeony trial, and there again are some new cultivars of botanic

garden provenance. Also new to me were some of the spurges which were on trial, and some attractive new dianthus cultivars.

Near to this trial ground is the Rose Garden, which all Soviet botanical gardens make such a point of. The Kiev example is, to me, more interesting than most Rose Gardens because far more use is made of climbers, *Wichurianas*, old shrub roses, and even species, than of the HTs and Floribundas which most of the botanic gardens in the USSR concentrate on. The plan of the rose garden is geometrical but curvilinear, the paths are paved and there are marble basins and fountains. The collection of Bourbon roses is a good one, and there are fine boskages of 'Hidcote' or something very like it, enormous old *R. moyesii* specimens; and the usual collection of HTs and Floribundas suitable for Kievan gardens. A big stone lily pond adjoins this Rose Garden, with terraces, seats and walks all shaded by quite pleasing specimens of *Salix babylonica*.

Perhaps the most charming really horticultural feature of this botanical garden, and its greatest distinction, is the collection of ornamental conifers. This is a skilfully landscaped section of little hills and valleys overlooking the Dnieper, and grassed. The views from it are of wide stretches of the river, with its steamers and yachts, and some of the bridges, and of three of the city's churches with their many golden, onion cupolas glittering in the sunshine. Planted skilfully in random clumps are a variety of silver furs—*Picea pungens* and *Abies concolor* being outstanding for their beauty of form and colour; of golden junipers; of *Thuyas*, pines and ginkos, with a number of dwarf, conical, fastigiate, and prostrate forms of many species. This little landscape of conifers is a complete success, interesting to the botanist, useful to the gardener, and pleasing to the eye of the visitor who is neither. Of course, most botanical gardens have some or all of these trees, but I know of no other where they have been combined like this in a single landscape garden.

YALTA

Below left One of the pergolas, and Italian cypress, in the Nikitsky Botanic Garden, Yalta. This great 'Italianate' garden is one of the USSR's two sub-tropical botanic gardens.

A bust of Lenin dominates the formal pool on the high terrace of Yalta's Nikitsky Sad.
Below right A green rose is one of the Nikitsky's rarities.

The great *Nikitsky Sad* outside Yalta on the Black Sea is very different from the gardens discussed above; but before we come to that something should be said in general about the satellite gardens which are grouped round, serve and are served by the big Central gardens of the principal SS Republics. A few words will do: for the most part they are either purely scientific or economic, and are, in fact, more like agronomic stations than gardens and as such hardly belong to this book. Others may be old botanic gardens of the educational-cum-ornamental type, and if they are not attached to or associated with a university or research institute, are apt to be rather neglected without being actually abandoned. A case in point is the old *Botanichevsky Sad* in Odessa. Odessa, like Kiev, is a city of many parks and gardens and avenues, and of fine sea views. But its old garden, perhaps because a new one is being planted and already becoming established, is a sorrowful place. Although some planting is still being done, and although there is a quite useful trial of garden perennials and shrubs; and although its great clumps of philadelphus bushes are magnificent it is for the most part a wilderness of fine old trees and overgrown shrubs. It has a few oleanders and even some true *Acacias*. It has some circles of clipped box enclosing specimen trees; and a little rock garden of cacti which must once have been enchanting. Its trees include *Thuyas* and *Cupressus* species, and some enormous *Robinia pseudoacacia* trees, for this white 'Acacia' is the symbolic flower of Odessa. The Odessa Botanical Garden is still a cool, shady and pleasant place to walk in away from the clamour of this thriving but gracious city. And it consoled us for its neglect by producing hoopoes for our delight. It forms a part of the vast botanical sciences network centred on the Museum Botanic Garden, to which, however, it is in no way subservient.

As for the *Nikisky Garden* in Yalta, for active work in botanical science, for conservation of the native flora, for work in the field of economic botany, and for sheer beauty, it takes very high marks indeed.

It was founded in 1812 by a Swede in the service of the Tsar Alexander I, Stevens by name. He chose the site on the Black Sea, five miles outside Yalta near the village of Nikitska, for its exposure and its magnificent topography. By so doing he set himself problems, for it was all rock and no soil. The land, a steep slope to the sea commanding glorious views over its vivid blue and of wooded, mountainous promontories, was cleared of thousands of rocks and stones, and then supplied, at immense cost in money and labour, with black earth from the fertile Ukraine. And since, like the whole Crimean litoral, Nikitska had a low rainfall, a huge reservoir was built to provide for irrigation.

The *Nikitsky Sad* itself covers 272 hectares (about 680 acres); but its directorate controls several satellite gardens and a nature reserve outside the garden. Two of the satellite gardens are nurseries which raise plants for sale, chiefly trees and shrubs, including fruit-trees. The total area involved is 900 hectares (2,250 acres), which makes this botanic garden complex one of the largest in the world. One of the satellite gardens which is purely botanic is at a high altitude, above the 400-metre contour because the flora of the Crimea is divided into vertical zones, each one different from the one below it. The January Minimum Temperature in the Central Garden is $-15°$C; in the high garden $-30°$C.

The Garden's laboratories, research institute and trial grounds are divided to serve a number of different 'services': the principal ones are Horticulture and Floriculture; Pure Botany (taxonomy) of the Crimean flora; Southern Fruit Crops (peaches, grapes, apricots, cherries, plums and olives); Sub-tropical Fruits (almond, persimmon, pomegranate, fig, etc.); Dendrology, including shrubs with trees; Economic Plants (chiefly aromatics for the scent industry); Entomology and Pest Control—biological methods of control receive priority; a special Department of Micro-climatic studies; Herbarium; Radio-biology laboratory; Phyto-physiological laboratory; Bio-chemistry laboratory; Library; and a studio for Plant Photography. Like all such institutions in the USSR, the whole complex comes under the general direction of the USSR Academy of Sciences. It seems a pity that in a country where botanic gardening is on such a grand scale, this old garden should have been allowed to decline.

The Central Garden employs 82 scientific workers, 180 technicians of various kinds, and 220 gardeners. The satellite gardens and nurseries bring the number of people working for the Botanic Garden to just over a thousand. There is also a Publications Department which publishes not only the Garden's scientific papers, but also popular botanical and horticultural works.

A principal object of all these departments and people is to find, study, and introduce from the wild, plants of the Crimean flora which may have economic value, either in themselves, or for some gene or genes which they carry and which can be bred into the genetical composition of new cultivars. In particular the Garden is interested in plants which are suitable for sub-arid steppe conditions; and in plants which may be useful in fixing soils subject

Fine trees in fascinating variety are the glory of the Nikitsky Sad at Yalta.

Below Nikitsky Sad: a rare, semi-prostrate weeping *Cedrus atlantica* var. *glauca*. Detail of foliage and cone.

to erosion on hillsides. A special mountain-top station attached to the Garden is in course of building, where scientists will be able to stay and study possible anti-erosion plants in the wild.

The Herbarium is a small one by Kew standards—91,000 sheets of which 51,000 are Crimean; but it is confined to plants of the southern USSR. It was started in 1904 during the directorate of Kusnetzov. Much of the material of the original collection was lost during the wars and revolutions of 1914–20, and the present collection was made between 1920 and 1964. It is rich in three families, as the Crimea is—*Compositae, Leguminosae* and *Gramminae*. This Herbarium is the basis of the standard *Flora Crimea* produced by the Nikitsky taxonomists, with its 2,300 species of steppe or sub-arid plants, and 2,100 alpines. It is also the basis of a revision, long in hand, of the classic *Florae* of the region.

More than half the plant species of the Crimea are 'Mediterranean' in character; the rest have affiliations with the flora of Central Asia. As it happens, Yalta lies on the line dividing these two and different geo-botanical regions, and the very fine geo-botanical map published by the Nikitsky makes this clear. Studies originating in the laboratories have led to field studies not only of anti-erosion plants but of legumes which may be good fodder plants, and of fruit-trees and shrubs which may be useful as stocks for the grafting of old cultivars.

The Nikitsky's practical achievements have been very considerable. Its staff have made a complete survey of the Crimean flora,* established its ancient kinship with the florae of other and often very remote geographical regions, and made comparative studies of the Crimean alpine flora with the florae of Corsica, Crete and the Dobrudja. The Assistant Director of the Nikitsky, A. M. Kornilitzin, a pupil of the great Vavilov, has made a study of the introduction of trees and shrubs into the Crimea from various countries, and on it based a flori-genetic method of plant introduction which is a further development of the botanico-geographical principle for selection and introduction of alien plants suggested by Vavilov. It has done work which has had practical consequences for the Soviet peach-canning industry, on the breeding of peaches especially for canning.† By collecting, studying and cross-breeding 947 varieties and cultivars of the almond over a period of thirty years, it has introduced into cultivation some new late-dormancy, late-flowering kinds suitable for the Crimean climate.‡ It has worked for 135 years on improving vines for the Crimean wine industry and on the introduction of new varieties. It has done much effective work in the field of pest control, and notably of fruit-tree mite parasites. It has bred new series of garden forms of chrysanthemum, carnation, cannas, phlox and snapdragon and worked on other species for the floricultural industry.§ It has done and continues to do fundamental work for the

*L. A. Rubitzov and L. A. Privalova.

†I. N. Riabov.

‡A. A. Rikhter.

§I. A. Zabelin.

essential oil, pharmaceutical and fruit-growing industries. It has done research work into tobacco-growing, two of its varieties (Drynbek 44 and Amerika 572) being now widely planted. It has made a study of all the apricot varieties and cultivars. Using its collection of 196 cultivars of *Ficus carica,* and pollen of *F. pseudo-carica* and *F. afganistanica* from abroad, it has produced the fig cultivars now used in the Crimean dried fig industry.¶ It has studied the anti-microbic action of plant secretions and shown how their antibiotic activity could be of use. It has studied the resistance to cold of several kinds of fruit trees; embryo-culture as a means of producing early varieties of fruit-trees; the family *Myrtaceae* for biologically active substances; made a study, now producing valuable results in practice, of salt-resistant and drought-resistant plants; suggested means of increasing the attar output of rose-petals . . . but a list three times as long would hardly suffice; and the point is made that the Nikitsky Garden makes important contributions to the agriculture and horticulture of the USSR and particularly of the Crimea.

The Nikitsky Arboretum is large and very pleasingly planted, threaded by winding, shaded gravel walks. Among its most interesting trees are an enormous *Taxus baccata* (Yew), a part of the primeval forest, and at least five centuries old; and a *Pistacia* which is 1,300 years old. And its most beautiful trees are its conifers: there is a whole vale of the three species of cedar in their varieties; a fine avenue of *Cupressus sempervirens* which is now sixty years old. Some specimens of *Sequoiadendron giganteum* have, in seventy years, exceeded 36 metres in height; splendid *Abies numidica* from the Atlas; many North American *Picea, Abies* and *Pinus*; a very big *Pinus montezumae*; some remarkable specimens of *Cedrus atlantica glauca pendula*; fine *Abies pinsapo* trees, and hundreds (three hundred in all) of other species of trees, shrubs and bamboos. From a score of points in the Arboretum, as

¶N. K. Arendt.

Every remarkable tree in the Nikitsky Sad is labelled with its scientific and vernacular names, age, place of origin, economic use and qualities, and the principal events in its history.

from elsewhere in this vast botanical garden, there are good views of the lovely Crimean mountains, their wooded lower slopes topped and backed by a great barren wall of limestone; and of the vividly blue sea and wooded promontories.

The collection, study and breeding of roses is an important branch of the Nikitsky garden's work. The Rosary has 1,600 species, varieties and cultivars and new cultivars are produced by large-scale hybridization. Among some of the best of these which may reach our own gardens eventually, since West European, including British, nurseries have been there to buy, are (names translated literally): 'Flame of the East', a crimson-scarlet Floribunda with very good resistance to heat and sunshine; 'Ayandak' (no translation), a very deep black-crimson which I particularly admired but which would win no prizes with us, having too few petals and a tendency to turn blue with age; a dark crimson HT called 'Crimean Night'; and a pretty, orange Floribunda called 'Yalta Lights'. The Nikitsky's rose-breeding aim is to find roses which are resistant to drought (M.A.R. is only 560 mm and mostly in winter); and the burning sunshine of the Crimean summer and autumn. The gardeners and hybridizers employed in this Rose Garden are for the most part women. Incidentally, all the scientific staff have cottages or apartments in the garden, with wonderful sea-views and their own swimming beach.

An interesting—and in early June attractive—feature of the *Nikitsky Sad* are the yuccas, mostly hybrids. But these were for the most part bred by Professor Rusanov of the Tashkent *Botanichevsky Sad*, and will be reverted to when we come to that garden. Some other features of the Nikitsky which I starred are as follows: the tremendous native *Sorbus domestica* trees; the pretty native *Cistus taurica* which yields a fixative for the perfumery industry; and a very remarkable specimen of *Zelkowa carpinifolia*. This is not really one tree; it was originally several which, planted too close, have grown together into one, forming a union of tissues not only near in the base but at numerous points, between branches, high up in the tree. The roots of this manifold plant occupy about 700 square yards. *Metasequoia glyptostroboides* has grown to about 40 feet in ten years. And incidentally, the beautiful *Libocedrus decurrens* seeds itself very freely.

Still on the subject of trees, the labelling in this botanic garden is a model for the labelling in others, or rather it ought to be. Trees of particular interest or of economic importance have a large metal label which gives a mass of information very clearly set out: scientific and vernacular names; place of origin; time of flowering and fruiting; type of soil preferred; frost and shade resistance; minimum temperature tolerated; method of propagation; commercial uses and quality of the timber; uses of bark, fruit, seeds etc.; date of planting—and so forth.

The Nikitsky Garden is not only topographically much more attractive than other Soviet botanic gardens, it has many more formal, strictly horticultural, design features; there are pretty belvederes, a charming little Baroque pavilion, a splendid water-staircase, a pretty little water-lily garden, the formal pool and its surrounding walks contained with a 'wall' of closely clipped *Quercus ilex*, and the pool overhung by a big *Salix babylonica*. In all the formal features masonry and evergreens are beautifully combined; or masonry and trees: the water staircase, for example, has a magnificent 'backcloth' of enormous *Platanus orientalis* trees, evergreen oaks, and *Laurus nobilis*. In the same spirit, miles of the walks are edged with clipped box.

Among the interesting trees outside the Arboretum in this botanic garden, and in addition to the great yew and the big pistacia and the manifold zelkowa already described, there are the *Poncirus* from China, which have been the source of material for the rootstocks now used almost exclusively in the citrus-growing parts of the USSR; and a five centuries old olive-tree which is carefully and lovingly tended and has an even more interesting

Cypress, pine and fir in the Yalta Botanic Garden. No garden in the USSR has a more splendid collection of trees, including a large plantation of *Sequoia* and *Sequoiadendron* species and varieties.

history. It was planted by a Greek settler in the fourteenth century; having become incorporated into various private gardens and at last into the Nikitsky, it also became the parent of several frost-resistant cultivars, selected over a period of years, from the seedlings of a mass-planting of its seed. There are large commercial plantations of this tree's hardy offspring not only in the USSR but also in northern Italy. Another splendid old tree is the largest specimen of *Platanus orientalis* I have ever seen, planted by Stevens in 1812; this tree is a native of Crimea. There is a small grove of *Quercus super* from which originated the commercial plantations of cork oak not, indeed, in the Crimea where the climate is too dry, but in the Caucasus region. And there is a whole valley, right down to the sea but best seen from one of the belvederes above, of *Sequoiadendron giganteum*, all raised from Nikitsky seed, still young, and which will grow in beauty and impressiveness for the next thousand years or so.

The Nikitsky has other large divisions. There are the trial grounds for plants of possible economic importance including fodder and lawn grasses; a big tree nursery, in addition to those maintained in the satellite gardens, for raising trees from seed gathered in the Garden; and a large botanical nature-reserve. This latter is a piece of primeval country, mostly juniper trees with associated shrubs and undergrowth, herbaceous perennials, annuals and a host of spring-flowering bulbs. Importance is attached to it because as the Crimea becomes more intensively cultivated, and as its population continues to grow, there is increasing pressure in the fewer and fewer pockets of wild country.

In most of the botanical gardens of the USSR, as of Germany, there is, in the more formal parts and especially in English eyes, a certain look of bleakness. And although the wild parts are full of interest and are, in terms of systematics, probably often superior to our own, it is almost easier to think of them as botanical nature-reserves than as gardens. Like its glorious alpine and maritime setting, equal to anything in the Mediterranean, the Nikitsky is an exception. It has a look of maturity, even of mellow age; and, in its formal parts of sophistication. It also has a look of richness which few temperate zone gardens outside of Italy or England achieve, and the elements of expectation and surprise. In many botanic gardens these refinements of design are overlooked. The Nikitsky Botanical Garden is not only one of the most interesting of the gardens described in this volume; not only one of the most active in the fields of botanical science and economic botany; it is also one of the most beautiful.

USSR

TASHKENT

Right One of the lakes, used for the cultivation of botanically interesting aquatic plants and marsh plants, in the Botanic Garden at Tashkent, in Uzbekistan.

The great distinction of the Botanical Garden of the Uzbek Academy of Sciences in Tashkent, capital of the Uzbek Socialist Soviet Republic, is its supremacy, among all the gardens we have seen, in the successful recreation of natural 'florascapes'. However, first a little about the Garden's scientific and economic work.

A botanical garden of about 25 acres—for the USSR a tiny one—was first planted in Tashkent in 1934; it was refounded in 1951 with an area of 80 hectares (about 200 acres). And the weight given to the sense 'botanical' rather than the sense 'garden' in its layout and stocking is undoubtedly due to the man who is now and long has been its Director, Professor F. N. Rusanov, a member of the Uzbek Academy of Sciences.

On the scientific side the Garden's laboratories are principally concerned with bio-chemistry. The Garden has a scientific staff of 35, and 70 gardeners. After biochemistry, the cytology and plant embryology laboratories are important. On the practical side it is involved, like so many Soviet botanical gardens, in the introduction into cultivation, hybridization and selection of potentially valuable plants from the wild; it concentrated on trees, and in the experimental arboretum has selected new poplar hybrids bred for speed of growth.

The Tashkent Garden also does work in the field of decorative horticulture; it has been particularly successful with tulips, many of which are native there, and has introduced some hybrids of its own into common use. Also with *Hibiscus*; and with *Yucca*, a genus in which Rusanov has specialized, with remarkable results. Over 150 hybrids have been raised, many are now in Soviet gardens, but only seven have, as far as I know, reached us in the West and they are, as yet, rare. The colour range of the new *Yuccas* is relatively great by reason of the intensification of red markings on the flowers which has been achieved; these markings are also very much more diverse than in any of the species, and another new diversity is in the forms of flowers; and, of course, hybrid vigour makes for great size.

Hybridization of species of *Hibiscus* has likewise yielded a whole range of new cultivars, and the Garden has published a book, by Rusanov, on this work, with colour plates of the best new cultivars—the soft pink 'Mozart', the white and crimson 'Ararat', the flesh-pink 'Pascha', the gorgeous cream and carmine 'Radno', the pure carmine 'Cosmonaut' and deeper carmine 'Moscow'; there are more, and all beautiful flowers on plants of good habit. They are beginning to embellish the gardens of those southern parts of the USSR where the genus is hardy.

A special interest of this Garden is in climbing and scandent plants, but in a specific context. In 1965 Tashkent was destroyed by the worst earthquake in its history, and 300,000 people were rendered homeless. A big programme of rebuilding was started at once and by 1968 85 per cent of the homeless had been rehoused in new blocks. This new city of a quarter of a million people was

built by other republics, each contributing materials and building labour according to its means, and the teams of workers camping on the spot until their part was accomplished. Naturally, the whole development looks raw and will continue to look like that for some years: but the Botanical Garden is co-operating in improving its appearance as quickly as possible by supplying large numbers of climbing plants to cover walls, verandas and balconies; and of trees to dress the streets. This has led to the starting of a programme of study, breeding and selection of climbers suitable for urban planting. To the best of my knowledge there is nothing quite like it anywhere else in the world. What the practical results will be is a question: but this kind of thing is in the spirit which maintains that even when the raw material is alive, what man needs man can make to suit his purpose.

The principal part of the Tashkent Botanical Garden, the wild garden-cum-arboretum (45 hectares), is ingeniously laid out. A central area is defined by a completely circular ring-road, from which the nine roads which lead the visitor through the nine principal geobotanical or other divisions, lead off and to which they return. Thus the divisions are completely separated from each other, and connected only through the ring road. This is how the Garden is now; but it is not finished, in the end there will be sixteen of these roads through sixteen divisions. The roads themselves are winding, so that a very large number of plants and plantings can be clearly seen from the road; there are, of course, small footpaths through the divisions as well. Professor Rusanov makes the point that to see the whole garden at present requires four and a half days divided into nine visits, one for each road; each road takes approximately two and a half hours to walk from end to end allowing time to look at plants. The principal divisions are the geobotanical—Uzbekistan with Central Asia; Soviet Far East with China; North America; European Caucasus; and some others, e.g. the Biology Laboratory's section, and a big Physic Garden, not a museum of plants but in use for the investigation of medicinal properties in native plants. A section is devoted to the economic plants of Uzbekistan; 20 hectares to the trial and selection of new introductions, chiefly trees and shrubs. There is a small garden demonstrating systematics, but, as I have said above, Soviet botanists are not much interested in mere taxonomy. A *Hortus* (Flora and Dendrology) of the Tashkent Garden is in course of production, in eight volumes, of which Vol. I is in the press, Vols II and III ready for press. The Uzbekistan and Central Asian section of the gardens has 350 species of trees and shrubs, as well as herbaceous and bulb genera. There are about the same numbers for the European Caucasus; and over a thousand for the Soviet Far East and China.

There is, of course, nothing new in planting a botanic garden after nature, but nowhere else have I seen this as well done as at Tashkent. Without impairing the natural look of the 'florascapes' Rusanov has contrived to place

Inflorescence of a hybrid *Yucca*, one of more than forty new *Yucca* crosses produced by Professor Rusanov in the Tashkent Botanic Garden.

groups or specimens of all the most interesting or beautiful plants near to the sides of his many miles of road. Modern Tashkent is a huge, thriving, noisy industrial city; the quiet, the deeply rural look, of its Botanic Garden is in delightful and very surprising contrast. In a sense, the 'ancestor' of this garden is Gravetye, for although it is strictly botanical, and has the necessary unsightliness of labels (labelling is impeccably maintained), it is an English 'Robinsonian' garden too, with the most floriferous species available within the rules doing duty as ornamental cultivars. I was fascinated by the completely 'natural', enormous, and in reality cunningly planted, boskages of wild roses—*rugosa, rubiginosa, moyesii, acicularis, wichuriana* and many more, for virtually all the West Chinese, Central Asian, East European and North American species are here. *Crataegus, Malus, Cotoneaster, Spiraea*—and *Lonicera* often already bearing orange or scarlet fruits in June—are among the many ornamental genera represented by numerous species. Some of the most attractive shrubs grouped near the roadsides belong to the genus we still, even in its shrubby species, call *Potentilla* but which the Soviet botanists call *Dasiphora*. The richness of the Central Asian and Far Eastern divisions are in part due to the fact that Rusanov himself has done a great deal of plant collection for the Botanic Garden, and so have several members of his scientific staff. With a kind of self-deprecatory wryness Rusanov points out a willow which, in the hurry induced by the fact that somewhere near the Chinese frontier he had a possibly hostile patrol at his heels, he 'collected' under the impression that he had found a new hypericum (the resemblance, I may say, is striking).

There is a very great difficulty in conveying the charm and interest of the Botanical Garden at Tashkent, in words or even in pictures. The photograph of an English country lane would look a little like one of the Garden's roads simply because detail would be indistinguishable. In spring the wealth of wild tulips and other bulb plants in the grass would hardly be apparent excepting in a series of close-ups; in summer a photograph would equally fail to convey the pleasure and excitement of seeing so many kinds of roses, so many honeysuckles, so many spiraeas, polygonatums, potentillas and scores of other genera in flower during an hour's walk. All is 'natural'; yet never in nature would you find a paradise as rich in plants as this, within so small an area.

Even in those divisions planted chiefly for the use of the various associated laboratories, the same good rule applies. I do not believe that Rusanov has much use for herbarium material; he seems to believe that useful studies can be made only of the living plant, and then only if that plant be ecologically 'satisfied', that is, in an environment and associations such as it would have in the wild. 'Tovarich,' he said, forgetting in his love of his subject that I have no right to that form of address, 'it is only the live plant which teaches.' He also believes that a genus can be known only after an exact and painstaking study of *all* its species; and that the man-made species designed in advance to have the attributes the botanical gardener has decided they shall have can be made only in the light of that knowledge. It was in search of that completion of knowledge about the genera which interest him that he travelled all over Europe, Asia and North America.

To return, for a moment, to some details: the fastigiate oak selected in this garden is represented by 120 cultivars with distinguishable differences; the two 'fossil' trees, *Ginkgo biloba* and *Metasequoia glyptostroboides* are here grown not as individual specimens, but in coppices which, given a couple of decades, will not only be pleasing to look at but may produce a range of seed in which the whole of the plants' genetical heritage will, under Garden conditions, get a chance of expression. That is something Soviet botanists are very enthusiastic about: plant as many of a species as you can and in one place; collect seed and raise it on as vast a scale as possible, and then be thorough in your study and selection from the seedlings. It is all a question of missing as few as possible of the combinations and permutations of genes, especially where naturally remote species are brought together.

This is necessarily a brief glance at the enormous subject of botanical gardening in the USSR, a sort of random sample of its 150 botanical gardens; and yet hardly random since Academician Tsitsin regards the five gardens I have described, with the Botanical Garden at Batumi, as the six most interesting of the many botanical gardens in the Union.

TROPICS

BOGOR

View across the lake in the *Hortus Botanicus Bogoriensis*, Java (Kebun Raya), towards the English Palladian palace built by Raffles and inhabited by Sukarno under house-arrest.

Key to Map
This garden is so richly planted that the key refers to families of plants dominating the area in question

1 Leguminosae
2 Palm collections
3 Euphoriaceae
4 Lauraceae
5 Moraceae
6 Anacardiaceae
7 Burseraceae
8 Myristicaceae
9 Annonaceae
10 Rubiaceae
11 Rutaceae
12 Taxaceae
13 Palmaceae
14 Icacinaceae
15 Pandanaceae
16 Liliaceae
17 Compositae
18 Bombacaceae
19 Malvaceae
20 Vitaceae
21 Hippocrateaceae
22 Combretaceae
23 Asclepiadaceae
24 Bignoniaceae
25 Acanthaceae
26 Verbenaceae
27 Aristolochiaceae
28 Flagourtiaceae
29 Myrsinaceae
30 Musaceae
31 Apocinaceae
32 Myrtaceae
33 Dipterocarpaceae
34 Guttiferaceae
35 Ochnaceae
36 Sapindaceae
37 Sapotaceae
38 Ebenaceae
39 Urticaceae
40 Araliaceae
41 Magnoliaceae
42 Meliaceae
43 Palace

There are over 500 botanic gardens in the world. Were I required to name the six best of the hundreds I have seen—best in point of beauty, extent, scientific service, economic service—I should without question include the Kebun Raya, that is Botanic Garden, of Bogor in Java, the erstwhile *Hortus Bogoriensis*; and the great Botanic Garden at Singapore. And neither of them would come sixth. This would be despite the fact that the Bogor Kebun Raya has suffered grievously from the depredations of militarist-politicians; and that the Singapore Garden seems likely to suffer increasingly from the same trouble. The present state of the Bogor Kebun Raya is remarkable, although it and its satellite gardens have had their scientific work brought to a standstill by withdrawal of financial support. The fact is that Indonesian governments have far too many soldiers, guns and tanks to be able to afford their people enough to eat, let alone to afford such luxuries as science. So the Gardens are maintained, and very well maintained, by the determination and goodwill of the staff, who devote to them not only their work but a good deal of money out of their own pockets. As for Singapore's Garden, cut off from the vast country it was designed to serve and be paid by, its future is doubtful.

THE HORTUS BOGORIENSIS

Although it was founded as a botanical garden in 1817, the *Hortus Bogoriensis* had still earlier antecedents as a garden of acclimatization. At present the complex of gardens to which it is central includes the Kebun Raya of Bogor (200 acres), one of the most beautiful and useful botanical gardens in the world; the Kebun Raya of Tjibodas, a mountain garden of fine trees and sloping lawns associated with 2,000 hectares of primeval forest reserved for scientific study; the Purwodadi garden on the drier side of Java for the trial and study of drought-resistant plants; and other satellite gardens at Lawans (East Java); Sibolangit (Sumatra) and Eka Karya (Bali). In all, well over 500 acres under cultivation and study.

THE KEBUN RAYA TODAY

Bogor's great garden comes as something of a shock if only because, if one has been visiting many botanical gardens, as I had been across the world before reaching Java, one is told that since the departure of the Dutch the place has been totally neglected and let down. The facts are different; as a garden, far from being neglected it is superbly well maintained. This is accomplished with an inadequate government grant by a group of mostly young, western-trained biologists, zoologists and botanists.

The Council for Sciences of Indonesia is a body directly responsible to the President; immediately under it comes the National Biological Institute; the Kebun Raya and the Herbarium Bogoriense are parts of a complex of institutes in the field of biology presided over (1967–8) by Dr Sumarwoto, under whom comes the Director of the Kebun Raya, a botanical scientist, Dr Sastrapradja. Trained in Europe and America, Dr Sastrapradja is devoted to the garden, its work and purpose; but having been called on by the Suharto government to undertake an administrative post connected with the National Biological Institute, in Djakarta, he is only able to spend one day a week working at the garden where, during the other five working days, his assistant, Dr Wurjantoro, takes his place. The team of young men and women, of whom these are only a few, all trained in their disciplines in America, Europe or the USSR was brought into existence by the foresight of Professor K. Setyodiwiryo, who directed the Institute in the post-war years and foresaw that with the withdrawal of the Dutch a new generation of Indonesian scientists would have to be trained to take their place. But Professor Setyodiwiryo was inspired to this plan and encouraged in carrying it out by the devotion and determination of Dr Kostermans of Leyden University, formerly of the *Hortus Bogoriensis*.

THE GARDEN AS IT LOOKS

The Kebun Raya at Bogor consists of 80 hectares (200 acres), of gently undulating land which has much topographical charm. A few of the trees and palms in this area are primeval; but for the most part the garden is planted, an artefact. The main walks and drives which criss-cross

The magnificent Torch ginger (*Phaeomeria speciosa*), one of the world's most spectacular plants. Each 'torch' is a waxy cone, often as much as 10 inches in diameter.

it are macadamized; the secondary and even very minor woodland paths are beautifully pebble-paved and the many stairways solidly and handsomely built of masonry. Even the larger and lesser drainage canals which form a grid over the whole garden and supply the several lakes and ponds are massively stone-lined and coped. All this fine and enduring construction work was done by Javanese craftsmen whose wage was 6 gülden per month.

The gardens contain 6,000 species of trees and shrubs and some 5,000 species of lesser flowering plants. In 1963 the National Biological Institute published the Garden's 36th Catalogue of its cultivated plants, a quarto volume of 240 pages excluding the index. A remarkable card-index system in the garden offices, giving a brief but sufficient history of every long-lived plant in the Garden associated with site charts, is still perfectly maintained and kept up to date; the oldest records are still complete; I made use of both while staying in the Guest House, inside the Kebun Raya. The woody plants, forest trees, flowering trees of spectacular beauty, many hundreds of shrubs, are all very well and expertly cared for. Much of this plant material is still of use to the authors of the great *Flora Malesiana*, in the production of which work Leyden University, the Bogor and associated gardens, and the Singapore Botanical Garden have been collaborating for many years.

One of the most striking features of this very beautiful garden is the long avenue of majestic *Canarium* trees, every one of which supports a gigantic tropical climber, often to the summit. Among these are the *Philodendrons*, one of Bogor's introductions to South-east Asia. Their enormous and beautifully cut shapely leaves make each *Canarium* trunk an object of visual delight, a vertical green garden. Nor are such climbing aroids the only plants supported by these trees, which are also hosts to such large epiphytic orchids as *Grammatophyllum speciosum*. Some of the inflorescences of this orchid exceed 6 feet in length. Many of the *Canarium* trees, forest giants, carry *Grammatophyllum* orchids as well as quite a range of the many hundreds of *Dendrobium* orchids native to the country. Other trees support whole ferneries of the stately *Asplenium nidus-avis, Platycereus* and other ferns.

As well as several of such avenues of huge trees draped from bottom to summit in epiphytes and climbers, trees are planted as specimens or as pure stands beautifully placed on the vast sweeps of pleasingly undulating lawns which slope gently or steeply to the lakes and ponds. Trees are natives or exotics from all over the southern hemisphere and the northern tropics, from Kashmir, from Central America and from the eastern Mediterranean. In one or two stands some of the largest *Canarium decumarium* trees carry big golden-yellow excrescences on the upper branches, the nests of wild bees. The bees in question are specific to that species of trees, and their wax is, or was, of commercial importance in the making of *batik*.

The collection of tropical palms at Bogor is a good one, with ninety-four genera, some represented by several species. The Hortus introduced some, notably the oil-nut palms since important in Indonesian agriculture.

Part of the Bogor Kebun Raya, as of the Singapore Botanic Garden, consists of an area of primeval rain-forest or jungle. It has a considerable fauna including five or six species of snakes, three of them poisonous, one of these being a cobra. (It is of interest that there has been no case of snake-bite among the garden workers and visitors for at least twenty years.) Nothing is planted into this wild area, and nothing taken out of it. But from the peripheral arboretum areas seedling trees are collected by the thousand, and supplied to the Government Forestry Department for reafforestation and similar projects.

The several sheets of water, one of which has a handsome fountain, are stocked with *Nymphaea*, *Nelumbo, Victoria* and other water-lilies and aquatic plants, including a fine range of reeds dominated by *Papyrus*. There are also a very large number of the genera and species of bamboo, including the gigantic ones which are of such importance to many industries in the East, even being used by the Chinese instead of steel to reinforce concrete. For economic as well as for purely botanical reasons bamboos have always been treated as important in the Kebun Raya flora. It is difficult to say which, in their beauty and impression of tensile strength, are the more impressive, the bamboos, or the palms, cycads, pandanus and rattans.

Ornamental horticulture is not neglected and the garden is kept colourful, especially along such principal walks as the Queen Astrid Walk and the Lady Raffles Walk, where spectacular plantings of canna and other vividly coloured tropical flowers are used, the scale being so large that garishness is somehow avoided. There are very cleverly contrived mass plantings of coloured-leaved aroids, crotons, coleus, etc., in abstract designs. There is a striking and curious contrast between the several style elements in this great garden, between the formal plantings with a very rich plant material beside rectilinear drives and walks, and the wildness represented by the burly, massive foliage, weird and gigantic inflorescences, and hypertrophied fruits of such trees as durian, jak and 'cannonball', beyond whose summits tower the cloud-draped Javanese mountains; contrast, too, between the 'English' romantic landscape gardening of wide sloping lawns and water-lily lakes, and the primeval forest alongside. In that respect, that is, in the range of horticultural styles using such a diversity of plants, Bogor is the finest garden we have seen, at once a botanical arboretum, a park of recreation, a nature reserve, and a botanical system garden.

If I had to name the most interesting, curious and florally lovely genus, though by no means the most spectacular, which I saw at Bogor, I should choose *Phaleria*, a member of the family *Thymeleaceae*; and notably *P. capitata*. The *Phalerias* are found in Java, the Moluccas, Borneo, New Guinea and Iran, and *P. capitata* comes from the Celebes. It bears its dense clusters of fragrant, white tubular flowers all over the trunk and main branches, well below the leafy parts, in a sort of scented white foam which hides the wood. At first glance it looks as if all the older woody parts of the tree are carrying a vigorous, snow-

Graveyard of the early European botanists—most of them Dutch or English—who worked in the great Bogor *Hortus* and died young of various tropical diseases.

Opposite left Avenues of trees at Bogor support gigantic lianas and climbing *Aroids*.
Right Modern Indonesian sculpture in concrete in the Kebun Raya.

white, sponge-like fungus parasite; closer inspection reveals the lovely detail of this extravagant inflorescence.

TJIBODAS

To reach the mountain satellite garden of Tjibodas you drive for an hour, climbing all the time, first in villages, between roadside bazaars, the shopkeepers selling great piles of *durian* stylishly bound into clusters of three with pandanus leaves; or of the pretty scarlet *rambutan* (tastes like lichee); or beautifully stacked or packed pineapples and coconuts. There is a cheerful population of graceful, slender people, porters carrying pole-loads and wearing the huge pandanus umbrella hats, tri-shaw boys with their brightly painted machines, housewives in the elegant Javanese dress. Higher, the roads are lined by terraced paddy fields, with banana plants in every gorge, corner or stream bed. It will be either raining or about to rain, and the mountain peaks ahead will be shrouded in cloud.

Tjibodas Kebun Raya comes as a surprise even greater than its mother-garden. It is an old 'English' landscape garden of hilly lawns and trees in dramatically beautiful country, with streams and a rock-and-water garden of South-east Asian plants. There is also a collection of orchids. Between 1944 and 1956 the garden was abandoned, or at least very badly neglected, and very great losses were sustained. But now this garden is one of the best maintained and most beautifully cared for that I have seen. Its enormous lawns are mowed—with small, sharp knives lashed to sticks, for want of proper tools; its trees are cared for; everything is order and neatness, and the head gardener a man whose brain contains a whole encyclopedia of Indonesian plant species.*

I think that the most beautiful feature of this garden is the grove (technically 'secondary vegetation' sprung up after tree-felling) of *Alsophila glauca*. Whether you stand on the hill sloping down to the little valley where these tree-ferns grow, or stand under them, their beauty of habit and the exquisite detail of their fronds exceed those of the fine tree-ferns in the New Zealand and Australian gardens. But this garden, about 4,500 feet above sea-level and with its mean annual rainfall of about 4,000 mm, is rich in all kinds of both native and exotic ferns. I starred in my notes species of *Adiantum, Aglaomorpha, Alsophila, Asclepia, Asplenium, Blechnum, Cheilanthes, Comogramme, Cyathea, Dicksonia, Diplazium, Drynaria, Hemitelia, Periploca, Pityrogramma, Polypodium, Polystichum, Pteris, Stenochlaena, Todea, Tylophora* and *Woodwardia*. I have seen no garden to compare with Tjibodas for ferns.

Second in beauty to, if far more impressive than, the *Alsophila* grove is the very grand avenue of *Araucaria bidwilii* and *A. cunninghamii*—one of the most handsome conifer groves I have seen anywhere. And the garden—it should perhaps be rather called an Arboretum—has a fine collection of conifers not all of them concentrated in the Pinetum section: there are some remarkably beautiful *Casuarinas*, eight species; fine examples of several species of *Agathis, Callitris* and other tropical and southern hemisphere conifers as well as cypresses of both groups, *Cryptomerias, Podocarpus* and pines, all well grown.

Unexpectedly, the *Rhododendron* in this garden are not, with a single exception, of much interest: the exception is the native epiphytic *Rhododendron javanicum*, which was in flower when I was in this garden in early December. The flowers are a rich orange and, as far as I could tell, the species is not very particular in its choice of host, for I saw it growing on both conifers and broad-leaved trees. (Deciduous trees will not do well at Tjibodas; the only one I saw was a cherry and it did not look happy. There are numerous roses but they seem to behave as evergreens.)

These two valuable botanical gardens, the natural woodlands preserved with them, and the satellite gardens already mentioned, have been very effectively rescued from neglect and restored to seemliness. Meanwhile the Herbaria and laboratories associated with them are being restaffed and their work restarted, including their share in the work of completing the *Flora Malesiana*, of which three volumes have now been published. Whether this good start can be followed up is a question; there is no doubt about the goodwill of the scientific and horticultural staff but none, either, about the fact that nationalist leaders in countries emerging from revolutionary situations prefer their friends, even in scientific posts, to men who, being neither friends nor enemies, are pure scientists indifferent to political ideologies provided they are given the means to get on with their work. The National Biological Insitute of Indonesia is a noble concept and looks fine on paper. Indonesian governments will find the money to bring it back to life; and to leave the management of it long enough in the hands of the present team to enable it to do, in the future, work as valuable as its begetter, the *Hortus Bogoriensis*, did in the past.

*It is an interesting fact that many of the working gardeners employed by the National Biological Institute in its Kebun Raya are analphabetics. The former Curator of the Bogor garden, Dr Soedjana Kassan, told me that some of these men have an astonishing and perfectly reliable memory for the names of plants, retaining many hundreds and always able to identify them correctly. I tested this for myself and was filled with admiration for some of the illiterate and erudite botanists I met.

PERADENIYA

The Sinhalese Botanic Garden at Peradeniya is remarkable for, among many excellencies, its great avenues of palms, both native and exotic.

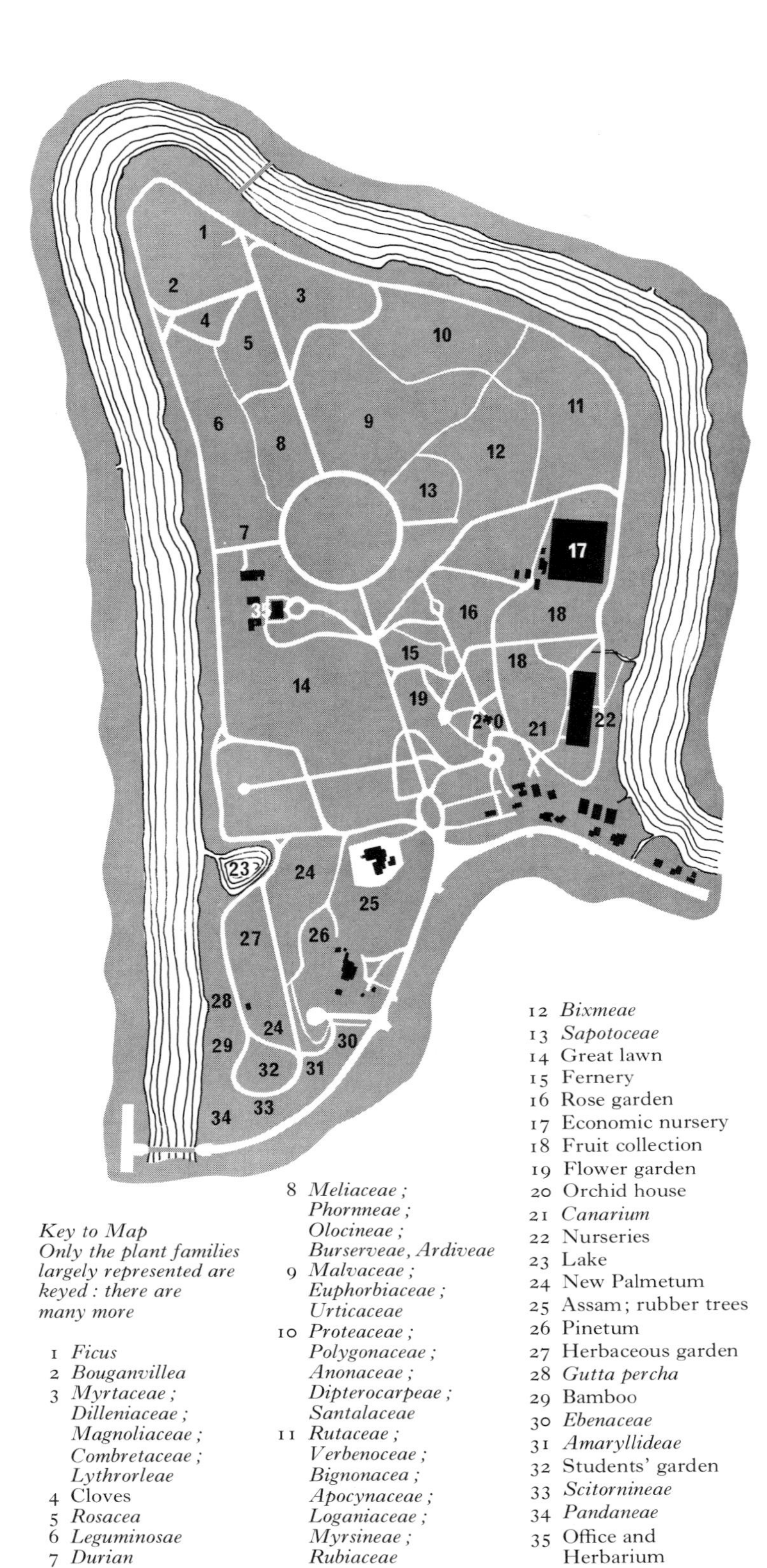

The first suggestion for a botanical garden in Ceylon came from Sir Joseph Banks when he was Director of the Royal Botanic Gardens, Kew. Nothing was done for some years until the idea was taken up by Sir Alexander Johnson, who urged it on the Prime Minister, Lord Liverpool. Johnson's suggestion was adopted and one William Kerr was appointed Chief Gardener and sent to Ceylon. Kerr began the laying out and planting of a garden on Slave Island and another at Caltura, where he made his home. The Caltura garden was specifically for the introduction and acclimatization of economic plants. But Kerr did not have time to accomplish anything, for he died within two years, in 1814. Presumably some progress was made under the new Superintendent, Alexander Moon, but the history of this earlier garden has been over-shadowed by the subsequent importance of Peradeniya. In 1821 Moon was told that the botanical gardens would have to be moved and he was instructed to seek a new and suitable site for them in the region of Kandy, the ancient capital of the Sinhalese kings which lies about forty miles inland towards the mountains, from Colombo. A month after receiving these orders Moon was ready with his report. Of the site he had chosen he wrote:

There are already a number of fruit and forest trees common to the Island, dispersed all over the Grounds, which will afford immediate shade and shelter to the more tender Exotic and Indigenous plants on their introduction and it is sufficiently spacious to admit of an extended Botanical arrangement, including experimental horticulture in general.

The place so described was Peradeniya. Moon's recommendations were accepted and a house was soon being built for him of traditional Sinhalese materials but in the new Anglo–Sinhalese style. He had included in his report an estimate of the 'establishment' he would require and this was allowed without quibble. This was not generosity; as will appear, the government intended the botanical garden to pay its way.

The area chosen was not one man's property and the authorities had to acquire piecemeal such plots as:

Agala Rotuweatte, extent 1 ammonam 2 pelah, of paddy containing 20 coconut trees, 13 Jak and 34 Kekuna and some coffee and fruit trees. Belongs to Delada Meligawa and is in charge of two Gamaheya who have paddy fields for their service.

And again:

Kandawatte, 1 ammonam 2 pelah . . . belongs to Natha Dewale and is inhabited by Kandawatte Appu who . . . performs the service of umbrella bearer at the Festival of Maha Parahara as well as supplying milk to the Dewale.

Another plot belonged to the 'Vidan of Peradeniya'. These and other plots were acquired by exchanging for them equivalent government properties, and it seems that everyone was satisfied. Some of the plots were burdened with traditional rents in kind, for example '20 coconuts a year to the Delada of Maligawa', and these Moon had to undertake to continue paying out of the new Botanic Gardens. Nor was that the whole of his financial undertaking. The government expected and required him to produce revenue: he was, for example, to send all the Gardens' surplus produce such as jak-fruit, coconuts, Kekuna sugar and coffee to the Revenue Commissioner in Peradeniya town, who was to sell it in the markets and put the proceeds towards the revenue. Moreover, if the

Ceylon, and consequently the Peradeniya garden, is remarkable for its orchids.
Above Vanda × 'Louise de Waldner'.
Below left A native *Dendrobium*.
Below right Amherstia nobilis.

Below Buttress growths of *Ficus elastica* in Peradeniya. This species was threatened with extinction during the 'Rubber Rush' following the discovery of vulcanization which made rubber plants valuable.

Superintendent did not need the whole area for botanical gardening he was to:

. . . attend particularly to realising as much revenue as possible and to apply as much of it as can possibly be applied to the growth of coffee . . . [and] it will be necessary you should consider and report how he [Moon] is to be supplied with labour at a cheap rate.

Such was the origin of one of the most beautiful and still one of the greatest botanic gardens in the world. The Peradeniya garden was 'inaugurated' in February 1822; but not until the appointment as Superintendent, in 1844, of George Gardener, a botanist of international reputation, did the Garden's real botanical career begin. It was Gardner who laid out and planted the lovely river walk and other fine walks and rides; it was he who planned and began to lay out a real garden of plants instead of a muddle of incomplete collections and market garden plots. But Gardener died in 1849, and it was left to Thwaites, his successor, to make the name of Peradeniya famous, first among botanists by his research work on the scientific side, and thereafter among gardeners by his progress with the work which George Gardner had planned and started. Even Thwaites, busy introducing such spectacular exotics as *Dendrocalamus giganteus* from Burma (1858), *Amherstia nobilis* from the same country (1860) and *Lodoicea maldwica* (*seychellarum*), the Double Coconut, did not complete the laying out of the Garden so that it did not attain its present most satisfying shapeliness until the next Superintendent, Trimmen, had taken up the work and brought the whole 145 acres into unity.

Not that the Garden remained unchanged exactly as Trimmen left it; there were many later minor modifications, and some major additions—I am still discussing the layout, not the collections. The once famous avenue of *Ficus elastica* trees, too closely planted by the Superintendent Watson in 1833, had to be felled in 1912–13, the ground prepared and replanted in 1914. Another avenue from the Main entrance down to the lake had to be replanted because the gigantic climber *Thunbergia grandiflora*, which had been planted to climb the trees, had killed them. In 1861 the Fernery, a very pleasing feature still, was added by the first Curator, William Cameron. Superintendents, Curators and, later, Directors, took immense pains with the Garden's superb sweeps of lawns, for the right grasses for the local conditions had to be found by trial and error. This work has never ceased, and lawn-grass trials are still maintained by the present Director, Mr Jayaweera. The largest of the lawns, a vast shallow bowl of land, covers 7 acres and is beautifully decorated with specimen trees. Both the Arboretum and the Palmetum were started early in the Garden's life and by 1922 the Arboretum had representatives of 48 botanical families. The Palmetum, with 138 species in 68 genera, is pleasingly extended all along the River Walk.

In the 1870s and 1880s Thwaites, followed by Trimmen, completed a planting of herbaceous subjects in systematic order; it is still useful to students of taxonomy.

The grounds had included a typical Sinhalese irrigation tank, some centuries old; in the 1890s this was converted into the lake as we have it now, and planted with the principal tropical aquatics—*Nymphaea*, *Nelumbo* and *Victoria* species. By the end of the same decade nurseries of both economic and ornamental plants had been established; and by 1902 the first experimental station. Of the Garden's several Palm Avenues the earliest, the Palmyra Palm avenue, dates from 1905; and the latest, the Royal Palm Avenue, from 1950.

Peradeniya's satellite Garden, Hakgala, high in the mountains, of which I shall also have something to say, was first planted in 1860 as a nursery for the newly introduced *Cinchona* trees. A second satellite garden was planted later in the century at Henaratagoda as a nursery for the rubber-trees, *Hevea brasiliensis*.*

PERADENIYA'S GREAT MAN

Henry George Kendrick Thwaites was at once the man who made Peradeniya famous and who did the people of Ceylon the greatest service which any one man has ever done them, using the Botanical Gardens as his means. He was one of the leading practical and academic botanists of his epoch; not only did he personally discover and collect and name twenty-five genera and many more species new to science; but, beginning in 1858 and finishing six years later, he described Ceylon's flora to the world of science in his *Enumeratio Plantarum Zelaniae*. Yet it is for his work in the field of economic botany that Thwaites is most highly honoured by those who know anything about it.

THWAITES AND COFFEE

Coffee was introduced to Ceylon by the Dutch; Gardner, Moon and Thwaites greatly increased the Island's output

*'Henaratagoda No. 2' was a famous champion tree of *Hevea brasiliensis*, and the parent of a high-yielding clone well known to rubber planters. Between 1908 and 1913 inclusively this single tree yielded, in 45 months, the record quantity of 392 lb of dry rubber.

Left The Palm Collection at Peradeniya is a fine one. This is a rare occasion, a *Corypha* in flower, which it does after thirty or forty years.

Right The roots like the fruits, of many species of *Ficus* provide endless pleasure for the eye and wonder for the mind.

by encouraging plantation, by supplying young plants from the Garden's nursery, and by introducing Arabian coffee through the Botanical Garden. Then, in 1869, a fungus parasite coffee-leaf disease was first noticed in the Island and within a few years was threatening the whole future of the Industry. Thwaites took two steps to deal with this crisis; he began introducing new varieties in the hope of finding a resistant one, with partial success with a Liberian strain in 1876, by which time most of the plantations were ruined; and he brought the well-known mycologist H. M. Ward to the Island to investigate the life cycle of the fungus and suggest a way to control it. Ward did establish the life-cycle of *Hemileiia vastatrix,* as he called the creature, but could suggest no remedy.* Meanwhile the coffee industry of Ceylon had collapsed, planters were ruined, and, as a consequence, the government was seriously embarrassed in its finances.

CINCHONA AND TEA

Meanwhile at Peradeniya Thwaites was not confining his fight to save Ceylon's economy to trying to save the coffee plantations. His first efforts to find a plantation substitute was concentrated on *Cinchona*. He had already obtained seeds of the quinine trees from Sir Clements Markham, who had his contacts in Peru. He now stepped up the propagation of plants and started a campaign to persuade planters that their salvation and that of Ceylon lay in planting *Cinchona*. He was not unsuccessful, for as early as 1865, 180,000 young *Cinchona* trees were distributed from Hakgala, and by 1876 the annual figure had risen to 1,224,000 plants. Like most scientists, Thwaites had far more foresight than the planters and businessmen; he did not believe in relying on a single crop; while pushing *Cinchona*, he was not only introducing variety after variety of coffee but turning his attention to another possible crop plant, *Camellia thea*; it was the one destined to become Ceylon's principal plantation crop.

Peradeniya had first introduced tea from China to Ceylon in 1828. When the wild ancestor of Assam tea was discovered, by a great-grandfather of the novelist John Masters, Thwaites introduced it to Ceylon. Throughout his term of office he was introducing other varieties of tea for trial in the Garden, and propagating and distributing young plants of promising kinds. Thus Thwaites and the Botanical Garden were unquestionably responsible for Ceylon's very profitable tea industry.

THWAITES AND RUBBER

I tell below the tale of how Brazilian rubber came East, as it is told in Singapore. There are some discrepancies between that tale and the story as it is told at Peradeniya. It will be recalled that from the 2,000 seeds of *Hevea brasiliensis* obtained from Cameta in 1873 Kew raised only twelve

*This was in 1880, and one cannot help wondering whether there was any contact between Thwaites and Ward on the one hand, and on the other with the workers who were trying to cope with another, newly pandemic, fungus parasite, *Phylloxera vastatrix*, which was rapidly wiping out the French and other European vineyards. Ward was probably the first mycologist ever to be called in to advise on an economic–agricultural problem.

plants. And that the six sent East were sent to Calcutta, where they died. The next Brazilian collection was made by Cross, who did not confine himself to *Hevea* but sent seeds of two other rubber-yielding plants, *Castilloa elastica* and *Manihot glaziovii*. (I saw both still growing in the Kebun Raya at Bogor.) All Cross's plants were tried by Thwaites at Peradeniya but the quality of the latex from all but *Hevea* were unsatisfactory. Of Wickham's Kew seedlings, Thwaites received 1,919 in 1876, and he seems to have discovered, independently of Ridley at Singapore, that only young seedling growth could be rooted as cuttings. By 1880 he was sending rooted cuttings to planters in Ceylon, South India and Burma. A year later the first seed was gathered from Ceylon-grown trees and three years later large-scale distribution of seed and seedlings began. There is no doubt that Peradeniya and Heneratagoda played a part as important as the Singapore Botanic Garden in establishing South-east Asia's great rubber industry, and it is no doubt significant that when the first World Rubber Exhibition was organized in 1906, it was held at Peradeniya.*

OTHER INTRODUCTIONS

Maintaining his policy of trying to persuade Ceylon's planters to diversity the Island's plantation crops Thwaites did not stop at coffee, tea, quinine and rubber in introducing new economic plants through the Botanic Gardens. Cocoa had been introduced by his predecessor in 1835, and by 1845 the Garden was offering young plants for sale. Worried by the swift spread of the coffee-leaf disease, Thwaites began pushing cocoa as a plantation

*It is a curious fact that only in Brazil, its homeland, has *Hevea brasiliensis* failed as a plantation crop. Even the great Ford Complex of companies was discouraged by the losses during commercial-scale trials. The difficulty is due to a parasitic leaf disease which seems never to have established itself in the Asian plantations. This is a killer. However, some resistant plants have been selected and it may ultimately be possible to grow rubber on a large scale in its homeland.

Left 'Cocos de Mer', so called because the fruits were found floating in the sea long before the palm was identified. The fruits take ten years to ripen.

Right Golden coconuts in the Peradeniya Botanic Garden.

crop in 1873, and at the same time he introduced and acclimatized the eighteen best cocoa varieties in cultivation in the West Indies. Cocoa has since been a fairly important plantation crop in Ceylon. But not all of Thwaites' Peradeniya introductions were successful: most curious is the case of *Coca*. Thwaites introduced and acclimatized this ancient Inca drug plant in 1870 and in due course there was some commercial plantings; cocaine is, of course, a valuable medicine as well as a dangerous drug. The refined alkali is far more potent than the leaf of the plant itself, and in any case the Inca governments had rigorously controlled the planting and use of *Coca*. The moderate chewing of coca leaves seems to have had no evil consequences. But the plantations in Ceylon were, of course, providing material for extraction of pure cocaine and the government of India soon took fright at the idea of growing *Coca* on a considerable scale and prohibited plantation.

Camphor succeeded at Peradeniya and was planted commercially; but a fall in the world price soon made planting uneconomic. Nutmeg was first brought to Ceylon by the Dutch; it was lost (possibly to a smut fungus), reintroduced in 1804, and from 1843 Peradeniya was selling young plants. It has since been a commercial crop in the Island, and one of the glories of the Botanic Gardens is a grove of old, enormous but still very fruitful nutmeg trees. Cloves and vanilla were two other Peradeniya introductions. The oil-yielding *Croton* was tried by the Garden but it never became established commercially for the demand for croton oil was never large; nor did the oil-palm, another Thwaites introduction, although the palm grows well in Ceylon. Thwaites tried cotton, but it failed at both Peradeniya and at Hakgala. Of the timber tree exotics now common in the Island, Peradeniya was responsible for two *Cedrelas*, a Mahogany, *Grevillea robusta*, a number of *Acacia* species, several *Eucalyptus* species and some useful conifers; work with these trees was chiefly done at Hakgala which is still a fine arboretum. There, too, Sandal, *Balsam* (*Toluifera pereirae*), and the most useful of shade trees for the tropics, *Pithecolobium saman*, native to Central America, were acclimatized and propagated. And finally three tropical fruit trees were introduced through and established by Peradeniya—Durian in 1850, Pawpaw in 1880, and Cherimoya in 1882.

BOTANY AT PERADENIYA TODAY

The influence and personal work of Peradeniya's present Director, Mr Don M. A. Jayaweera, ensures that the Garden is still a centre for botanical work; and there is a lot of it to be done, for by no means the whole of Ceylon's flora has been described and classified. By way of example, as a plant collector Mr Jayaweera has found and described ten new species of orchids; he has, in fact, made a special study of Ceylon's orchids, spending six years in the collection of 150 species, another eight years in their study. His major work on them is now in the press and meanwhile he has ready for the press another useful work, an exquisitely illustrated (with fine drawings made by one of his subordinates) account of Ceylon's 750 medicinally useful plants. A work of this kind is always important; for one thing the Ceylonese make use of their own as well as of Western medicine; and secondly, traditional medicinal herbs have repeatedly yielded valuable drugs on proper investigation, the case of digitalis being classic. Among papers published by Mr Jayaweera as a result of work done on plants grown in the Botanic Garden and in its associated herbarium are a number on the genus *Mussaenda*; a number on the genus *Duabanga*, including the description of a newly distinguished species which he has named *taylori* in honour of Sir George Taylor; and a study of the pollen morphology of the *Sonneratiaceae*. In short, although Peradeniya may no longer have an economic role, its scientific role is still important.

IMPRESSIONS

Ceylon is an island whose beauty transcends all expectation. This is not only in the splendour of its mountain and river scenery and in the perfect marriage of wild rocky country and agricultural artifice in the thousands of acres of hill country tea plantation; but also in the manner in which its richly lush jungle country fulfils the most romantic dreams of what such country should look like. And it is equally so in the Royal Botanical Gardens; one hears and reads of the beauties of Peradeniya, but they far surpass expectation, and from the moment one enters the main gates the eye is delighted and the mind engaged by the spectacle of plants at their most colourful, their most

A general view of tropical foliage and flowers in Peradeniya.

strange and ingenious in form and habit, their most majestic in size and bearing, disposed with art over a piece of country with every topographical advantage, skilfully and lovingly cultivated.

There are three avenues of palms: I have referred above to their planting. The specimens of the sixty-four-year-old Cabbage Palm (*Oreodoxa oleracea*) Avenue have slender, stately, perfectly vertical trunks swollen at the base, and crowns of gigantic leaves. The Palmyra Palm (*Borassus flabellifera*) Avenue is older. Planted in 1887, it is a beautiful ornament of Peradeniya, but it was introduced as a food plant and is much grown in Northern Ceylon; the fleshy embryo of the germinated seeds is used as a vegetable, the nuts are eaten, the sap yields toddy and a form of sugar called *jaggery*. Then there is the young Royal Palm Avenue, which should be at its stately best by about AD 2050.

Among the most spectacular of the flowering trees the following are especially remarkable: *Amherstia nobilis*, large to very large trees with pinnate leaves and huge, pendent, scarlet inflorescences of pea-flowers; *Delonix regia*, 'Flamboyant' with leaves like a gigantic 'mimosa' and crimson-scarlet flowers; *Tabebuia guayacan*, with its great clusters of mauve trumpet flowers—these are only three of the scores of tree species which relieve the Garden's major theme of diverse greens with mighty domes and pyramids of vivid colour.

The form and majesty of many forest trees is even more impressive than the brilliant colours of the kinds which have spectacular flowers. Among the grandest of these are the enormous specimens of *Duabanga taylori*, whose girth at 5 feet from the ground often exceeds 40 feet; the *Ficus* in many species, including the Bo Tree, *F. religiosa*, made sacred by its shading of the Prince Siddhartha, the Buddha, during a part of his long meditation; *F. benjamina*, one superlative specimen of which has a regular domed head 54 yards in diameter; *F. krishnae*, whose leaves are curiously shaped like Krishna's cap. Then, *Hura crepitans*, the Sand Box Tree, various mahoganies, *Celtis cinnamonea*, and the lovely *Tabebuia rosea*. In another planting of figs two are outstanding; *F. retusa* for its complex of adventitious roots which have become supplementary trunks; and *F. parasitica* for sheer bulk of timber.

Many trees in this Garden support immense climbers and lianas, notably in the Liana Drive, a tree-lined winding road through tropical jungle, where the trees support such bright-flowered climbers as *Bignonia unguis*, with fantastically hooked tendrils and canary-yellow flowers; the purple-flowered *Securidaca volubilis*; the 'Nagadarana' (*Bauhinia anguina*), whose seeds are said to cure snake-bite; *Calamus*, the climbing Palm; the huge-leaved, golden-variegated *Pothos aurea*, which, for some reason, is called 'Colombo Agent'; the red-and-gold *Thunbergia mysorensis* whose flowers are like slipper orchids. The wonder and diversity of this Drive are indescribable.

Among the most curious plants at Peradeniya I have starred notes of the following as particularly striking. Beside the very beautiful Lake Drive, *Antiars toxicaria*, the 'Upas Tree', source of a Javanese ordeal-poison and once believed to kill people and animals by its venomous exhalations; in practice the mortal poison called *Ipoh* used for poisoning blow-pipe darts is extracted from the bark, and such darts are still used for hunting in Java, their range, which I tried for myself, being remarkable; *Ladoicea maladivica*, the Double Coconut or Coco-de-Mer of the Seychelles—the trilobed fruits of this extraordinary dioecious Palm, often found floating far out to sea, were known and used for centuries before the Palm itself was identified. The fruits of *L. maladivica* take ten years to ripen and the palms themselves must live for many centuries, for although only one new leaf is produced each year, specimens are known up to 100 feet tall. Then *Parmentiera cerifera*, whose fruits, like fat yellow candles in shape, are borne all up the trunk and on the main branches, growing directly out of the bark.

Art and nature together have provided the Peradeniya Garden with exceptionally beautiful vistas. From some parts there are fine views of the Nantana Hills, but it is the River Drive which is so rewarding in this respect. This follows the River Mahaweli almost completely round the Garden, and on one side it is planted with magnificent clumps of the giant bamboos *Gigantochloa aspera* and *Dendrocalamus giganteus*—many canes exceeding 120 feet in height—and on the other with a collection of palms. At the beginning of the Drive is a good rockery of tropical succulents topped by stands of conifers including the 'Kauri Pine', *Agathis robusta*, and the 'Bunya-Bunya Pine' *Araucaria bidwillii*. At the point where the River Drive intersects the Joinville Drive you have, on the one hand the River Drive continuing between a sort of gateway formed by two colossal Jak-trees some of whose fruits have exceeded 70 lb a piece, and on the other a commanding view across the 7-acre lawn and its mighty *Ficus benjamina* tree, to the Gannaruwa Hills.

A very pleasing feature of Peradeniya is the Flower Garden, a good example of formal, almost stylized planting for colour, more horticultural than botanical; a curving path between mass plantings of *Coleus* and such ornamental variegated grasses as *Phalaris arundinacea* leads to the Octagon House. Backing the path-side plantings are beds cut in the lawns, of familiar herbaceous perennials, the *Gerberas* and *Cannas* being particularly fine. The Octagon House is made not of glass but of wire netting on a metal structure in the form of an octagon, entirely grown over with the small-leaved, evergreen, climbing fig, *Ficus repens*, so that inside it is in deep shade. In this shade flourishes a collection of *Anthurium alocasia*, and also various *Philodendrons*, *Marantas*, *Dieffenbachias* and other plants with decorative leaves. Peradeniya's main collection of *Anthuriums* contains the best specimens of the genus I have seen anywhere, with a remarkably wide range of colour in the flowers. Another house is devoted to the Garden's orchid collection which is of more botanical than horticultural interest.

As in other tropical gardens, many of Peradeniya's trees and shrubs carry a load of epiphytic plants, including

In all the Far Eastern botanic gardens the cultivation and study of bamboos is encouraged by the economic importance of this family of plants in the East.

Bromeliads, of course, but by far the most curious being the giant orchid which I also saw at Bogor, *Grammatophyllum speciosum*. One specimen growing on a Brazilian ironwood tree, *Caesalpinia ferrea*, produces 6-foot, erect racemes of yellow and brown flowers. Among those plants which make a provision of nitrogenous food for themselves—for example those which form symbiotic relationships with colonies of ants—one of the *Asclepidae*, *Dischidia rafflesiana*, the Malayan Leaf-Pitcher plant, is represented at Peradeniya by a good specimen growing over a red-flowered 'Frangipani' (*Plumeria rubra*): this *Dischidia* has both normal leaves and leaves which, folded into a tubular pouch, become full of water in which insects drown themselves by accident; the plant, a climber, then forms adventitious roots from the stem and grows them into the 'pitchers', from which they thus feed the plant on decaying animal-matter, or its end-products dissolved in water. *Plumerias*, incidentally, are much-used host plants for epiphytic orchids of the genera *Vanda*, *Arachnis*, *Dendrobium*, *Epidendrum* and some others.

Reflecting one of the fine walks in the Jardim Botanico of Rio de Janeiro, Peradeniya has a row of *Couroupita guinanensis*: it so happened that whereas I saw this tree in flower in Brazil, in Ceylon the trees were laden with their huge, woody, spherical fruits which give the species its vernacular name of 'Cannonball Tree'. This is only one of the hundreds of species in this Garden which make one marvel at the infinite diversity of forms, the endless ingenuity of devices adopted by plants in solving their one problem, survival of the species.

It is not practical, short of devoting a book to this single garden, to describe more than a very small fraction of the tens of thousands of beautiful trees, shrubs and groundling plants at Peradeniya or in any other great garden; I have more starred notes of the 'Daffodil Orchid', *Ipsea speciosa*, with its golden yellow flowers and a sap so aphrodisiac that, according to Sinhalese legend, a prince murdered his sister the princess, with whom he had fallen passionately in love as a result of an inadvertently taken a dose of it, to save them both from the crime of incest; of *Eugenia malaccensis* with its gorgeous crimson flowers; of the evergreen *Fagrea fragrans* tree, its myriads of white flowers distilling a sweet fragrance; of the gigantic 'Maiden Hair' ferns in the Fernery; of the formal beauty of freshly opened nutmeg fruits when the 'mace' is still a rich crimson; of the grand scarlet-and-green fruited 'Rambuta' trees (*Nephelium lappaceum*) whose fruits are as delicious as those of the related Litchi of the extraordinary buttressed trunks of *Canarium commune*.

Yet perhaps the most admirable attribute of Peradeniya's Botanic Gardens is not the range and diversity of its plants but its unity. A gardener who was not in the least a botanist, or a garden lover who did not know one plant from another could wander for days with ever renewed pleasure inside this fine work of art without troubling at all about the materials of its composition.

SINGAPORE

Singapore is an island about the size of the Isle of Wight at the tip of the Malay peninsular. It was formerly a part of the Federated Malay States and Singapore city its greatest city. It owes its importance to Stamford Raffles, who, unable to continue in his intention to make Djakarta the most important port and *entrepot* in South-east Asia, conferred that destiny on Singapore instead.

The foundation of its botanic garden is not by comparison an early one since the city itself did not exist until Raffles called it into being. The first European botanical garden in the tropics was Pamplemousse in Mauritius (1735), which had a hand in introducing nutmeg, pepper, cinnamon and other useful plants from the East Indies and which played an important part in starting the sugar industry in Mauritius. The British started a botanic garden at St Vincent in 1764; it was for that island, with a cargo of young bread-fruit trees to grow cheap food for slaves, that Bligh was bound in the *Bounty* when her crew mutinied in 1787. The Jamaica Botanic Garden was founded in 1774; Calcutta's in 1786; Penang's in 1796 and the *Hortus Bogoriensis*, now the Kebun Raya of Bogor, in 1817. Trinidad was next in 1819, Peradeniya in 1821; and at last Singapore in 1822, at Raffles' own suggestion and only three years after his foundation of the city itself. (Raffles had already started the second Penang botanical garden in 1820.) Raffles was advised in this project by Nathaniel Wallich of the Calcutta Botanic Garden; nutmeg, cloves and cocoa were the first economic plants to be planted at Singapore. But this first Singapore garden only lasted until 1829, when the Government of India, in one of its fits of economy, closed it down.*

It was a new Agri-Horticultural Society, Singapore's third, which in 1859, got a government grant and the loan of convict labour and 60 acres of land, part primeval forest, to start a new botanical garden. This was to be supported by membership subscription. Eleven acres of this primeval forest have been preserved and form part of the Garden today. The woods are inhabited by python and cobra; and also by monkeys which roam all over the Garden and occasionally bite the children who tease them.

The laying out of the Garden, still much what it was from the start, was entrusted to a neighbouring nutmeg planter, Laurence Niven, who turned out, as luck would have it, to be a gifted horticultural artist. The Society and its Garden were soon flourishing, and in 1866 an additional 25 acres were bought to extend the Gardens, the very handsome house now occupied by the Directors was built, and the big lake was dug and filled. All this expansion put the Society heavily in debt and they even had to go to the government for help in paying Niven's salary arrears—he had been appointed Supervisor of the Garden. A second government grant was applied for some time after this, to enable the Society to maintain the Botanic Garden; it was accorded on condition that the Society show collections of economic plants, for educational purposes, in their Garden. But the day had passed when the Society could make ends meet, and in 1874 its officers asked the government to take over and maintain the Garden in the public interest. This was agreed to; and Sir Joseph Hooker at Kew was asked to recommend a man to fill the post of Superintendent. He must be 'a practised as well as a systematic botanist and able to travel in the Malay peninsula not a little for the purpose of investigating its vegetation'.† Hooker sent one of his own men, James Murton.

Murton's youth was doubtless an advantage on his collecting trips by which he added a great many new species to the Garden's collections; and also, perhaps, in providing the energy with which he set up and maintained connections with other botanical institutions all over the world and by which he enriched the Garden's collection of exotics. But it was a handicap when it came to exercising authority over older men and his career at the Botanic Garden was a stormy one. However, that did not prevent him from having a hand in the planting of the associated Economic Garden of 102 acres: this was started by the government and the University of Malaya in 1879 for the trial and acclimatization and propagation of such potentially valuable crop plants as *Cinchona*, coffee, numerous species of Eucalyptus, tea, maize, sugar and, most important of all, rubber.

Murton's conduct in authority led to his dismissal in 1880; there are hints that the manner of his private life gave offence and perhaps even caused some kind of scandal. His successor was another Kew man, Nathaniel Cantley, who had been assistant Superintendent of the Mauritius Botanic Garden. Cantley had all the Singapore plants properly labelled, added considerably to their number, surveyed the timber resources of the Malay peninsula, a most exacting task, and planted potential timber trees from Malay forests in the Economic Garden for trial. He started the Singapore Herbarium; he nursed the first rubber plants received from Brazil by way of Kew; and finally he introduced a number of European vegetables to Malayan horticulture, through the Botanic Garden. He also initiated and supervised the planting of street trees in the city, providing the young stock from the Garden. He died in 1888, presumably of exhaustion. He consistently overworked, for his health had never been good and the climate did not suit him. With the appointment in his place of Henry Nicholas Ridley the Singapore Botanic Garden entered upon its epoch of greatness and Malaya received the man who was to make her fortune.

INDIARUBBER AND PARARUBBER

East Asia has a good source of latex in that majestic native tree, so widely planted as an ornamental all over the tropics, *Ficus elastica*. The latex from this tree yields indiarubber. The East Asians never discovered the use of latex to make rubber: that discovery was made by the

*For the foregoing and some of the following facts I am indebted to Purseglove, J. W. in *The Gardens Bulletin* (Singapore, 1959). No publisher is named owing to a Minister's hostility to any celebration of the Garden's centenary in that year.

†Burkhill, I. H. *Gardens Bulletin, 11, 2*, 55–72 and *3*, 93–108, 1918.

Courtyard garden in the Singapore Botanic Garden. On the left, stems and leaves of the 'Jade Vine'.

Black (Australian) swans on one of the water-lily lakes in the Singapore Botanic Garden. The lake scenery in this garden is exceptionally beautiful and very much in the manner of 'Capability' Brown.

Overleaf The Jade Vine, *Strongylodon macrobotrys*, the most beautiful plant in the Singapore Botanic Garden. It is an enormous climbing plant of the family *Leguminosae*; approximately life-size. The inflorescences are up to 2 feet long.

South American Indians, who got their latex from a small tree, *Hevea brasiliensis*. When rubber first became commercially important following the discovery of vulcanization there was a rubber-rush in Brazil, the people going by thousands into the forests to tap the wild *Hevea* trees whose latex yielded pararubber which was to become the only commercially important kind (unless you count the product of the Gutta-tree, guttapercha, as rubber). In Malaya there was a lesser but similar rush into the forests; but in order to get the latex from *Ficus elastica* the trees were cut down and therefore killed, and the government of India became alarmed at the prospect of the whole species being exterminated. Action was taken by the authorities.

At the request of Sir Clements Markham at the India Office, James Collins was instructed to carry out a survey of the world's latex-yielding plants so that a species suitable for plantation could be chosen. There are quite a lot of these, but of them all *Hevea brasiliensis* was considered to be the most promising and, again at the request of the India Office, Hooker at Kew undertook to send a collector to the Amazon basin to collect seeds. Several seed collections were made but either the seeds failed to germinate or the few seedlings which were raised and which were nursed in the Calcutta Botanic Garden and in a selected nursery in Sikkim, died. Next H. A. Wickham, later knighted for his services, was sent to Brazil where he collected 70,000 *Hevea* seeds. With the full co-operation of the Brazilian government and the help of the Jardim Botanico in Rio de Janeiro, this collection of seed was rushed to England in a chartered ship.* The seed reached Kew and was sown; germination was only 4 per cent but that yielded about 3,000 seedlings. Most of the seedlings were then sent in Wardian cases (small portable greenhouses), as recounted above, to nurseries prepared by the Peradeniya Botanic Garden at Heneratgoda in Ceylon; eighteen plants were sent to the *Hortus Bogoriensis* in Java; and fifty seedlings were sent to the Singapore Botanic Garden, where they died. But a second consignment of twenty-two seedlings sent to Singapore were successfully established in the Botanic Garden. Later, nine were transplanted into the private garden of the Resident, Sir Hugh Low, at Kuala Kangsar and a permanent site for the Botanic Garden seedlings was found in the Economic Garden. There Murton found it possible to propagate by cuttings taken from very young growth; cuttings had always failed before, apparently because taken from mature material. Murton's discovery made it possible to increase the number of plants much more rapidly and, incidentally, to propagate only from high-yielding plants.

Meanwhile the known methods of tapping the *Hevea* trees were either killing them or yielding inferior latex. Ridley, not long after his appointment, set about finding a better way and after repeated experiments discovered a method (excision and bark-paring) which confined the wounding of the tree to a single cut which, reopened at suitable intervals, yielded more and better latex without harming the tree. By 1891 some of Ridley's rubber had reached London where it was declared to be excellent. By 1899 rubber in commercial quantities was being produced by Ridley's method in Malaya. Ridley continued his experiments in the Botanic Garden and at Kuala Kangsar, improving tapping techniques, working out optimum planting distances for plantation work, and the general care of the plantation trees. By 1910 the technique of cultivating *Hevea brasiliensis* for maximum yield of the best quality rubber without damage to the trees had been perfected in the Singapore Botanic Garden. It was none too soon, because meanwhile a disease was destroying the Malayan coffee plantations and new plantation crops were badly needed.

Ridley urged and continued to urge that rubber could and should replace coffee. He was ignored, laughed at and even officially reprimanded for 'wasting his time'. At last, in 1896, a Chinese planter, Tan Chay Yan, convinced by Ridley's arguments, planted 40 acres of Ridley's young *Hevea* trees at Bukit Lintang. It was in honour of this pioneer planter that in 1966 Mr Burkhill, present Director of the Singapore Botanic Garden, named that Garden's best prize-winning orchid hybrid, *Vanda* × 'Tan Chay Yan'.

*A widely accepted story is that Wickham smuggled the seed out against the will of the Brazilian government. A complete refutation of this charge will be found in *Borracho do Brasil*, 1913. No objection whatever was raised by the Brazilians to the export of these *Hevea* seeds.

The breeding of hybrid orchids, especially *Vanda*, from native Malaysian species has been a principal concern of the Singapore Botanic Garden. Commercial growing of these hybrids is important to Singapore's economy.

RIDLEY'S WORK ON OTHER ECONOMIC CROPS

Ridley did not confine the botanical garden's work in economic botany to *Hevea brasiliensis*. By 1889 he had satisfied himself that cocoa would be a profitable plantation crop. The original *Theobroma* trees were introduced from Mexico and he had studied their cultivation in the Economic Garden. Moreover it was he who first urged the commercial planting of the oil-palm, another introduction from the Americas. Seeds of all the suitable economic crop species were produced by and distributed from the Botanic Garden: over seven million *Hevea* seeds as well as tens of thousands of seedlings were sold to the planters (Burkhill, *op. cit.*). Furthermore the Botanic Garden became, despite the rival importance of the Peradeniya Botanic Garden, the world centre for technical advice on rubber cultivation.

PURE SCIENCE

But economic botany by no means consumed all of Ridley's remarkable energy. As a field botanist he travelled and collected new species in the remotest parts of Malaya, in Borneo, Java and Sumatra, the Cocos and the Christmas Islands. He was the first to describe and determine literally thousands of species hitherto unknown to science. He wrote the first accounts to be published of Malayan timber-trees, fibre plants, esculent, drug and dye plants. He also wrote what is still a standard work on *Spices* (1912) and a five-volume *Flora of the Malay Peninsula* (1922–5). As an administrator he was responsible for such valuable measures as the Coconut Trees Preservation Ordinance, which saved an important food plant from extinction. He became Director of Forests, Straits Settlements. He founded the *Agricultural Bulletin* of the Malay Peninsula; he published, through the Botanic Garden and its Herbarium, five hundred scientific papers and books; and in 1930, long after his retirement, as a monument to a long life of service to mankind, came his *The Dispersal of Plants throughout the World*. Nothing could exhaust him, and on his hundredth birthday on 10 December 1955 he said how glad he was to have lived to see the Singapore Botanic Gardens 'the best tropical botanic gardens in the world'. He died on 24 October 1956.

POST-RIDLEY

Although Ridley never paused in his work he retired as Director of the Botanic Garden in 1912, and I. H. Burkhill, formerly of the Calcutta Botanic Garden, took his place. Like Ridley, he was interested in economic botany and he wrote his *Dictionary of the Economic Products of the Malayan Peninsula* during his term of office. Still the standard work, it was reprinted in two big volumes in 1966. Burkhill also made a thorough study of the genus *Dioscorea*, the yams, both on field expeditions and in the Garden. His successor was his erstwhile assistant R. E. Holttum (1925–49), later Professor of Botany at the University of Malaya; he specialized in ferns and monocotyledons and he discovered that orchids cross-breed in the wild to produce new 'species'. It was Holttum who initiated the breeding of free-flowering orchid hybrids.

MONKEYS AS PLANT COLLECTORS

From 1929 to 1946 Mr E. J. H. Corner, now Professor of Botany at Cambridge, was assistant Director of Singapore's Botanic Garden. Professor Corner was responsible for a very remarkable innovation in the science of plant-collection and field botany, the enlistment of monkeys, *Macacus nemestrina*, as collectors and their enrolment, *de facto* if not *de jure*, on the staff of the Botanic Garden. He had observed that the Malays trained these monkeys to climb tall palm trees and gather coconuts for them. With the help of a skilled Malay monkey-handler Professor Corner trained eight of these so-called pig-tailed macacs to gather wanted plant material from the inaccessible heights of enormous forest trees. He could direct them quite specifically to the piece of material—a twig, a flower, fruit or epiphyte—which he had picked out with his binoculars and wanted for his work. One monkey in particular, perhaps a genius among its kind, knew eighteen words of the Malay dialect in which it had been trained and was, moreover, able to find at the tree-tops fruits and flowers which had been shown to him on the ground. But monkey-training for botany was not Corner's only, or even his principal, contribution to the work of the Botanic Garden. His specialities were mycology and trees. His admirable *Wayside Trees of Malaya* was published in 1940, his *Natural History of Palms* is a valuable book, and he still, at Cambridge, continues his work on the genus *Ficus* in Malaysia.

THE JAPANESE PHASE

In 1942 the Japanese army took Singapore. Very shortly after the fall of the city Professor Tanakadate of Tohoku University arrived to take charge of the Botanic Garden, with the happy result that the military did not get their hands on it or their trampling feet inside it. Holttum the Director, his assistant Corner, and Mr C. X. Furtado, the chief botanist and specialist in the International Rules of Nomenclature, had been interned; Tanakadate had them released and instructed them to get on with their proper work in the Botanic Garden. Thus Holttum continued as Director in all but name until, in 1942, Professor Kwan Koriba of Kyoto University was appointed Director. Holttum continued to work at his special subjects. Like Tanakadate, Kwan Koriba left his English subordinates to continue their proper work while he himself took the opportunity to do a useful piece of research into the growth behaviour of certain Malayan trees.

> It was particularly fortunate for the Singapore Botanic Garden that the Japanese scientific offers and the British staff who remained were able to preserve the Herbarium and the Library as part of the cultural heritage of Malaya and that no loss whatsoever was suffered in them. . . .

After the Japanese retreat the Garden came under British Military Administration. Once the war was over

Another aspect of the Singapore Botanic Garden's superb lake scenery. Not even the exotic foliage can make this anything but an 'English' landscape garden in the grand manner.

the old staff reassembled and by 1954 there was, with some newcomers, a full establishment once again. The Gardens were again being improved, plant-collecting was resumed, the first two volumes of Holttum's Revised *Flora of Malaya* were published; and J. W. Burkhill, son of the former Director and now Director himself, began to make the first serious study of Malayan seaweeds.

The realization of Malaysian independence brought with it a movement to 'Malayanize' all the country's institutions, and presumably this process will continue in Singapore, although it has separated from Malaysia and, with the population mainly Chinese, the process is more likely to be one of 'Sinalization'. The officers at present serving under Mr Burkhill are Mr A. G. Alphonso, Curator and a very active plant-collector; and Dr Chew Wee Lek, who, after three years at Cambridge, is Assistant Director. It is the separation of Singapore from Malaysia which has put the future of the Garden in doubt; members of the government, including the Prime Minister, recognize the value of their Botanic Garden as a scientific and as an ornamental institution. The fact does remain that the Garden is an institution which was designed to serve, and therefore to be supported by, the whole of the Malay peninsula, that is, a country the size of Britain, and not just by a country the size of the Isle of Wight. Apart altogether from the Botanic Garden's value as a centre for South-east Asian plant taxonomy, as a great work of horticultural art, as an institution for botanical research and experiment, and as a school of botany, it still could have an important function in economic botany and could, therefore, be of use and profit to the industry of South-east Asia; and that might, perhaps, be a good reason to hope that support for its work might be expected from outside the tiny frontiers of Singapore island. An example of what I mean by this may be in place here: J. W. Burkhill's study of Malaysian seaweeds has reached a point where the commercial value of some of them has been established; the era of 'marine farming' is about to begin. So much so that even some politicians are aware of the fact, and in the Singapore budget debate of 1967 the Member for Sembawang, Mr Teo Eng Siong, included the cultivation of certain seaweeds among his suggestions for the bettering of the country's economy. In the same debate, more conventionally, another Member urged the further expansion of the prosperous orchid-propagation industry—one which Singapore owes to its Botanic Garden which is still working in that field.

THE BEAUTY OF SINGAPORE'S GARDEN

The moment of greatest pleasure for me during my visit to the Singapore Botanic Garden was my first sight of *Strongylodon macrobotrys*—the fabulous 'Jade Vine'. This is one of several climbing plants growing over a handsome pergola which decorate one side of a formal sunken garden centred on an oblong lily pool and fountain and backed by the beautiful little slatted fernery. This Jade Vine is a vigorous, very large liana of the *Leguminosae* family, with characteristic pinnate leaves, evergreen, of course, large and glossy, deep green. The inflorescences hang down, are over 18 inches long, and consist of very numerous and large, narrow-winged pea-flowers with an unusually large keel; their clear pure colour is an indescribably and very lovely shade of jade-green, although inflorescences in shadow showed rather a deep, metallic, greenish-blue. I have seen this extraordinary colour in only one other flower, that of certain Chilean *Puyas*.

The Jade Vine is not the only spectacular beauty of this part of the Garden: covering a pergola at right-angles to the first and growing all over some very large trees in the background are specimens of *Mucuna novo-guinnensis*, a colossal leguminous climber with big pinnate leaves and enormous racemes of vividly scarlet pea-flowers. But the whole of this sunken garden area is a place of striking visual delights: there are other prettily-flowered climbers on the pergolas, the stairways to the sunken garden are decorated with coloured foliage plants in big pots.

Excellent use is made of varying levels of the Garden by terracing and lawns which dramatize the fall of the land. There are good pergolas in several quarters of the Garden and some of the flowering shrub collections, notably of bougainvillea, are planted formally.

However, for the most part the Garden is a park-like landscape garden of fine trees and palms, wide, sloping sweeps of lawn, lakes and shrubberies. There is also the 11 acres of primeval forest already mentioned. It is curiously impressive, retaining in the midst of artificial gardens in the skirts of a great and teeming city the brooding stillness broken by small movements and small cries, the undisturbed woodlands which one would hardly expect to survive the proximity of artefacts and urban noise.

In no part of the Singapore Botanic Garden is the topographical beauty more evident than in Palm Valley: here the land falls away into a bottom from the drive which skirts it, rising beyond in a gentle slope to a wooden background where tall forest trees make an interesting skyline. The valley is planted to a collection of bamboo genera; and to palms second only to the Fairchild's and Sydney's.

Probably the most important family of plants in the Singapore gardens is *Orchidaceae*, for this is a world-famous orchid garden not only for the collection of species—collecting still continues and the Curator does a good deal of it himself—but for hybrids and hybridization. New hybrids are distributed to commercial growers and every year thousands of unflowered seedlings are sold off to make room for new crosses. Orchids are grown in rows in the open, under slatted roofs; and, in the early stages, under glass. Although native Malayan orchids form the most important part of the collection, there are also many exotics, of course. Seedling are raised in flasks, the usual practice now, by the hundreds of thousands, but grown on in lesser numbers. Orchid growing on a commercial scale is important to Singapore.

INDIA

CALCUTTA

The Botanic Garden in Calcutta is remarkable for its collections of water-lilies and for the quality of its landscaping in the English style of the eighteenth century. Britain's imperial greatness spread the artistry of men like 'Capability' Brown and Humphrey Repton throughout the five continents.

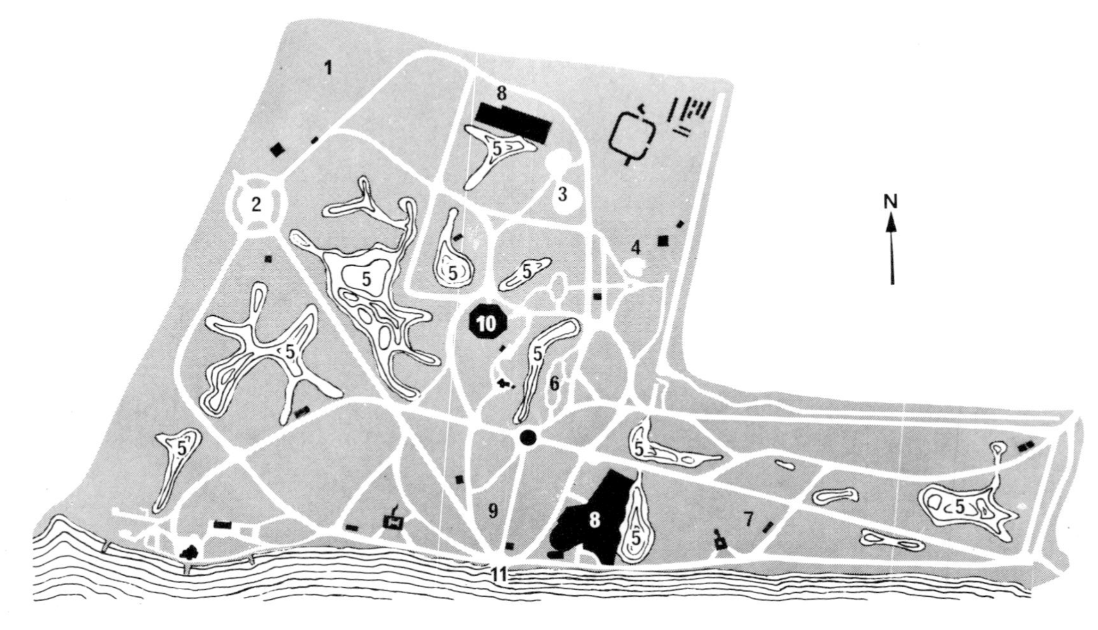

Key to Map

1 Economic garden and experimental ground
2 Great Banyan tree
3 Lotus garden pool
4 Flower garden
5 Lakes; water-lilies
6 Orchids
7 Pandanus plantations
8 Nursery
9 *Roystoneas*
10 Large palm house
11 Water gate

As I have noted elsewhere, the earliest so-called botanical garden in the tropics was probably Pamplemousses, in Mauritius. But strictly speaking it is not, and never has been, a botanical garden: these eighteenth-century official gardens in Europe's tropical colonies were started and maintained as nurseries for the acclimatization of potential crop plants from overseas, and later some of them acquired botanical status by falling into the hands of a botanist curator or director, who developed a scientific side and founded an Herbarium; others, as they grew in beauty, became simply recreational. Such was the case of the Royal Botanical Gardens, Pamplemousses, still a beautiful and an interesting garden but without either Herbarium or greenhouses.

The Calcutta garden is in the category of gardens started for economic reasons and then growing into scientific institutions. But the origin of what is now known as the Indian Botanic Garden is probably unique, for it is to be found in the sensibility of administrators who were deeply shocked by the loss of life due to starvation following a series of crop failures in the province. It occurred to some of them that there might be other crop plants than those commonly cultivated, or cash crops which could be sold for food, which might be grown in Bengal. One of these men, Colonel Robert Kyd, an officer in the East India Company's army, suggested the establishment of a garden of acclimatization in which newly introduced crop plants could be tried; the same garden, Kyd suggested, could also be used to make trial plantings of the spice plants in which the Company was trading; and to raise teak trees for timber to repair the Company's ships. On 31 July 1787, eleven years after the founding of Pamplemousses, Kyd's suggestion was approved by the Court of Directors.

Kyd was instructed to find a suitable site and he chose an area on the bank of the Hooghly called, at that time, Mugga Thana. Mostly unreclaimed jungle and subject to serious flooding, a part of it was worked by a few peasants; they were persuaded to leave by suitable compensation; a dyke was dug round the site to prevent flooding; it was fenced, and the first plantings of fruit trees, spice plants and some timber trees were made. Thus the foundation of the Calcutta Botanical Garden is usually dated 1787; and Robert Kyd is considered to be its first, albeit amateur and honorary, Superintendent. Kyd remained Superintendent, carrying out the original economic purpose of the Garden, for six years, dying in May 1793. In November 1793 the first salaried Superintendent of the new Garden was appointed; this was Dr William Roxburgh, who happened to be one of the greatest botanists of his time, with the result that from the day of his appointment the Calcutta Botanical Garden was turned from its original, economic purpose, to the service of scientific botany. Roxburgh had been the Company's botanist in Madras and was already an expert on the flora of India.

It was Roxburgh who made the rest of the world interested in the botany of India's immensely rich and varied flora, at that time new to science and to horticulture. It was also Roxburgh who brought the Botanic Gardens to life and made them world-famous, bringing both live plants and dried material for the Herbarium from all over India so that students could study them in one place. He worked closely with the botanist Koenig in south India, and through him was in touch with Linnaeus and Retzius and consequently with the latest advances in systematic taxonomy. The East India Company proved very liberal in its attitude to the change from economic to scientific work which Roxburgh insisted on, and financed the work of the Botanical Garden very adequately.

As fast as he planted the Botanical Garden, Roxburgh described the plants he was collecting, and in 1814 he issued the first considerable catalogue of Indian plants ever to be published, his *Hortus Bengalensis*. By that time he had been forced by illness to retire to the Cape of Good Hope, where he died in 1815; his other publications, based on his work in the Botanic Garden, were a *Flora Indica* and his *Plantae Coromandeliana*. The 1814 catalogue described 3,500 species. There is, however, still

In Calcutta's Botanic Garden. *Left* A *Bougainvillea* cultivar. Tropical gardens have a greater range of colour in this genus than we are used to in Mediterranean gardens.
Top right One of Calcutta's vast range of water-lily species, varieties and cultivars.
Bottom right *Wagatea spicata*, a woody climber which is native to India, flourishes in the Calcutta Botanic.

Below In most Far Eastern botanic gardens muscle-power, human or animal, is still more commonly employed than the power of machinery. Ploughing for a new planting in the Calcutta Botanic Garden.

unpublished but cherished in the Botanical Garden library and Herbarium an even more remarkable example of Roxburgh's scientific industry during his term as Superintendent: this is a 35-folio volume record of the plants he had collected in India, with very remarkable colour plates drawn under his supervision by Bengali artists. The cost of publishing this work would be enormous, but the present Director of the Botanic Garden, Dr Mitra, has made a start on a project to reproduce the work in facsimile so that every botanic institution in the world will be able to possess a copy.

The naturalist Buchanan Hamilton succeeded Roxburgh when the latter's health broke down but did nothing remarkable in the Garden, and the next advances were made by the Danish botanist Nathaniel Wallich, who was appointed Superintendent in 1817 and held the post for thirty years. Wallich was a man of tremendous energy and robust health. He collected plants himself all over India and even outside her frontiers; he added very largely to the living collections, and he also added many thousands of sheets to the Herbarium, usually sending duplicate sets to one or more European Herbaria. He compiled the enormous plant list which is still one of the treasures of the Botanical Garden library and is known to botanists all over the world simply as 'Wallich's Catalogue'. But during his reign, although the number of specimens in the Garden was more than doubled, nothing was done to transform what was simply a scientifically valuable collection of plants into a garden proper. Falconer, who succeeded Wallich, and Thomson, who succeeded Falconer, were both strict scientists, the latter especially famous for his collaboration with Sir Joseph Dalton Hooker, who visited the Calcutta Botanical Garden in 1858 and again in 1860, in the latter's *Flora Indica*.

For a while, during the term as Superintendent of Thomas Anderson who succeeded Thomson in 1861, the Calcutta Botanical Garden once again played an important part in the economic botany of India. Anderson was the first Conservator of Forests in Bengal; but he was also the man principally responsible for introducing *Cinchona* as a plantation crop into India. He used the Calcutta Botanic Garden as his garden of acclimatization, and it was while superintending the first commercial plantations of *Cinchona* in Sikkim that he died of a sickness contracted in the course of his work. His assistant, another of Hooker's collaborators on the *Flora Indica*, C. B. Clarke, was Superintendent of the Botanic Garden until 1871 when, with the appointment of Dr later Sir, George King, the Garden got its first horticulturist and garden designer.

King's task, as he saw it, the task of making a work of horticultural art out of a scientific collection of plants, was to some extent facilitated by the damage which had been done by the disastrous cyclone of 1864. This storm was accompanied by a colossal tidal bore up the Hooghly, a wave so huge that it dumped two large ships inside the Garden and laid nearly the whole area under 6 or 8 feet of water, destroyed at least a thousand trees and many thousands of shrubs and herbs. About half the total

Two aspects of one of the world's most remarkable trees, the great Banyan in Calcutta's Botanic Garden. There are one thousand trunks and the walk round the tree is about quarter of a mile. It is still growing vigorously.

collection was wiped out, and three years later a further 750 trees were destroyed by another, lesser, cyclone. So when King took over the Garden it was almost total ruin and part of it had already reverted to jungly swamp.

The site was perfectly flat; it had none of the topographical features which help to make the beauty of a garden. King seized the opportunity presented by its ruinous condition to set about creating them artificially. For example, he got rid of the new swamp by excavating a number of artificial lakes, and he made use of the large quantity of soil from these excavations to give the Garden several different levels, to endow it with small hills, slopes, little valleys. If any garden designer ever created his landscape from the ground up, King did so at Calcutta. As for King's replanting, it was on more or less phytogeographical lines but always with a proper regard for design.

Although King is remembered in the Indian Botanical Garden as its chief horticultural designer, he was also an excellent botanist; it was he who, in 1887, started publishlishing the *Annals of the Royal Botanic Gardens, Calcutta*; still more important, in 1890 he initiated the Botanical Survey of India. Today this is the Authority presiding over all the botanical work in India, including the work of the Botanic Garden; it is directed by one of the most famous and venerable figures in the history of botany, Father Santo Pao, SJ.

Sir George King was, then, the maker of the Garden as we see it now. Several of his successors have been distinguished botanists and gardeners, among them, notably, Sir David Prain, later Director of the Royal Botanic Gardens, Kew; Dr K. Biswas, the Garden's first Indian Director; and the Kew-trained Dr D. Chatterjee. But none of them made any changes in the Garden as such; they respected, used and maintained it. The present Director, Dr Mitra, a botanist but also the first landscape gardener to become Director since King, uses the Garden in his teaching course in landscape gardening design as a superlative example of its kind. It is pleasant and curiously reassuring to find the art invented by such Englishmen as Hoare of Stourhead, 'Capability' Brown, and Humphrey Repton being taught in India two hundred years after their time, inside a belated masterpiece of that art.

THE INDIAN BOTANICAL GARDEN NOW

The Indian Botanical Garden is one of those in which there is not the slightest difficulty in identifying the outstanding feature of interest: it is the great Banyan Tree. This specimen of *Ficus bengalensis* is certainly one of the largest living creatures in the world and it may well be *the* largest in terms of sheer bulk of tissue. There is no record which enables us to fix its age, but it cannot be much less than two centuries old, and it may have been a young tree, native to the site, when Kyd first started the Garden. This species of *Ficus*, like some others, drops aerial roots from its branches, which, rooting themselves when they reach the ground, in time form subsidiary trunks. Thus, as the branches grow longer and longer horizontally they are not only supported by a series of pillar-like trunks but are provided with still more nourishment so they continue to grow indefinitely. In the case of the great Calcutta tree this process has been assisted and hastened by care and training, for new aerial roots in selected suitable places are not left dangling and tangling and only slowly reaching the ground, as in the wild, but are tied to and drawn down by bamboo poles which support them until they are well rooted. When I visited this mighty plant in 1968 it had about one thousand such auxiliary trunks. The dimensions of the tree are extraordinary: it is not remarkably tall, for that is not in the nature of the species; it is about 100 feet at the highest point. The latest recorded measurement of the circumference of the whole complex

is well over a quarter of a mile, and the canopy of the tree covers 4 acres of ground. The great Banyan is still growing vigorously and there seems no reason why it should not continue to do so indefinitely.

The Palm Houses in the Garden are interesting. In the climate of Calcutta such houses are not, of course, necessary to exclude cold but to exclude excessive sunshine; glass would be quite inappropriate. They are elegantly shaped metal structures prefabricated in England in the nineteenth century, covered with wire netting, and overgrown with a number of interesting and pretty climbing plants whose foliage provides the shade for the collections of rare palms, ferns and moisture or shade-loving plants inside. Of the creepers which thus provide this damp coolth the most remarkable is the tremendously vigorous *Porana panniculata*, whose mass of white flowers so completely cover the roof of one Palm House that, from a suitable distance, it looks for all the world like a snow-clad mountain peak. Inside the house the charming layout is that of a formal maze and the collection of shade-lovers is a rich one. The interior of this *Porana*-covered house has, by the way, become a resting place for certain migratory birds, and provision of clean water is always made for them in gigantic bivalve shells from the Indian Ocean.

The Indian Botanic Garden reinforced my impression that the *Leguminous* trees provide us with more, and more spectacular, ornamental kinds than any other family. I have here in mind the Garden's specimens of *Saraca declinata* and *S. indica*, the first a mass of rich orange flowers, the other scarlet; and of its avenue of *Brownia* and *Amherstia*.

The Palm Collection, very prettily planted, is not a large one. About forty species are grown, but it includes some fine specimens of an interesting rarity, the branching palm *Hyphaene thebaica*; branching monocotyledons are an exception to the rule for their kind. One of the Garden's best vistas is through the Palmetum.

A special flower garden is devoted to the cultivation, during the Bengal winter, of European summer garden flowers, chiefly for the guidance of private and institutional gardeners. Remarkably good results are obtained with many of these exotics in the very difficult climate of Calcutta, but some—for example roses—are far from easy to grow, and with these the Botanical Garden does useful work in selecting the varieties best able to tolerate the local conditions.

The big propagation nursery is one of the best in Asia, and although it is maintained chiefly to supply the Botanical Garden's need for replacement plants, the surplus of young plants is available for sale to the general public and other institutions.

The pleasantly planted garden of succulent plants, *Euphorbiaceae* and *Cactaceae*, has one especially interesting and attractive plant belonging to the genus *Pereschia*, a cactus which has true leaves; it is a tall, woody shrub with exceptionally beautiful flowers.

Below The natural buttresses of trees of the genus *Ficus* are of inexhaustible interest in tropical botanic gardens. They develop as a means of stabilizing the tree where rooting is relatively shallow.

Sunset over the Entebbe Botanic Garden. This East African garden, on Lake Victoria, Nyanza, is one of the most beautifully sited in the world.

NEW WORK

Much new work is being undertaken in the Indian Botanical Garden. Doctor Mitra is interested in high-altitude alpine plants and has himself made collecting expeditions, chiefly in Sikkim, up to 19,000 feet. The cultivation of high-altitude alpines, even of genera common in gardening such as *Primula*, is seldom easy and work done on them in this Garden may lead to the selection or breeding of easier cultivars to enrich our rock-garden flora. New shrubberies are also being planted in the Garden to replace old and ill-conditioned ones, and again the chance is being taken to experiment with the less familiar genera, which while being a service to horticulture is also serving a botanical purpose. In much the same spirit, the big collection of Indian grasses, while primarily for botanical purposes, serves horticulture by helping the gardening staff to select species suitable for lawns, never easy of cultivation in tropical climates.

A Students' System Garden is well maintained and used for the teaching of taxonomy. Not far from it a garden is maintained for field experiments with market-garden crops, and some ornamental plants. The vegetable trials are particularly important in a country the majority of whose 350,000,000 people live chiefly on vegetable food. The Indian Botanic Garden even works rather outside the horticultural field in conducting trials of cereals as well as garden crops.

One of the most beautiful spectacles in this Garden is provided by the enormous number of diverse water-lilies in every sheet of water—there are several ponds as well as the lakes excavated by Sir George King. The number of species and varieties grown is larger than in most botanical gardens because the scientific staff have in hand a programme of research in the morphology of these plants, which may be followed by a programme of breeding to produce improved garden varieties. Other trials, of economic rather than horticultural or botanical importance are very much in the tradition of the Garden which introduced quinine, rubber, vanilla, baobob, carob, clove, cinnamon, mangosteen and other crop plants to Indian agriculture, are in hand. Of these the most unusual and interesting is the trial of Balsa, *Ochroma lagopus*, whose wood was used by the ancient Peruvians for their sea-going rafts and has half the specific gravity of cork; it is valuable in the manufacture of life-saving apparatus, in accoustics and in heat insulation;* another trial which could have considerable economic consequences is that of *yerba-maté* which, surprising though it may seem, does compete successfully with coffee and tea in at least two Old World countries. And yet another is that of a millet with a 2-foot spike (the normal is about 1 foot or less), which may prove suitable for arid conditions.

If the trials of esculent and other economic crops contribute nothing to the beauty of the Indian Botanic Garden, they do, by making the Garden a productive unit in the country's agri-horticultural complex, help to justify, to a population painfully far from rich, the four or five lakhs of rupees spent annually on upkeep and on pure science. This Garden is, in fact, one of the few still making a useful contribution not only to science and ornamental horticulture but to the economic life of the country. It would be monstrous if it were not so, for one approaches the Garden through some of the most terrible slums, some of the most atrocious scenes of abject and squalid poverty which the world affords. In a way which made a powerful impression on me the Indian Botanic Garden is a microcosm of much of modern India: science and art are present at the European level; the means are not. I saw no machine tools in use; but three hundred people work in the Garden and maintain it well for the heartbreakingly miserable wage which the Garden can afford, doing with their hands and the simplest tools work which, if the Garden were as advanced on the practical and technological side as it is on the purely scientific and artistic side, would be done by a score of men with machine tools. In that case 280 people would starve for want of the pittance they do earn.

At all events, and at whatever cost, and in the face of difficulties we simply do not encounter in Europe, much less in America, the Indian Botanic Garden is making its contribution to civilization, which is more than can be said of the botanical gardens of some rich southern hemisphere countries which contribute nothing to tomorrow. The Indian Botanic Garden is an ornament to Calcutta: but the work which its scientific staff is doing makes it very much more than an ornament.

*See De Wit, H. C. D. *Plants of the World* (London, 1967).

ENTEBBE

Opposite Calliandra, a genus belonging to the *Leguminosae*, is used as an ornament in tropical gardens of both hemispheres. The name is derived from two Greek words—*kallos*, beauty; *andros*, stamens—the stamens being the spectacular organs in this genus. They are American plants and introduced to East Africa by Entebbe. *Bottom left* Flowers of a climbing *Nepenthe* on the pergola at Entebbe. *Bottom right* A typical aspect at Entebbe.

Late in the nineteenth century the Government of the Protectorate of Uganda in East Africa, decided that it ought to have a botanical garden. Somebody in the administration had noted that it was through such gardens that the tropical colonies of the European powers as well as Britain had introduced and acclimatized commercially valuable plantation crops and had also improved their range of subsistence crops. It was primarily for a similar purpose that in 1898 Mr A. Whyte was authorized to found and plant a botanical garden at Entebbe on Lake Victoria, the chosen site being at a rather high altitude, and only a small fraction of a degree off the Equator. Entebbe was never a scientific garden in the full sense, that is, it seems not to have had even the beginnings of a herbarium or a taxonomic department. It was started as an economic garden and it grew into a recreational one, still 'botanic' as will appear, but without a scientific staff.

Nearly all the crops currently grown in Uganda were introduced from some other country. Cotton was introduced to East African agriculture long before the Botanic Garden was started; but Whyte and his successors were responsible for the introduction, acclimatization, propagation and distribution of, for example, rubber (*Hevea brasiliensis*), coffee, tea, tobacco, cocoa, pineapples and sisal. A single one of these might justify the Garden. And even when, in the 1930s and thereafter, advanced experimental work on economic plants was taken over by the Department of Agriculture's new Experimental Stations, the Botanic Garden still continued its work of introducing and trying possibly useful plants, as well as, increasingly, ornamentals.

In fact with the decline of the Botanic Garden's importance in the field of economic botany, its importance in the field of ornamental gardening rose. It remained 'botanic' until fairly recently in the sense that the plantings were done with botanical species and that there were collections of native plants properly labelled. But that such East African towns as Nairobi and Kampala are among the most lavishly beflowered in the world is largely due, in the first instance, to work done at Entebbe. The streets, squares and roundabouts of Nairobi, for example, make far better use of *Bougainvillea*, in a score of species and varieties and wide range of colours, than any city I have seen, and one can find the origin of this at Entebbe.

For some years the Uganda Botanic Garden ran courses in horticulture for boys leaving secondary school; the pupils were trained in both theory and practice and were then found jobs as municipal and institutional gardeners all over the country; hence the quality of municipal gardening in the cities of Uganda and Kenya. For a time this was a great success; but the course has been discontinued because it has now become very difficult to place the graduates in work, and because there is a severe shortage of funds. In the field of botanical work, seeds are collected and cleaned, from the Botanic Garden's flora a seed-list printed, and exchange arrangements maintained with gardens overseas and elsewhere in Africa.

As a visitor one's impression of the Entebbe Botanic Garden is one of a great park of hilly, undulating lawns, fine broad walks, a variety of magnificent trees planted with art; and fine collections of shrubs and herbs. The great beauty and the remarkable serenity of this Garden is as much enhanced by its siting on Lake Victoria, with a wonderful view of that topographically lovely freshwater sea, as are those of Sydney's Botanic Garden on the shore of Botany Bay.

There are two well-maintained herbaceous borders planted with tropical herbs and shrubs, separated by a lawn and a series of pergolas overgrown with a variety of especially interesting climbing plants. Among the most spectacular of these are *Aristolochia labiosa*, with its big, speckled brown pitcher flowers; *Thunbergia mysorensis*, whose flowers are scarlet and yellow; *Allamanda cathartica*, with large, butter-yellow trumpet flowers and tremendous vigour of growth; the charmingly delicate *Antigorion leptopus*, whose lacy flower racemes are both pink and white; the crimson-flowered *Lonicera sempervirens*, an evergreen honeysuckle; and some good *Passifloras*.

Much colour is provided by the good *Hibiscus* collection, most of which is concentrated in a single shrubbery so that species and varieties can be easily compared; by the very numerous and colourful *Bougainvillea* collection—the range of colours is from white through the yellows and oranges to scarlet, crimson and a range of purples, of forms from prostrate to arboreal; and by the *Plumeria*, 'Frangipani', collection. The shrubs of this genus are rather gaunt but the flowers are lovely and their scent overpowering; they are, of course, used as Temple Trees all over the Far East and, with jasmine, their flowers are commonly offered to both the Buddha and to the Hindu gods.

I have no difficulty in picking out the plant which distinguishes the Entebbe Botanic Garden—it is by far the most spectacular, the most interesting and the most novel of flowering trees (novel, at least, to me). This tree grows out of the middle of the rockery of arid-soil succulents which do not do well in this fairly high-rainfall garden. *Monodora myristica*, the 'Calabash Nutmeg', is a native of Uganda and a breathtaking sight when in flower. Here perhaps I may quote from H. C. D. de Wit (*op. cit.*), who says:

> The seeds of *Monodora myristica*, a medium-sized West African jungle tree, yield spices and drugs. In *Monodora*, the carpels have been completely fused into a spherical structure with a lignified wall to which the numerous seeds are attached. The seeds of *M. myristica* contain an oil that tastes of nutmeg and is, in fact, used as a nutmeg substitute. In the West Indies, to which they were brought by African slaves, they became known as Jamaican, American or Mexican nutmeg. The flowers of the twenty or so species of *Monodora*, all indigenous to West Africa, are particularly large and attractive and may be so numerous as to hide the branches and leaves from view. The bizarre form of its curly and pendent petals has earned *Monodora crispata* and related species the name of 'Orchid Tree' though its smell is said to be repulsive.

Three comments: Uganda is in East Africa but the

In tropical gardens the above-ground roots and buttresses of trees and *Pandanus* make an interesting contribution to the visual beauty of the scene.

The Entebbe Botanical Garden is above all an arboretum and a palmetum. The Garden introduced and acclimatized many American and Asian exotics to East Africa.

magnificent specimen of *M. myristica* at Entebbe is labelled as native to Uganda; maybe the range is wider than Professor de Wit allows. In his plate of *M. crispata* the 'crisped' petals are shown closely hugging the central pouch, but in *M. myristica* as I saw it the petals stand elegantly out, like an overskirt of stiff silk, from the pouch and there is no scent, repulsive or otherwise. Finally, immensely numerous though the flowers be, they do not 'hide the branches and leaves' because they are pendent on long stalks. The frilled and scalloped petals are about 6 inches long, triangular, speckled scarlet and gold. The fruit of fused carpels is almost as large as a tennis ball and as hard as an oak-gall. Certainly this lovely *Monodora* is the star of the Entebbe Garden even though much more familiar beauties as the *Erythrinas,* the *Couroupitas*, the gorgeously flowered *Cassias* and the aptly named 'Flamboyants' (*Delonix regia*) are there to compete with it.

Among the many large and well-grown trees of this park-like botanical garden one of the most interesting natives is the 'Mpewere' *Piptadenia africana*, a member of the *Mimosa* family. There are several enormous and stately specimens in the Garden; they have bipinnate leaves and a tremendous number of creamy flowers which turn coppery red as they age.

One part of the Botanic Garden is given over to primeval bush, which is very impressive to walk in: a drive divides this from forest and while on the one hand there are exotics planted into the wild and it has been used for trials of exotic under-storey plants, the other side is native and unplanted. Not far from this 'botanic nature reserve' there is an old rubber plantation, a relic of the Garden's economic botany days; the trees in it support a collection of pepper vines. One finds a large number of economically valuable plants all over the Garden, evidence of its useful and profitable past. Thus there is a plantation of cocoa dating from 1903 and whose trees support the climbing *Vanilla* orchid; *Eugenia aromatica* (Clove) from the Moluccas; *Eugenia jambolana* (Jambolan) from Java; *Litchi chinensis*; *Artocarpus heterophyllus*, the inevitable Jak-fruit tree; *Psidium guajava* (Guava); *Persea americana* (Avocado Pear) and others including mango, mangosteen, soursop, pawpaw, lime, banana and plantain. As well as spices and fruit trees one comes across coffee, tea and tung; the coffees are both Arabian and the native Ugandan kinds.

The Palm collection is neither large nor diverse but it includes some beautiful species, such as *Borasus flabellifer*, tallest of the East African native Palms; *Elais guineensis*, the Oil Palm; *Ravenala madagascarensis*, the water-storing 'Traveller's Palm'; *Raphia monobuttorum*, the native raphia Palm, the 'Ivory Nut Palm', *Phytelephas macrocarpa ; Caryota mytis*, the 'Fishitail Palm' from Malaysia; and some Australian *Archontohoenix alexandrae*.

Although the Entebbe Botanic Garden is fascinating in detail and endowed with some outstandingly interesting and beautiful plants, its striking attributes are its fine horticultural quality, the beauty of its lakeside topography so admirably exploited in the manner of its planting and its grouping of trees; and, remarkable among the botanical gardens I have seen, its delicious peace and quiet which are emphasized rather than disturbed by the calls and movements of many birds and small animals, and the restless flutter of Uganda's lovely butterflies and, at twilight, her great night moths.

Has the Entebbe Botanic Garden a future? I do not know; it certainly could have. I met the retiring (1968) Curator, Mr Kateregа, and his successor, Mr Mubi, and learned that the Garden is being extended into parts never cultivated. But it is regarded principally as a recreational park much used by the civil servants and their families from the nearby government offices. I do not know what hope there is of making more botanical use of the Garden, of giving it a laboratory and of starting a national herbarium; or of restoring some of its old economic importance by giving it a technical staff and some assignments in economic botany. That, at all events, would be the proper course.

RIO DE JANEIRO

One of the marvels of tropical botanic gardens like Rio de Janeiro's are the arboreal genera such as *Couroupita* and *Artocarpus*, which flower and bear their fruit directly on the trunk or main branches.

Although the idea of systematic gardens of plants may well have reached Europe from the Americas, yet when Europeans began to plant such gardens in the Americas some centuries later they had certainly forgotten the debt which they were thus unwittingly repaying. One of the earliest of such botanical gardens in the New World was Rio de Janeiro's; and like a number of other great botanical gardens, it had its origin in a royal garden.

Very early in the nineteenth century the Portuguese royal family fled from Lisbon to Rio in order to escape the fate of the Spanish Bourbons who had been trapped into captivity by Bonaparte. Thus the centre of the great Portuguese Empire shifted from the metropolis to the Brazilian colony. Unlike the Spanish royals, the Portugese were not entirely useless creatures, and on their arrival in Brazil they founded a number of new industries under crown patronage, the 'Fabricias e Fundacoes do Reino' administered by the 'Junta da Fazenda e Arsenais'. One of these Crown factories was located in the grounds of the house and sugar-cane mill belonging to one Rodriguo de Freitas, a few miles from the waterfront centre of Rio de Janeiro, whose heir sold the property to the Crown Junta. In 1808, not long after his arrival in Brazil, the Prince Regent, Don Joaõ VI, saw this estate, was much pleased with its beauty of topography and verdure and, on 13 June of the same year, decreed that the land surrounding the old de Freitas mill be prepared for the planting of a garden to be devoted to the cultivation and acclimatization of economically useful plants from the Portuguese East Indies. The garden was to be called the Real Horto—Royal Garden; it is now Rio de Janeiro's Jardim Botanico. It was at that time, and for some decades remained, the principal channel through which East Asiatic plants of commercial and medicinal value were introduced into South America; and as such its role has been extremely important.

Early in its career of service to both botanical science and to horticulture and agriculture the Real Horto was enriched by a supply of new plants in a manner which must be unique in botanical history. In 1808 a frigate of the Portuguese navy, the *Princeza do Brasil*, was wrecked off Goa. One of her officers, Chefe de Divisaõ Luiz de Abreu Viera y Silva, was picked up by the brig *Conceicaõ* bound for Brazil by way of the Cape of Good Hope. She was taken by the French—Portugal being Britain's only remaining ally against Bonaparte—and Luiz de Abreu, made a prisoner of war, was interned on the Ile de France, a French West Indian possession. The island had, at that time, a famous *Jardin d'Acclimatation*, the 'Jardin Gabrielle', where a large number of useful plants introduced by the botanists Poivre and Menouvilles were cultivated. In 1809 Luiz de Abreu, a man of wide reading and interests as well as great courage and ingenuity, not only contrived to escape from internment, suffering great dangers and hardships, but to carry with him seeds, and even some roots, of the best plants in the 'Jardin Gabrielle'. He went to Brazil, presented his booty to the Prince Regent, and saw roots and seeds planted in the Real Horto.

INTRODUCTION OF TEA

Thus from the beginning and during all its early history Rio's Botanical Garden was more concerned with economic than with taxonomy or scientific botany. Portugal, like Britain, had in this respect the advantage of a great overseas empire. Thus as early as 1812 the garden received an important consignment of Asiatic plants from the Marechal Manoel Marques in Portuguese India; and in the same year and at the instance of Luiz de Abreu, the *Senador* of Macau, Rafael Bottado, sent to Brazil the first seeds of *Camellia thea* (at that time the *Thea viridis* of Linnaeus), thereby introducing the cultivation of tea into first Brazil and later other South American countries including, in 1940, Argentina. More Asiatic plants reached the garden every year from the East. In 1817 the cultivation of sugar-cane in Brazil was greatly improved and increased in production by the garden's initiative in getting the Governor of Cayenne, Maciel da Costa, to send for trial plants of the superior cane variety called *cana de Caiena*. The Real Horto's first epoch of successes ended in triumph when Prince Joaõ was crowned king of the United Kingdom of Portugal and Brazil and marked his continued interest in the garden by renaming it the Royal Botanic Garden and attaching it to the new Museo Nacional. That was in 1819, and when Don Joaõ returned to Lisbon, his recovered capital, then his son, Don Pedro, who remained in Rio de Janeiro as Viceroy, took the garden under his protection.

THE IMPERIAL EPOCH

The Jardim Botanico's second epoch was one of tribulation. It began quite well when the Prince Viceroy, its patron, listening to complaints that the garden's work had been hindered rather than helped by the bureaucratic administration of the Museo Nacional, restored its Director's independence, making him directly responsible to the Ministry of the Interior. But when Don Pedro became involved in his struggle to separate Brazil from the mother country and make it an independent Empire under his own rule, the Botanical Garden was not merely neglected, it was more or less destroyed. And when, under the new Imperial Government, the Carmelite botanist Frei Leandro do Sacramento was appointed Director of the Botanical Garden he found it, as he says, 'em deploravel estado de abandon'. More than half of the plants were dead and the rest overgrown. Tropical gardens very quickly succomb to the alarming exuberance of natural vegetation if they are neglected. But Frei do Sacramento was not only a man of courage and persistence, he was a considerable scientist, a Fellow of the (English) Royal Society, a member of the Munich Academy of Sciences and of the Ghent Royal Horticultural Society, one of the most advanced societies of its kind in the world. He restored order to the Jardim Botanico, extended the area under cultivation, planted avenues of trees, adopted the Linnaean system (belatedly) in arranging his new plantings, changed the topography of the garden by making hills with the soil taken out while digging artificial lakes and ponds to

Left Anthurium species, of the family *Araceae*, are among the beauties of all tropical botanic gardens, and have become familiar in the northern hemisphere as excellent house plants.
Below left Inflorescence of the 'Travellers' Palm', so called because wholesome drinking water can be tapped from the leaf stalks. It is not a palm but a member of the family *Musaceae*, related to the banana.
Below right Flower-bud of a species of *Philodendron* in the Rio de Janeiro Botanic Garden.

Fruits of a *Ficus* species in the Rio de Janeiro Botanic Garden. The open ones have shed their seeds.

accommodate aquatic plants, and began the policy of seed exchanges with other botanical gardens, including that of Cambridge. When Frei Leandro died in 1829 he had made Rio's Jardim Botanico into one of the three or four really important scientific institutions of its kind in the tropics. Moreover he left so much planned and even started that his successor could do nothing, during his term, but finish what do Sacramento had begun. He was rewarded for this fidelity by the flowering, during his term and for the first time in Brazil, of the Royal Palm grown from seed stolen from the 'Jardin Gabrielle' by Luiz de Abreu. This plant, probably about 250 feet tall now, is the mother not only of the entire and magnificent Royal Palm Avenue which is the Garden's most spectacular feature, but probably of every Royal Palm now growing in Brazil.

The next Director to make any real improvement in the Garden was Senador do Imperio Candido Baptista de Oliveira. He drained the still marshy areas of the garden into ditches which he used to feed the canals, which he had handsomely constructed of stone and ornamented with balustraded bridges, canals which run through the garden over a clear bottom of boulders set in concrete, cooling the air, and filling the whole garden with the pleasant sound of running water. To supplement his water supply, Oliveira built a handsome aqueduct to carry water into the garden from the steep hills which form its background.

Although, as we have seen, the Garden has already done very important work in economic botany, little scientific work had been done under the two directors who succeeded Frei do Sacramento. But in 1859 a naturalist, Alves Serraõ, commonly known as *Frei Custodio*, Brother Custodian, undertook the re-identification and systematic classification of the Garden's specimens, for Frei do Sacramento's work in that field had never been completed and was, in any case, out of date. But after only two years of this work he resigned when it was decided to make the Jardim Botanico a dependency of the Instituto Fluminense de Agricultura, a sort of semi-official, semi-academic body comparable with the Royal Society of Agriculture in Britain but with much political power. Frei Custodio was convinced that it would be impossible for him to do the work he had to do if he were submitted to the whims and fancies of the Instituto administrators. The Secretary of the Instituto was appointed Director in Serrao's place, and the scientific staff became subordinate to him. The Instituto's governing body had decided to found and maintain, with Government funds, a School of Agriculture and Model Farm—this was the period when English farming technology was making an impression on farmers and governments all over the world—and for that purpose they brought in an agricultural specialist from Austria, Dr Karl Glals. One of the sites he inspected for his farm and school was the *fazenda* Macacos, whose land marched with the Botanical Garden; he decided that it was ideal for his purpose; and as a final result in due course found himself directing not only the farm and school but the Jardim Botanico as well. But as that was not really within his field of work, the garden declined in importance, becoming

little more than a recreational park for the citizens of the capital. Typically, Glals's only contribution was an artificial grotto. His immediate successor planted the Frei Custodio walk, as it was called in memory of the garden's greatest director, with trees of *Terminalis Cattapa*; and one other short walk now called the rua das Arecas.

The garden was not restored to botanical importance until, having been divorced from the Instituto Fluminense by the Minister of Agriculture, it was put in charge of its first great major botanist.

THE GREAT RODRIGUEZ

Joaõ Barbosa Rodriguez began his term as Director by reforming the systematic classification of the garden's specimens. Next he dredged and canalized the river Macaco which was apt, in some seasons, to flood a considerable area of the garden. He cleared and planted new ground, naming the new plantations and walks after former directors. He built the garden's first greenhouse and planted an arboretum. Rodriguez also began the building up of an Herbarium which the garden had lacked, making scientific work there very difficult; and the collecting of a botanical library, now a considerable one, both of which were paid for out of the privy purse of the emperor Dom Pedro II. Rodriguez also published the first full list of the garden's plants, the *Hortus Fluminensis*. Thus, when he died in 1909, Rodriguez had not only fully restored the garden; he had enlarged and improved it.

The Botanic Garden in Rio de Janeiro, like Peradeniya in Ceylon, is remarkable for avenues of Palms. This one, of 'Royal' Palms (genus *Roystonia*), is probably the tallest in the world.

Of the Directors who came after him those who did most to consolidate his work were A. P. Leaõ (1915–31), a medical botanist of distinction; Joaõ Kuhlmann (1944–51), a taxonomist; and Paulo de Campos Portos. During Leaõ's term the Herbarium was much enriched with material collected in Amazonia by Adolfo Ducke. Campos Portos annexed to the Jardim Botanico a small nature reserve of importance, the Reserva Florestal do Italiaia, and he also enlarged the library; he was also responsible for starting the garden's review *Rodriguesia*. Campos Portos had served the garden, in various capacities both within and outside its establishment, for forty years when he became its director in 1951; between that date and 1960 he started and encouraged the study of experimental (as distinct from taxonomic) botany, a policy which was followed by his successor F. R. Milanez. An important impulse was given to the study of the internal morphology of plants not only by the fact of Dr Milanez's special interest but by the gift to the garden by the Rockefeller and Ford Foundations of an electron microscope. In other respects also the garden's laboratories are well equipped; its scientific library of 4,000 volumes and 1,500 periodicals is adequate; and in its 54 hectares about 7,000 native and exotic species are now cultivated.

THE JARDIM BOTANICO TODAY

To the northern visitor who is not used to tropical vegetation the most striking, sometimes even amazing, feature of the Rio de Janeiro botanical garden is the variety and magnificence of foliage, notably in that part of the garden devoted to the flora of Amazonia. In this respect the *Aroids* are remarkable, so wonderfully fanciful are the variations on the theme of arrow-head leaves and spathe-and-spadix flowers. It is little wonder that the only completely original artist working on gardens design in our time, the Brazilian architect Roberto Burle Marx, has found inspiration for his work in forms of Brazilian foliage and that, applying genius to that material, he has created masterpieces of abstract landscape gardening. Most stately and beautiful in their habit are the variously divided, gigantic leaves of the big *Monsteras*; and the vigour and growth of the huge *Philodendrons* which climb big trees to the summit is a source of wonder. In that garden it is impossible not to be repeatedly surprised and filled with admiration by the endless variety, complexity and ingenuity which plants have shown in devising means to a single common purpose, survival and propagation of the species. One's wonder persists, everything in the garden tends to heighten it—not least, a charming and unique distinction of Rio's botanical garden, the gigantic and vividly coloured butterflies which move among the flowers.

Among the strangest of the trees are those composing the *Couroupita guianensis* avenue planted by Campos Porto. (Several important or interesting arboreal species are represented by avenues rather than single specimens.) These trees, with their flowers like huge and brightly coloured sea-anemones growing out of the trunk and branches far from the canopy of foliage, so held my admiration that I returned to them time after time. Even so, of all the tree-lined walks the great nave-like avenue of Royal Palms which forms the spine of the garden is the most remarkable. Probably there is nothing quite as fine in any garden in the world. The slender straightness of these palms, topped each at two hundred feet by enormous leaves, repeated with perfect regularity across the whole

Below Laid out when French styles in garden art were dominant, the Rio de Janeiro Botanic Garden is formal in design and its architectural elements are reminiscent of a great French garden, as are its straight paths and stately avenues. Its works of sculpture too, are in the late eighteenth- and early nineteenth-century French taste.

garden, is indescribably impressive.

Among the most splendid and most curious of the garden's big trees are the various figs—*Ficus elastica* with its immense spread of magnificent leaves; *Ficus clusiaefolia* with its stylized fruits; others of the genus with grandly buttressed or weirdly manifolded trunks. Competing with the figs for sheer grandeur and surpassing them in the enormity of their fruits are the 'Jak-fruit' trees, *Artocarpus integrifolia*, the bag-shaped greenish-yellow fruits weighing up to 20 lb apiece. For leaves, and for the decorative quality of fruit clusters, the shrub-shaped *Ficus roxburghii* are among the most interesting plants. And here, as at Les Cèdres, I enjoyed especially the huge clumps of the gigantic bamboo *Dendrocalamus giganteus* towering to well over a hundred feet.

Among the smaller flowering trees and shrubs, the various *Hibiscus* species provided the loveliest colours—the collection of this genus is a large one; then the *Mimosas* and related legumes, some with crimson flowers; and the *Euphorbias* planted alongside the canalized river Macaco. But in this garden I returned time and again to admiration of the foliage rather than the flowers, to the endless variety of forms, the endless variations on the theme of green.

A singularity, from our point of view, of tropical botanical gardens is the growth of plants on plants; epiphytic orchids are, in certain gardens already described, an obvious case in point, but in Rio the *Bromeliads* are more interesting. Millions of North Americans and Europeans have become familiar with some of these, such as *Vriesia* and *Bilbergia* species, as house plants. But to see hundreds, many in fantastic flower or even more fantastic fruit, growing clustered up the trunks and along the branches of big trees, is new. The only European garden where this has been achieved in the open garden is Les Cèdres.

Below Jak-fruit, *Artocarpus communis*, and *right Couroupita guianensis*. Single fruits of *Artocarpus*, a genus related to Mulberry, may weigh as much as 80 lb. The vernacular name for this *Couroupita* is 'Cannonball Tree'.

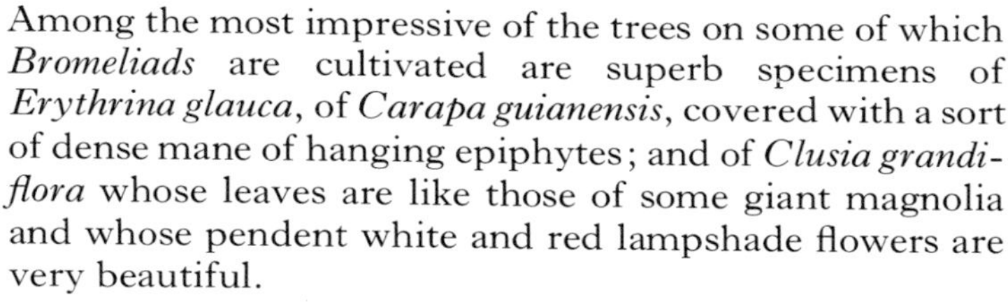

Among the most impressive of the trees on some of which *Bromeliads* are cultivated are superb specimens of *Erythrina glauca*, of *Carapa guianensis*, covered with a sort of dense mane of hanging epiphytes; and of *Clusia grandiflora* whose leaves are like those of some giant magnolia and whose pendent white and red lampshade flowers are very beautiful.

THE LAKE

The small artificial lake in one corner of the garden is pleasingly landscaped, made interesting by reason of its aquatics, and impressive by the partly encircling stands of the enormous 'Travellers' Palm', a strelitzia-like or banan-like plant about 40 feet tall, with leaves 8–9 feet long—*Ravenala madagascarensis*. The plant gets its vernacular name from the fact that when pierced at any point in its stems it yields a good supply of fresh water.

The garden contains a number of statues and busts of former directors and royal patrons and the several sections of the garden are named after them, as are the principal walks. There are few buildings, but a charming feature near to the lake, and one which is full of interest, is the *Orchidaria*. It is an octagonal greenhouse which has a glass roof but sides of open trellis-shaped wooden slatting painted a pretty blue. When we saw it it was filled with Anthuriums, and a nearly complete collection of all the Maidenhair ferns. Adjoining this is the long, open-sided, roofed structure which houses the garden's orchid collection; and near this again is the charming stone pergola which supports the collection of *Bignoniaceae* genera. Likewise in this quarter of the Garden is the Mother Palm, mentioned above, from whose seed have sprung most of the Royal Palms in Brazil. The collection of Palms is a fine one and very pleasingly planted; and most attractive is the bright red young growth of the palm *Cyrtostachys renda*. Trees, palms and shrubs are all planted in a well-tended turf of a small, broad-leaved plant which does duty for grass and is remarkably tolerant of deep shade and high temperatures. And the whole garden—walks, turf, trees, shrubs, waterways, herbaceous plants and bulbous genera—is well and lovingly tended.

NOTE: The Jardim Botanico of Rio de Janeiro is especially rich in certain families of plants among the large number represented. There are twelve genera and many more species of *Bignoniaceae*; about ten genera and more species of the *Araceae* in which Brazil is naturally rich; a large collection of tropical composites in nine genera; thirty-seven genera of tropical *Leguminosae*; about twelve genera of *Palmae*—these are a few examples taken at random, for, short of publishing the Jardim's 300-page *Hortus*, one cannot really give an idea of its wealth in tropical American plants.

SOUTHERN HEMISPHERE

CHRIST-CHURCH

In 1847 a number of worthy persons devoted to the Church of England formed an Association in London, called the Canterbury Association. Its object was to found a Church of England Colony in New Zealand, and the prime movers were one Robert Godley; the Archbishop of Canterbury; and Lord Lyttleton. As a result the town of Christchurch, in South Island, was founded, and when, in 1850, the first party of settlers arrived, they found the town pegged and taped for them by the Association's surveyor. From the very start an area of 500 acres was set aside to be preserved as an open space, a 'lung' of the city. This area was called Hagley Park after Lord Lyttleton's seat in Staffordshire. An Association Ordinance secured the new Hagley Park 'for the recreation and enjoyment of the public'; however, the Superintendent of the Colony had power to put roads through it; and also 'to make plantations and gardens'.

1 Recreation area
2 Aquatic plants
3 Native New Zealand plants
4 Cockayne memorial garden
5 Rock garden
6 Fernery
7 Greenhouse complex
8 Rose garden
9 Archery lawn
10 Clematis collection
11 Herbaceous collection
12 Pine hill
13 Azalea garden

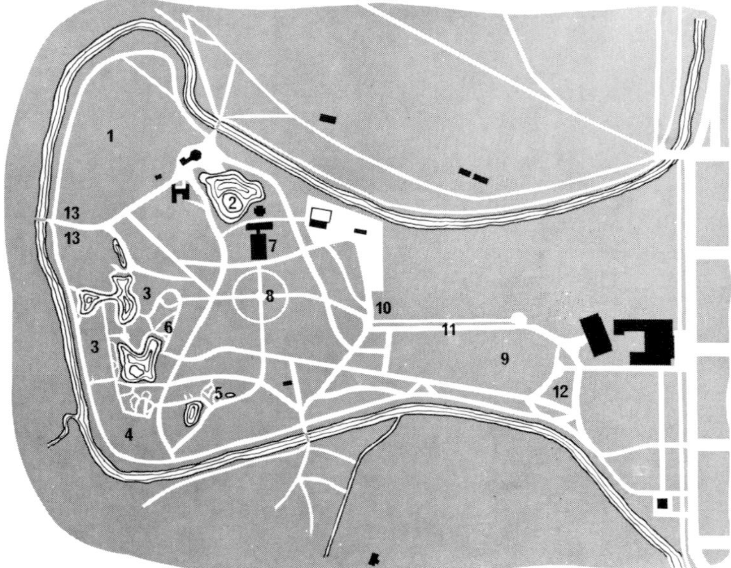

It was in 1863, as far as I can determine, that a group of Christchurch citizens made a move to start a botanical garden; chiefly with a view to the acclimatization and propagation, under proper care, of useful and beautiful exotic plants. They turned for advice to the Government Gardener in charge of Domains (Crown Lands), one Enoch Barker, trained as a gardener in several of England's great gardens, who had arrived in the Colony in 1859. With Barker's help the Botanical Garden was founded and Barker planted its first specimen tree, the 'Albert Edward Oak' which commemorates the marriage of the then Prince of Wales and Princess Alexandra of Denmark. When the Canterbury Horticultural and Acclimatization Society's objects and purposes were formally approved by the Provincial Government; and when a Commission was appointed to 'promote the cultivation of the Government Domain in connection with the objects of the Acclimatization Society', only then did the planting of the Botanic Garden begin in earnest. A fund of £1,000—say about ten or fifteen thousand of our debased currency—was furnished by the Provincial Government.

Although the Commission and the Acclimatization Society working more or less together did good pioneer work in the way of trenching, draining and otherwise preparing the soil, and in beginning plantation, the Gardens did not enjoy their first great epoch until the period presided over by the Armstrongs, father and son, as Government Gardeners of the Domain.

John Francis Armstrong, who began his working life as a gardener's boy in England at the age of twelve, emigrated to New Zealand in 1862 and was appointed to the Domain of Christchurch in 1866. There, for

Above right Once again in a remote land the English landscape style turns up—this time in New Zealand. But then the first makers of the Botanic Garden in Christchurch were English.

sixteen years he and his son, Joseph Beattie Armstrong, worked to lay out and plant the Botanic Gardens of 75 acres and the much larger Hagley Park beyond the Gardens boundaries. But they did much more than that, for, carrying out the original purpose of the Acclimatization Society, they introduced, established, propagated and distributed no less than four thousand species, both useful and ornamental, by way of the Botanic Gardens, to the farms, plantations and gardens of New Zealand. The Armstrongs raised European trees in the Gardens—the Canterbury district had no native trees—in extraordinary numbers: between 1870 and 1882 the Gardens were the source of 763,034 trees for general planting throughout the Province. That was by no means all they did, for they were keenly interested in the native flora, Joseph Armstrong in particular. He collected plants throughout the Canterbury region; he established a fine collection of living specimens in the Botanic Garden and of herbarium material in the associated Museum; and he sent seeds of New Zealand to many botanic institutions in Europe receiving in exchange seeds of Old World plants and American plants to add to New Zealand's floral wealth. Both he and his father also found time to write for numerous journals, learned and unlearned, publishing their discoveries about the flora of South Island, and especially the Canterbury region.

The Armstrongs were singularly ill-rewarded for their very valuable work. For example, Joseph's herbarium material formed a unique and very large collection of plants new to the botanical world; Julius Haast, later Sir Julius von Haast, head of the Christchurch Museum, sent so large a part of it to European herbaria and museums without asking Joseph's permission, taking all the credit for collecting it himself, that when Joseph, dying in 1926, left his collection to the Botanic Gardens, it was found to contain only 900 specimens of the many thousands he

In many genera of Australasian flowering trees and shrubs the stamens, rather than the petals, are the spectacular organs of the flower. This is most familiar to northern hemisphere people in the *Acacias* (the florist's 'mimosa'); and it is the rule in *Banksia* (*top left*), *Kunzia* (*top right*) and *Eucalyptus* (*bottom*), all in the Melbourne Botanic Garden.

intended to bequeath. Haast, in fact, played Elias Ashmole to Armstrong's Tradescant. Before that time, however, both father and son had resigned from the Garden's service on being required by the Board—which had taken over the management of the Gardens in 1876—to concentrate on various unbotanical labours in the Gardens.

Much of the trouble between the Armstrongs and the authorities was due to shortage of money; they were expected to do half a dozen different jobs because hard times made it impossible to find the money to pay enough gardeners. As a result, when a new Curator, Ambrose Lloyd Taylor, took over he found the Botanic Garden and the Domain both neglected; and as there was still very little money allocated to him his 18-year term of service was one long struggle and frustration, though he did in fact succeed in clearing and planting more land and in restoring and improving the Garden as a whole. Before he reached New Zealand Taylor had had about the best training a gardener could have had: his father had been a gardener at the Duke of Bedford's Woburn; the boy was sent to the Royal Botanic Gardens, Kew, as a student gardener; his career had included terms as head gardener to Lord Rothschild; and to the Duke of Devonshire at Chatsworth. Only a man trained hard and dedicated to his trade could have accomplished what Taylor accomplished at Christchurch with the wretched means given him. His plantings in Hagley Park are now stately avenues; it was he who, with the help of local nurserymen, replanted the Botanic Gardens after they had been devastated by fire; he even had himself sworn-in as a constable in order to deal with hooliganism and vandalism in the Gardens; at one time he was acting as the Botanic Gardens Secretary because the authorities would grant no money to pay the salary of a professional secretary.

The next Curator to do good work in the Christchurch gardens was another Englishman, James Young, who took over in 1908 when the money famine was ending. He had been trained in horticulture in England and in forestry in Australia. He created the rose garden in the Christchurch Botanic Gardens, built and stocked the two major glass-houses—the bigger one, in which tropical plants are grown and which is called the Cunningham House was modelled on the big Palm House at Glasnevin (see p. 58). In one part of the Gardens were a number of shingle-pits, for the subsoil at Christchurch Botanic Gardens is an ancient river bed. Young transformed them into the ponds which are now the most attractive feature of the Gardens and he also made the Bog Garden in the same quarter.

Young's successor was the first New Zealander to take charge of the Gardens, James MacPherson. MacPherson redesigned the Rose Garden, making it a circle defined by clipped evergreen hedges and one of the prettiest Rose Gardens we have seen in any Botanic Garden; he also made the Rock Garden as it now is. He greatly expanded Young's plantings of New Zealand native plants; he made a collection of *Magnolia*, another of the azalea series of *Rhododendron*, and he underplanted the park and woodland areas with daffodils.

Although the Botanic Gardens and their Curators and Directors played their part in the introduction of ornamental and probably also some salad plants to New Zealand commerce, they did not, on the whole, make much of a showing in the field of economic botany. It is possible, however, that they had a good deal to do with the introduction, acclimatization, propagation and large-scale distribution of New Zealand's most important exotic timber tree, *Pinus radiata*.

CHRISTCHURCH GARDENS TODAY

The parsimony of the Municipality responsible for the Botanic Gardens in Christchurch has deprived the Garden of any scientific importance it might, and probably would, otherwise have had. It now has no scientific staff. It is a fairly safe rule that when a Botanic Garden falls into the hands of a town council rather than those of a national government department or a university, its budget will be so restricted that its scientific and often also its economic work will stop. But the Christchurch Garden is still horticulturally one of the most beautiful in the world.

Elements which contribute to its quality are the gently sloping and undulating topography—few gardens have such a pleasant fall of ground; the river Avon vistas; excellent landscaping; the wide range of plants chosen for their horticultural attributes rather than as botanical specimens; and the splendid, healthy growth of plants in the rich, acid soil over the perfect drainage provided by the old shingle beds which lie at various depths from the surface. Perhaps the most striking quality of this Garden is, indeed, the rate and healthiness of growth: *Pinus radiata* at just over a century old stands 134 feet, straight as a poplar and with a fine, spreading head. A specimen of *Arbutus menziesii* must surely be the biggest in the world. The old *Rhododendron arboreum* hybrids have made enormous growth yet remain very floriferous. And such exotics as *Protea cynaroides* are grown, and, indeed, flourish. Mr L. J. Metcalf, the Assistant Curator, has recorded that this gorgeous South African shrub will, in his garden, stand 16 degrees of frost without showing any sign of damage, which it certainly will not as a rule.

The Garden still plays a part in the development of good garden cultivars from native species; the most interesting and in many ways the most pleasing of the plants I saw there were in the collection of *Leptospermum* hybrids and selected sports. This whole group of cultivars was very much finer than any of the *Leptospermums* one sees grown in European gardens. The Hebe collection also possesses some exceptionally fine garden plants. It is perhaps in this field of work that the Garden's controllers will allow its staff to make a practical contribution; a good start has been made; the Christchurch garden is not only one of the prettiest, it is horticulturally one of the most instructive we have dealt with in this book, and if it can no longer contribute anything in the field of pure science or of economic botany, at least it can be used for the advancement of ornamental gardening in New Zealand and the southern hemisphere generally.

MELBOURNE

Opposite Both the South Yarra (Melbourne) and Perth Botanic Gardens have representative collections of Australian plants. This *Cephalotus follicularis* (enlarged four times), Australian Pitcher Plant, is insectivorous, using the same trap device as the *Nepenthes*.
Below An aspect of South Yarra, Melbourne, the most beautiful botanic garden in the southern hemisphere.

The Royal Botanic Gardens in South Yarra, Melbourne, have the distinction of being the most beautiful of their kind in the southern hemisphere and perhaps in the world. They are inferior to the Royal Botanic Gardens, Kew, and to a number of other botanic gardens, in size; much inferior to Kew in the number of species grown, and in the quality of their scientific work; they are not as beautifully sited as the Royal Botanic Gardens, Sydney; one can, without difficulty, compare them unfavourably with half a dozen European botanic gardens, not to mention many American and Russian gardens, in particular ways. But in the matter of sheer beauty they are superlative, and it is worth crossing the world to see them.

For a country of only 12,000,000 people and less than two hundred years of history, Australia is rich in good botanic gardens. Adelaide has a garden of 240 acres, specializing in native Australian plants and notably in those from arid regions; it has a fine collection of the remarkable West Australian plants, and its work in studying drought-resistant and salt-resistant plants is of value far beyond Australia. The small (46-acre) Brisbane botanic garden specializes in succulents, has a good collection of palms, and an arboretum of the genus *Eucalyptus* with about five hundred species. At Canberra a small acreage of what will one day be one of the largest botanic gardens in the world has already been planted. This garden will be confined to natives, and at the time of my visit the attractive series of pools and waterfalls had already been completed and there were impressive plantings of *Melaleuca, Callistemon, Grevillea, Hakea, Telopea, Acacia, Boronia* and *Banksia* species, as well as a whole rain-forest section. In this primarily scientific garden, landscaping and garden design are being respected. Hobart has a small botanic garden chiefly devoted to Tasmanian natives. Perth has a thousand acres of nature reserve as well as a new but quickly growing garden of West Australian plants. In Queensland, Toowoomba has what must be the largest 'botanic' garden in the world, if you count as such its forty-four square miles of botanical reserve. As for the Royal Botanic Gardens, Sydney, we shall come to them once we have dealt with the history and beauties of South Yarra's garden.

For a reason which will presently appear, Melbourne Botanic Gardens in South Yarra were not always the first in point of beauty. Writing of the Australian botanic gardens which he visited in about 1870, Anthony Trollope had this to say:*

> The public gardens of Sydney deserve more than a passing mention. . . . Those in Melbourne in Victoria are the more pretentious and, in a scientific point of view, no doubt the most valuable. I am told that in the rarity and multiplicity of plants collected there they are hardly surpassed by any in Europe. But for loveliness, and that beauty that can be appreciated by the ignorant as well as the learned, the Sydney gardens are unrivalled by any that I have seen. . . .

**Australia and New Zealand* (London, 1873). Reissued in part by Nelson (Melbourne), 1967.

Clearly, a hundred years ago Sydney's botanic gardens were superior to Melbourne's as a garden if not as a scientific institution. Of Melbourne's garden at that time Trollope goes on to say:

> . . . the Melbourne gardens are the most scientific but the world at large cares little for science. In Sydney the public gardens charm as poetry charms. At Adelaide they please like a well-told tale. The Gardens at Melbourne are as a long sermon by a great divine, whose theology is unanswerable but his language tedious.

Trollope was by no means alone in his complaints of the excessive devotion to botany and disregard of gardening in the Melbourne Botanic Gardens. The fault, as we shall see, was the great von Mueller's; but so perfectly was it corrected by his successor Guilfoyle that once the latter's work was done even Trollope himself would have been forced to prefer Melbourne's garden to that Sydney.

Lake scenery is one of the beauties of Melbourne's Botanic Garden.

South Yarra, Melbourne: another aspect of a garden made in the English landscape manner but with Australasian plants. The tree on the extreme right is a *Melaleuca*.

In 1842 the Chief Surveyor of the Settlement at Melbourne marked out an area for a botanic garden, although the colony itself had only been established for a single decade. This first choice, in the heart of the city, was dropped, and in 1845 Governor La Trobe suggested instead the Garden's present site, making use for his purpose of a committee of the town council which had been set up to consider the business. La Trobe's site is in a bend of the River Yarra, and was certainly chosen so that the Governor could be sure of a pleasant outlook from Government House and security against encroachment by the city's growing industries. A professional, John Arthur, was appointed first Superintendent of the Garden which was at first only 5 acres in extent. But La Trobe, like so many English and Irish gentlemen of the day, a good amateur botanist and and gardener, kept a close eye on the new Botanic Garden and provided it with plants and seeds at his own expense. He did even better than that: he firmly excluded the Town Council from any say or interference in the administration of the Garden, thereby saving it from the worst blight that can afflict any botanic garden, municipal pettifogging. It is very doubtful if the Garden would have had the great future which was in store for it had it not been so well protected by Governor La Trobe against the city fathers.

John Arthur seems to have been little more than La Trobe's garden boy. He died after four years in office and was succeeded by a Scot named John Dallachy, who made the first great extension of the Botanic Garden by excavating a swamp to make the great lake which is still one of the most beautiful features. Moreover during Dallachy's term of office the number of exotic species in the Garden was raised to 5,000, of natives to 1,000. Furthermore Dallachy succeeded in making the Melbournians aware of their Garden: according to a Report made to the Victorian Parliament in November 1851, eight hundred visitors came to the Garden every Sunday and the Melbourne Horticultural Society was holding its Shows there. Dallachy not only managed the Botanic Garden very well, he went out on plant-hunting expeditions and he never failed to send one of every plant he collected for Melborne to Sir William Hooker at Kew.

Such, then, were the beginnings of Melbourne's Botanic Garden; and the man who was to make it world-famous while provoking Trollope's complaints about its excess of science, had meanwhile arrived in Adelaide in 1847. He was a young German doctor of philosophy, very learned in European botany, Ferdinand Jakob Heinrich Mueller. Within a few years of his arrival in Australia and despite its remoteness from the world centres of learning, he had made himself a world reputation as a chemist, as a botanist and as an explorer—a reputation extending so far beyond Australia that it was Sir William Hooker who, in 1872, persuaded Governor La Trobe to create the post of Government Botanist, for Mueller's especial benefit and for the advancement of science. It was, then, as Government Botanist that Mueller, in the intervals of his exploring and botanizing expeditions all over the new continent, slowly took over control of the Botanic Garden as Dallachy, no match for the determined German, slowly lost control. By 1857 Mueller had poor Dallachy working away from Melbourne as one of his plant collectors; while he himself had got just what he wanted and needed for his work, complete control of a publicly financed botanical garden where he could grow the material he and his subordinates were gathering, and study it from the taxonomic and morphological points of view. And for the next sixteen years what counted at South Yarra was not horticulture but pure botany. Hence Trollope's complaints; I suspect that he must have met the didactic scientist at dinner one night and been thoroughly bored by him. However, during those sixteen years:

Dr Mueller worked furiously at systematic botany. The whole of his working hours were devoted exclusively to botanical work, both systematic and economic. No family ties, no social pretensions were ever there to interrupt his phenomenal output of botanical papers, books and memoranda. And the gardens became, with humourless teutonic thoroughness, a living textbook of systematic and economic botany, with garden beds for pages. In seventeen years Dr Mueller had travelled 27,000 miles on horseback and foot. He had examined more than 1,000,000 botanical specimens. He had become honorary or corresponding member of two hundred scientific societies. But, as a contemporary writer put it, '. . . the gardens themselves did not feed the hunger of the eye'.*

Mueller meanwhile had been suitably honoured. He had been created a baron with the right to the particle *von* by the King of Wurtenburg; he had been knighted by Queen Victoria and variously distinguished by almost every European government; he had been elected a Fellow of the Royal Society. But when as a result of public complaints very much in the manner of Trollope's he was given a new Herbarium to work in at the Garden gates but quietly edged out of the control of the Garden itself, Mueller became unforgivingly embittered, and during the remaining twenty-three years of his life and work at the very threshold of the Botanic Garden he never set foot inside it again.

**Melbourne's Garden*, Crosbie Morrison (Melbourne, 1946).

It was on the solid foundation laid by Von Mueller that his successor, as great an artist-gardeners as the German was a scientific botanist, created the beauty of this superlative garden as we see it now. The new Superintendent was an Englishman, William Robert Guilfoyle, who saw how he could make the best of the Garden's gently sloping and undulating topography and how Von Mueller's collection could be used as material in the making of a great landscape garden, without in any way deprecating its value as botanical material. He belonged to the new English school of what I have elsewhere called 'paradise' gardeners and he seized his chance to create a work of garden art. The task took him thirty-six years, in which time he increased the extent of the Garden by 40 acres; by the time he retired in 1909 he had made such a garden as no subsequent Director has wanted to change. Guilfoyle was Australia's greatest horticultural artist as Von Mueller was her greatest botanical scientist, and, like him, a world figure whose fame was by no means confined to his own continent.

Of the successors to these two great botanical gardeners, John Cronin (1909–23) was a practical horticulturist rather than a botanist; his contributions were in the fields of pruning techniques which he improved in the course of a series of experiments in the Garden; and of hybridizing notably of dahlias and watsonias. As a botanist, he planted a special area of the Garden with Australian native plants. William Laidlaw, who came next, was a Scot and a plant

The genus Eucalyptus has over 500 species confined, with a few exceptions, to Australia. This painting of *Eucalyptus tetraptera*, the 'square-fruited mallee', is by Margaret Stones, whose first volume of a projected five-volume *The Endemic Flora of Tasmania* (Ariel Press, 1967) showed her to be of the company of great botanical illustrators.

pathologist, but he died within a year of his appointment. Frederick James Rae (1926–41) seems to have been more remarkable as an administrator, a public figure, than as a botanical gardener, but the Garden flourished in his care as it has done in that of a series of later Directors whose task has been to maintain the Melbourne Botanic Gardens up to the high standard set by their great predecessors, and to keep steadily advancing the work of completing the study of Australian and Southern hemisphere florae in both the Garden itself and in the Herbarium.

THE MELBOURNE GARDEN NOW

As I have said already, the South Yarra Garden is chiefly remarkable for the manner in which by a very skilful disposition of lawns, the layout of paths, the use of water, and above all the placing of specimen trees and groups of trees, it presents itself as a long series of landscape pictures—vistas, panoramas, prospects and perspectives, all of quite exceptional beauty. The trees, which are one of the principal materials of these compositions, are magnificent, the growth attained in about a century is very remarkable for growth rate in this Garden excepting during severe drought; it is about double the average. To me the most beautiful, as they are the most interesting trees, are the natives, not only among the *Eucalypts* of which there is a large collection, but especially the 'Paper Barks' —various species of *Melaleuca*; and the Eastern Pacific *Araucarias*. I do not think the Norfolk Island Pine quite so beautiful a tree as its Argentinian relative *Araucaria imbricata*, but it is still one of the loveliest of the conifers, as is *Callitris columnaris*, the graceful and stately 'Cypress Pine' of Queensland, of which there is one quite perfect specimen in the Melbourne Garden. Also very striking are the superbly grown specimens of the rather oak-like *Grevillea robusta*, of *Cinnamomum camphora* and of *Araucaria cunninghamii*. For tree-lovers the Melbourne Botanic Gardens are a paradise: consider, for example, the familiar *Cupressus macrocarpa*, which here attains a height and girth far exceeding anything known in its own country or elsewhere, a size to astonish American specialists who are familiar with it at home.

And all this is still leaving the spectacular flowering trees out of account: I have already mentioned the tree species of *Melaleuca* because, prettily though they flower, their great merit is in their habit, shapeliness and in the pale beige papery bark. But beyond comparison the most eye-catching trees in this botanical garden are the *Erythrinas*—*E. caffra*, the 'Kaffir Coral Tree'; *E. secundiflora* with its big racemes of flaming orange flowers; *E. indica*, the 'Indian Coral Tree'; and, of course, *E. crista-galli*, which so impressed us in Argentina's Botanic Garden, and as a street tree in Buenos Aires. The colour played on in all these is red, and is in beautiful contrast harmony with the lavender blue of the *Jacarandas* which are also a feature of this Garden. In passing—a large number of the trees in the Melbourne Botanic Garden are the homes of myriads of the loudest cicadas I have ever heard, and their tremendous shrilling will always be associated with this in my memories of the garden.

Among other native flowering trees which are well represented are *Banksia*, with the huge bottle-brush flowers and curious cylindrical fruits; *Lomatia*; and *Telopea*, so close in looks and habit to the related *Embothrium* species of South America, and the Tasmanian *Telopeas*.

Forest trees and flowering trees splendidly grown and in great variety provide, then, the grand flora of Melbourne's garden; another texture is worked into the pattern by the tree-ferns which are native—one can see them at their best in the wild in the Maquarie Pass south of Sydney, and also in the gorges of the Dandenong Hills near Melbourne. No northern temperate zone garden, and no tropical garden but those at high altitudes, can make such use of the sweetly curving giant fronds and fiddle-handle buds of these graceful plants. Only in one other botanical garden did I see Melbourne's tree-ferns surpassed: Tjibodas, in Indonesia (see p. 198). At the time of my visit to Melbourne the tree-ferns were suffering, however, from the severe drought of 1967, as indeed was the whole State of Victoria, where the sheep were dying like flies and lambs selling for sixpence. It is deplorable that in a country where experience has shown that lethal droughts will occur every few years, practically nothing has been done by successive governments to provide for these conditions by adequate water-storage, water-rationing, and the use of untapped water resources; but then, of course, *dolce far niente* is very much the Melbourne style. The Palms also, of which Melbourne has a collection inferior only to those of Fairchild, Sydney and Peradeniya, were showing signs of distress. So, likewise, were many of the gigantic *Ficus* trees, those mighty plants which embellish Melbourne's parks as well as her Botanic Gardens.

Colour is provided in the Melbourne Garden not only by the great number of flowering trees but by thousands of flowering shrubs, and, of course, herbs. Many of the shrubs are natives, for example, *Callistemon* in numerous species, the gorgeous 'Bottle Brush' bushes; the shrubby *Grevilleas* of which there are now some fine garden cultivars which could probably be grown in south-west England; *Acacia*, *Boronia* and many others. The best flowering exotic shrubs are in the genera *Camellia* and *Rosa*. To some extent all the orders of plants, from annual and perennial herbs up to large trees, are planted in geobotanical groups; but the rule is not strictly kept and is often broken for the sake of good garden display or composition.

Although much good work is still being done in taxonomy at Melbourne, especially in the collection, determination and description of native plants, and although the Herbarium is a large one and constantly being added to, the emphasis is now very definitely on horticulture rather than botany—despite the kind of beginning which Von Mueller gave to the Botanic Gardens and perhaps because of what Guilfoyle made of Von Mueller's beginnings.

SYDNEY

Sydney's Botanic Garden is one of the most gloriously sited in the world, near the point where the great botanist Sir Joseph Banks landed to initiate the study of the Australian flora. In the background, the Sydney Harbour Bridge and the Opera House. The flowering tree is a *Jacaranda*.

To go from Victoria to New South Wales was to move from an area of total drought to one of comparative lushness where rainfall had been 'normal'. The grass was green, the foliage of trees and shrubs glossy with health, and because the sun had not been so merciless so early in the season, there was more colour. But I know of no botanical garden in the world so beautifully sited as Sydney's, out on a peninsula in the grand bay of which Trollope wrote a century ago.

> I despair of being able to convey to my reader any idea of the beauty of Sydney harbour. I have seen nothing equal to it in the way of land-locked sea-scenery . . .

Trollope had seen Dublin Bay, the Hudson River, the Bay of Spezia and Bantry Bay. I cannot go with him quite so far; he had not seen Rio de Janeiro and I have. But I have not seen a botanical garden, and only one garden of any kind (Illnacullin, off Glengariff) so prettily sited as Sydney's on the foreshore of Port Jackson.

The Gardens on this site were originally planted to supply the Governor of the Botany Bay colony, his household and military staff, with vegetables and fruit. The site has the distinction of being the first piece of land in Australia to be brought under cultivation; as early as 1788 Governor Phillips was reporting to London '. . . a farm of nine acres in corn'. Also planted there were the first vegetables brought as seed from England; and the first fruit trees, from South Africa. The farm there became 'official' in 1792 when the Farm Cove area of the present Botanic Garden site was named Governor's Farm—that was still during Governor Phillips' term of office. And at about this time Phillips set aside a much larger area—Phillips' Domain so-called—for future use as Gardens and Parks. Within this Domain lay the land now occupied by the Botanic Gardens, though the Gardens themselves were not inaugurated as such until 13 June 1816, when the Colonial Botanist, Charles Fraser, was appointed Supervisor, later Superintendent.

Governor Maquarie, who succeeded Phillips, tried to have the Botanic Gardens shifted; he wanted to retain the whole area as part of the grounds of Government House. But for some reason he was forced to give up this idea although George Caley, Sir Joseph Bank's plant collector, had started a private botanic garden in 1800 at Paramatta, and it was still flourishing in 1810 and might have provided the substitute which Maquarie wanted. Besides Caley there were other proto-botanic gardeners in Australia, men whose work came to nothing in their lifetime but who set the example to be followed in the future: Joseph Gerrald, a Scottish deportee had started planting in the present Botanic Gardens area in 1794: and in the same region one Henry Dodds, who had come out to Australia as Governor Phillips' personal servant, began a sort of model farm and farming school for the colonists. Sir Joseph Banks' interest in Australian plants had made it certain that there would be a botanic garden at Botany Bay, and it was doubtless under his influence that first Caley and then Fraser began botanical planting and study.

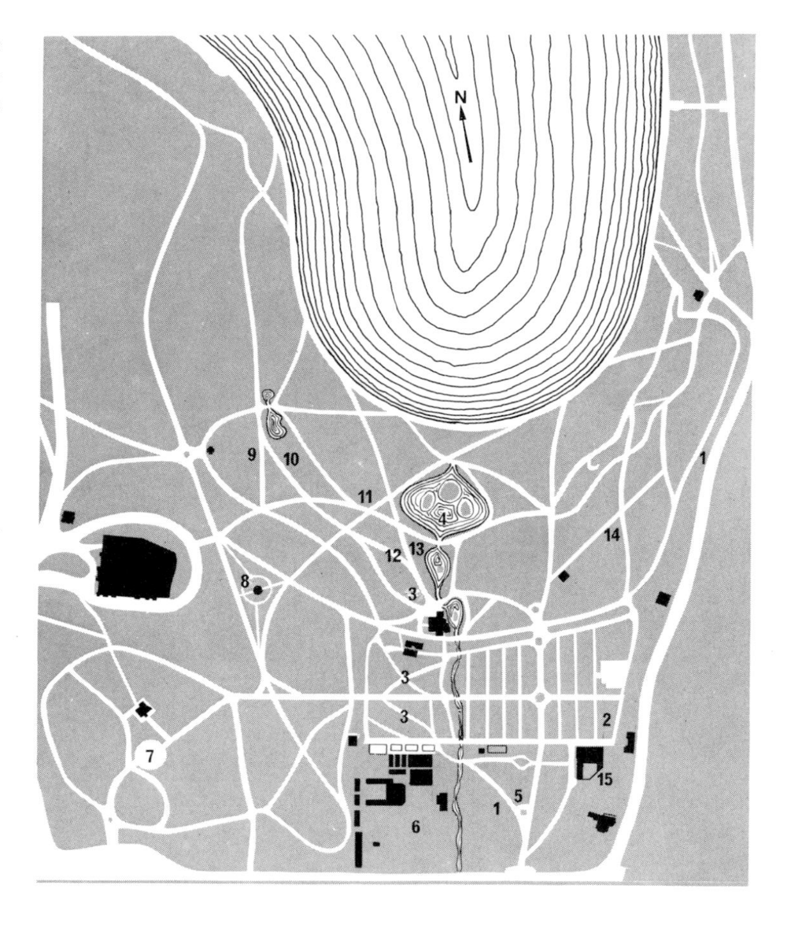

Key to Map

1 Australian native plants
2 Succulents
3 Palms
4 Ponds
5 South African plants
6 Nursery
7 Sunken garden
8 Rose garden
9 *Leguminosae*
10 *Myrtaceae*
11 Medicinal plants
12 Conifers
13 Far Eastern plants
14 Cycads
15 Herbarium and general offices

In a brief account of the Botanical Gardens* Mr R. H. Anderson, a former Director and Chief Botanist said:

> The early development of the Gardens was stimulated by the great interest shown in Australian plants by botanists and nurserymen in other parts of the world. Most of the important botanic gardens and similar institutions considered it most desirable to have a 'New Holland' collection of plants and this led to a brisk exchange of seeds and plants. The Sydney gardens provided a centre for such exchange and gradually valuable collections were added to the existing plants.

The early Superintendents were all active plant collectors for, despite the beginning made by Banks, a colossal task of plant discovery, description and determination faced Australian botanists. Charles Fraser accompanied Oxley on his three exploring expeditions between 1817 and 1819 and a large number of specimens of the New South Wales flora were collected. Fraser also col-

*Published some years after it was written in *Australian Parks*, Vol. 3, No. 3, Feb. 1967, and brought up to date by Mr K. Mair, the present Director.

lected for the new Gardens in Tasmania, New Zealand and Norfolk Island. Richard Cunningham, Superintendent of the Gardens from 1833 to 1835, was killed by a war party of Aborigines when, as a member of Mitchel's second expedition to the Darling River, he wandered away from the main party in search of new plants. Alan Cunningham, who succeeded him (1837–8), had been with Oxley and Fraser in 1817 and he spent the years 1817–22 collecting plants along the Australian coastline as botanist to the *Mermaid* Survey ship; on other exploring expeditions he collected for the Sydney Botanic Gardens in New South Wales, Queensland, New Zealand and Norfolk Island. The next four Directors (the title was changed *c.* 1847) stayed in the Gardens and Herbarium, sorting, determining and describing, cultivating and propagating the enormous amount of material which had been collected. The next plant-collecting Director was Charles Moore (1848–96), who explored and collected in New South Wales, the New Hebrides, the Solomon Islands and New Caledonia. All the Directors followed the policy of sending newly found Australian plants abroad in exchange for exotics, until such exchanges were being maintained with 185 foreign institutions. The establishment of both native and exotic plants in the Sydney Gardens was never easy. The soil was poor; there was a serious shortage of suitable water; and there was one difficulty peculiar to Australia, for according to Anderson (*op. cit.*), 'In 1852 Moore lamented the loss of most of his workmen who joined the gold rush of that year'.

By 1896, when Moore was succeeded after nearly half a century of service to the Gardens by J. H. Maiden, it is probable that although many Australian plants remained undescribed the back of the botanists' great task in Australia had been broken. Since then the Directors and staff of the Royal Botanic Gardens, Sydney, have been engaged in the work of completing, in the Gardens themselves and in the National Herbarium, the task of systematizing the flora of the continent. Thus Maiden (1896–1924) was responsible for the *Critical Revision of the Genus Eucalyptus*, and in our own time there was Anderson's *Trees of New South Wales*; other publications which are standard works are, by way of examples, Rupp's *Orchids of New South Wales*; Maiden's *Forest Flora of New South Wales*; and Blakeley's *Key to the Eucalyptus*.

THE GARDENS TODAY

Although the Sydney Gardens are not perfectly flat they depend for topographical beauty, as I have said, on their situation, but that is so remarkably lovely that want of internal features is much less important than it would otherwise be. And in the variety of foliage forms and colours the Gardens rival those of the Jardim Botanico in Rio de Janeiro, while in the grouping of forms and colours it is little inferior to Melbourne. The general plan of its 65 acres is one of vast lawns, groups of trees, specimen trees, shrubberies, rockeries and flower-beds, and among

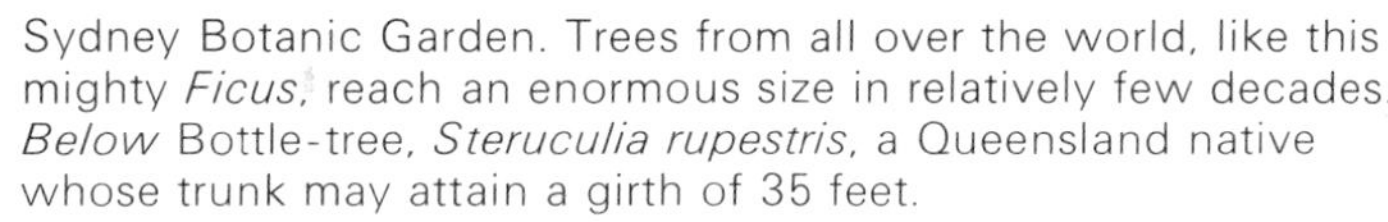

Sydney Botanic Garden. Trees from all over the world, like this mighty *Ficus,* reach an enormous size in relatively few decades. *Below* Bottle-tree, *Steruculia rupestris*, a Queensland native whose trunk may attain a girth of 35 feet.

them are the broad, curving walks and straight drives.

The Sydney Gardens have excellent and very colourful plantings of a very wide range of flowering herbs of much more horticultural than pure botanical interest. The four parts of the Gardens (growth over a hundred years has been in four sections) are still clearly defined. The Middle Garden is the oldest and it is chiefly planted with trees and shrubs; the Upper Garden, opened as part of the Gardens in 1876; the Lower Garden, laid out by Fraser *c.* 1833; and the more recent Garden Palace Grounds.

A number of the Gardens' finest trees are native to the site and older than the Gardens. Near the Maiden Memorial Pavilion there is a group of very ancient *Casuarina glauca*, the so-called Swamp Oaks, though they are, of course, conifers and nothing to do with oaks. Elsewhere there are old plants of *Eucalyptus tereticornius* and *E. bothroides* which are survivors of the forest primeval. Also among the best specimens of big trees are some of the Gardens' memorial plantings—a *Martinsuella imperialis* planted in 1868 by the then Duke of Edinburgh; a *Melaleuca viridiflora* planted in 1881 by the Duke of York, later King George V. In the same year the then Duke of Clarence planted a *Livistona australis* which is now a splendid tree, as is a very fine *Araucaria columnalis* planted in 1901 by the Duchess of Cornwall and York.

As well as the forest trees remarkable for beauty, size, form, grace or simply rarity, there are the more spectacular if less imposing flowering trees. It is, I think, for these that Sydney's Gardens are chiefly memorable in the months of November and December—for the flaming crimson-scarlet *Creocallis wickamii*, a member of the *Proteaceae* rather like *Embothrium*; the even more spectacular *Metrocideros excelsa*, a mighty dome of raspberry red flowers in striking contrast with the lavender of *Jacarandas*; *Stenocarpus sinuatus*, the 'Wheel of Fire' tree; the scarlet-flowered 'Ilwarra Flame Tree', *Brachychiton acerifolia*; *Brachychiton discolor* with its large flesh-pink flowers of waxy substance; and, inevitably, the many representatives of the genus *Erythrina*.

Sydney has one superlative collection, very handsomely planted, its Palmetum, probably the second finest in the world. So dense is the Palmetum that to walk in it is to move into an older, pre-arboreal world of palms, tree-ferns and cycads, a world of prehistoric plants.

The Gardens are rich in good native and exotic shrubs. There is no point in listing the familiar ones which are in every botanic garden where they will grow; some less familiar which I enjoyed in Sydney were *Ochna serrulata* with its curious rabbit-faced crimson flowers; the ferocious, leaf-less *Colletias* entirely composed of huge, iron-hard spines, yet bearing a froth of pretty flowers; the big, shrubby Begonias; and the curious *Monodora myristica*, whose bright salmon-pink young foliage is in vivid contrast with the dark green of the mature leaves.

The Royal Botanic Gardens, Sydney, are a monument to the origins of Australian horticulture and agriculture; botanically a good teaching and research institute; and horticulturally an example of very good gardening.

KIRSTENBOSCH

The identifying feature of the great National Botanic Garden of South Africa at Kirstenbosch is Table Mountain. In the foreground, *Proteas*, noblest members of the remarkable Cape flora; middle-ground, *Leucadendron*, the Silver Tree peculiar to Table Mountain.

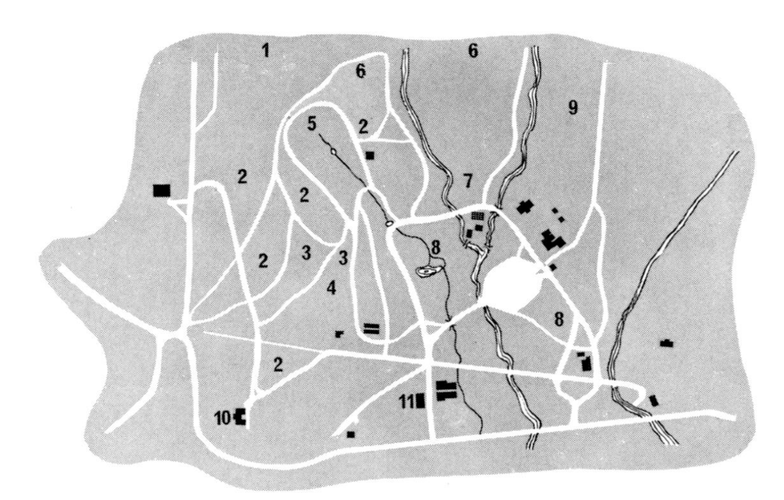

As it is planted as a wild garden it is not possible to key the thousands of species grown and only a few special areas can be clearly distinguished by generic names.

1 Leucadendrons
2 Proteas
3 Succulents
4 Rock garden
5 Cycads
6 Heaths
7 Arboretum
8 Lawns
9 Oaks
10 Herbarium
11 Office and nursery

The South African National Botanical Garden, at Kirstenbosch in Cape Province, is at all times of the year one of the most beautiful gardens in the world; by reason of the special flora of the Cape, it is also one of the most distinguished; and finally it is the mother garden of no less than nine new botanic gardens which are being planted in various parts and different climates of South Africa, as a direct consequence of the enthusiasm and energy of the Director of Kirstenbosch, Professor Hedley Brian Ryecroft, a specialist in the family *Proteaceae*.*

The Botanical Garden's site, only a few miles from the heart of Cape Town and in the lee of Table Mountain, has an interesting history. Until about the middle of the seventeenth century it was untouched, untrodden bush; or perhaps, if any human being has ever set foot on it he was one of those 'bushmen' who were the aboriginal men of South Africa and whose rock drawings were the first artefacts in the country's human history. But in 1657 Leendert Cornelissen, in the service of the Dutch East India Company in Kaapstad, Cape Town, left the Company's service to become a *vrijtimmerman*, a free sawyer, or timber-merchant. As such he sought and was leased a grant of forest discovered and chosen by himself at a place where he could fell and saw timber. Before he made the grant the Colony's governor, the great Jan van Riebeck, visited and inspected the site for himself. The grant was, in fact, a 'loan' of land, made on condition that Cornelissen exploit it to the Colony's advantage. The place became known as Leendertsbos.

Leendert Cornelissen's workers were slaves; not Bantu, there were none there, they were slowly migrating into the Cape at about the same time as the Dutch. Van Riebeck has some entries in his journal which indicate that even slave labour did not make the free-sawyer's life very easy. Thus, for example, on 28 April 1658:

> Guinea and Angola slaves are inclined to desert especially now that 14 slaves have run away last night from the free-sawyer's Leendert Cornelissen and . . . who treat their slaves better than other freemen do . . .

and again, on 25 October 1658: 'Today another slave belonging to the sawyer Leendert Cornelissen deserted . . .'. One savours the word 'deserted' as applied to slaves. But it is the strength of such men as Van Riebeck, or of our own John Hawkins a century earlier, or of today's ultra-Afrikaaners, that they know they have the divine right to enslave the black man.

The only useful timber tree on Leendertsbos must have been 'Yellow-wood', and fairly soon this was all felled, sawn and used as beams and planks in house-building and ship-building. What exactly happened to the holding immediately thereafter is not clear. In order to acquire the name Kirstenbosch it must, at some time before we next hear of it, have passed into the hands of some free burgess named Kirsten, although the name does not emerge into Cape history until a century after Cornelissen's time. The name Kirstenbosch first turns up in 1795, when the British occupied the Cape for the first time and an inventory of the Dutch East India Company's property was drawn up and handed to the new bosses. By then there was a house on the place, and both house and land reverted to the Batavian Republic in 1803, only to go back into British hands during what the Afrikaaners call the second British Occupation. The British government of the time adopted a policy of selling the freehold of Crown Lands to raise funds to pay their inflated Civil Service. Half of Kirstenbosch was sold to the Colonial Secretary Henry Alexander, and the other half to his assistant, Colonel Christopher Bird. Among other things which Bird did on his part was to build a charming little pool, to contain a natural spring on the hillside, using the small Batavian-style but Cape-manufactured bricks. The pool was bird-shaped, to signify its maker's name, and it is now, in a tree-shaded and ferny glade, one of the prettiest and certainly the oldest artefact in present-day Kirstenbosch. Bird sold out to Alexander for £1,000, and Alexander built himself a pretty house on the estate; one of his house guests, in 1816, described the local scenery as romantic, which it is; and the estate as well wooded, which means that the forest felled by Cornelissen had regenerated.

When Alexander died two years later and his widow decided to take her children back to England, the government advanced her money against Kirstenbosch and later sold it, apparently in lots. At all events, there were several owners, chiefly of the Van Reenen, the Eksteen and the Cloete families, which were all interconnected. And it was the Anglo-Dutch Charles Duffy Hendrik Cloete who first did any considerable laying-out and planting of a park or garden at Kirstenbosch; his plantings included an avenue,

*They are as follows: Worcester in the Karoo, 240 acres; Betty's Bay, Caledon region; Cape Flats; Orange Free State, near Bloemfontein; Drakensburg and Easter Free State, Harrysmith; Nelsprint Botanic Garden; Natal Botanic Garden, Pietermaritzburg; Natal Sub-tropical Garden; Darling Floral Reserve.

oaks and other exotic trees some of which still survive although most European trees grow themselves to death in the Cape climate and their wood is coarse, soft and shaky. Orchards and vineyards were also planted; small parts of the estate were detached to make freeholds for members of the Cloete family. Then, in 1895, what remained was bought—153 *morgan*, about 400 acres—by Cecil Rhodes, apparently with the object of preserving it for the nation. Rhodes paid £9,000 for the estate and left it to the State in 1902. It was thus that Kirstenbosch, or most of it, reverted to public ownership and became available for the planting, in due course, of a National Botanical Garden.

FOUNDATION; AND ITS BOTANICAL ANTECEDENTS

In one sense the ancestor of the National Botanic Garden was the Company's Garden, now called the Town Garden, in Cape Town. This was first cultivated as a farm where Van Riebeck tried to grow his grain. But Cape Town is a windy city; when there is no gale from the north-east there is a high wind from the south-west. The Company's corn was repeatedly 'laid' until cereal cultivation was shifted to more sheltered sites inland. The Company's Garden was then used to grow vegetables and some fruits, for the Dutch were very early aware of the antiscorbutic quality of fresh vegetables and fruits. However, in the second half of the eighteenth century, the Company's Garden became at least partially 'botanic', if only because the man sent from Holland to be Company's Gardener, J. H. Auge, was not only recommended by the great Boerhaave, who had trained him, but came out to the Cape in the company of the illustrious botanist, Thunberg.

Of the fruit trees originally planted by the Company's gardeners before Thunberg's time, two are still flourishing, a huge old pear tree divided into four trunks near the base and now a fine ornament to a lawn; and a *Morus niger*, which was shedding its load of mulberries when I saw it in late January. Both are well into their third century. Of Auge's plantings—one determined, the others presumptive—there still remains a big clump of *Strelitzia alba*; a *Cycas circinnalis* from Madagascar; a fine old giant *Ginkgo biloba*; a cork-oak, introduced as a parent of trees to yield corks for the wine industry; and possibly one or two others. This very pleasing little Town Garden, now associated with the botany department of Cape Town University, has some other very fine old trees, an immense 'Norfolk Island Pine', an *Araucaria bidwillii*, some well-grown 'Paper Bark' *Melaleucas*, *Agathis*, specimens of the native timber-tree, that 'Yellow Wood' or *Podocarpus latifolius* which attracted Leendert Cornelissen to Kirstenbosch. *Grevillea robusta* grows well here and there are two huge, tree-like specimens of *Aloe bainsii* and a wonderfully floriferous and fragrant *Rothmania globosa*. There is an excellently planted display-house or orchids, bromeliads and other tropical exotics, a pretty, formal rose garden, a collection of such palms as can be grown in Cape Town,

A majority of the 90-odd species of the genus *Protea* is confined to South Africa. This picture shows not two flowers but many hundreds, for the true flowers, tubular in shape, are densely packed in the centre, the surrounding rays being not petals but brightly coloured bracts. Scores of *Protea* species grow in the Kirstenbosch garden.

Dr Ludwig Pappe, one of the three founders of Kirstenbosch's Botanic Garden.

and a pleasing garden of scented herbs for the blind.

Such is the Town Garden today; its origin, the Company's Garden, was the ancestor of South Africa's botanic garden.

In the days when South Africa's rich and fascinating flora was still being discovered and described, the scientific work was all done by European, often British, botanists working at home with herbarium material sent to Europe by collectors. As R. H. Compton, a former Director of Kirstenbosch now working as a botanist in Bechuanaland, says in his *Kirstenbosch, Garden for a Nation*:

> The monumental work, 'Flora Capensis' was brought to completion by the Royal Botanic Gardens, Kew, after more than a century of study, for the most part of botanists who had no first-hand knowledge of South African plants in the field.

But in a country colonized by the two European peoples most forward in botany and horticulture and whose flora was so rich and new, it was as inevitable as it was desirable that the centre of study should shift to South Africa itself. But here again Compton has something pertinent to say:

> In the early days of the development of a country such as South Africa—and such days lasted well on into the second half of the nineteenth century—the intellectual and cultural aspects of life take a low place in popular estimation and funds, private and public, are rarely forthcoming for activities without evident financial advantage. Generally, cultural and scientific enterprise is left to those whose tastes lie in those directions, and who are able through inheritance or success in business to cultivate them from their own resources, or whom in receipt of an income from some other source, are able to devote their spare time and energy to pursuits of this kind. Much of the great botanical achievement in South Africa has been brought about through devoted and highly skilled amateur effort . . .

But as more and more professional botanists took over from the amateurs it became clear that a botanic garden, laboratory and herbarium were needed.

Kirstenbosch is thus one of the very few botanic gardens, if not the only one, which came into existence in order to facilitate the study of a particular flora and not for the introduction of economic plants or the study of botany in general. Its ancestor, the Company's Garden, had its origins in economic need; but its predecessor—the botanic garden started in 1848 by the plant collection botanists Ludwig Pappe, Karl Zeyher and Peter MacOwan, in Cape Town—had the same purely scientific and local purpose. This earlier garden, badly sited and without a Herbarium, was not a success. Dr Harry Bolas, the Cape's greatest amateur botanist and collector whose private Herbarium, left to and neglected by the National Botanical Garden as a result of which it is now housed and valued by the University, writing of the Pappe-Zeyher-MacOwan garden, said:

> There is a great want of a proper representation of Colonial plants which ought to be the first to be attended to. There should be a botanic garden of a national character.

Nothing was done at the time, and when, in 1891, the Cape Town municipality assumed responsibility for this garden it had, according to Compton, become nothing more than a recreational park. The complaint that 'Cape Colony' was 'the only British possession without a botanic garden' then came, in 1895, from Kew. Kew's *Bulletin* had in fact been drawing attention to this for some years and warning that some Cape species were in danger of extinction. However, not until after the South African War and the 1910 Act of Union could anything effective be done about it.

The man who was soon to make Kirstenbosch was an Englishman. Here, briefly, is an account of Harold Welch Pearson's antecedents: born Lincolnshire, January 1870; won the Darwin Prize at Christchurch, Cambridge and graduated with First Class Honours in both parts of the Natural Sciences Tripos; visited Ceylon and worked at Peradeniya on a travelling scholarship; back at Cambridge as Assistant Curator of the Herbarium; elected Frank Smart Student in Botany Gonville and Caius College; 1899, University Walsingham Medallist; 1900, Assistant Director, Kew, under J. Thistleton Dyer; to Cape Town in 1903 as Harry Bolas Professor of Botany, and there created the botany department and attracted students from a zero beginning. Pearson was elected a Fellow of the Royal Society for his work on the morphology of the genus *Welwitchia*.

This was the man who, at the Annual Meeting of the

Below The spring, popularly but incorrectly known as Lady Anne Barnard's Bath, one of the oldest features of Kirstenbosch.

Opposite: Top left Protea barbigera, the Woolly-bearded Protea, showing the bracts still closed round the flowers. *Top right* One of the lovely Cape Heaths: of the 500 species of *Erica* known to botany, 470 are South African and grown at Kirstenbosch. *Bottom Encephalartos altensteinii* in the Cycad Amphitheatre, which has every Cycad genus known to science.

Overleaf The higher, wild garden at Kirstenbosch where the trees, shrubs and herbaceous plants are all natives. The Cape flora is probably the richest in the world for any place of comparable area.

South African Association for the Advancement of Science in 1910 devoted his address as President of Section C to 'A national Botanic Garden'. He discussed the nature, work, purposes, site, financing* and development of the kind of institution he wanted and he concluded:

> The South African Botanic Garden cannot be merely an economic undertaking; it must also be an expression of the intellectual and artistic aspirations of the new nation, whose duty it is to foster the study of the country which it occupies, to encourage a proper appreciation of the rare and beautiful with which nature has endowed it.

*Pearson had done his homework very thoroughly; for example he was able to point out to the meeting that Peradeniya received one two-hundred-and-seventy-fourth part of Ceylon's revenue; he calculated that on that basis the Union government should grant the projected South African National Botanic Garden £32,000 per annum. Call it a quarter of a million of our debased pounds, or about half a million rand.

Pearson, a pure scientist but a good, shrewd lobbyist in the cause he had so much at heart, must have judged the political leaders of his country—British or Afrikaner, they have never been remarkable for intelligence or breadth of mind—and found them wanting; for it was to his friends in the business community that he turned for support. In an article following up his address he dwelt on the functions of the kind of Garden he envisaged, and if he emphasized the importance of using it to make trials of potentially valuable crop plants it was chiefly because he knew the kind of men he was trying to interest, for he never showed much interest in economic botany in practice and he wrote more about the advancement of botanical science, about botanical education, and about the conservation of rare and beautiful species. Finally, it was Pearson who, having prospected for possible sites, suggested Kirstenbosch, which by then had been left to the nation by Rhodes but remained unused. Pearson thought that this would be particularly appropriate since the adjoining and much greater part of Rhodes legacy of land to his country, the Groote Schur estate, was to be the site of the National University with which he considered that the Botanic Garden should properly be associated.

Pearson not only accomplished all this but at last in 1911 he succeeded in convening a public meeting of notables, presided over by the Chief Justice, Baron de Villiers de Wynberg, and this meeting adopted Pearson's motion that a botanic garden be established in the Cape Peninsula, setting up an executive committee to found a National Botanic Society. It is clear that Pearson was counting less than ever on government support. His meeting got a good press but still more time elapsed before the desirability of establishing a botanic garden was moved in the House of Assembly. The motion was supported by the Prime Minister, Botha, and the government agreed to contribute financially, though on a minute scale; £2,500 for a house for the garden's Director; and £1,000 a year towards upkeep, a very far cry from that one two-hundred-and-seventy-fourth of the revenue.

PEARSON AT WORK

The funds available for the Kirstenbosch Botanic Garden were and have since remained inadequate; Directors have done great things there, and they have done them with small means. Pearson's determination and energy had to take the place of money. His body lies buried under a memorial cross in the Garden and this is appropriate, for he killed himself, dying of overwork in his early forties, in the making of it. Not that it was his intention to take all upon himself; he was too modest for that, he applied to Kew for help and Kew agreed to send him W. J. Bean, at that time Assistant Curator; but by then it was half-way through 1914, war intervened, Bean was unable to leave, and Pearson was obliged to depend on himself. In the matter of garden design he was a product of the English 'romantic' school and it was natural to him to make Kirstenbosch a wild woodland garden about a more or less formal nucleus, rather in the manner of Harold Peto

Top left Protea cynaroides, the 'King Protea'.
Bottom left Disa uniflora, the unique Table Mountain orchid.
Right, top to bottom Protea neriflora; Serruria florida, the 'Blushing Bride'; *Protea barbigera*; and *Protea mellifera*.

but without that artist's passion for Italianate garden buildings. An initial attempt to reconcile botanical system with ornamental effectiveness was given up as hopeless; as Compton (*op. cit.*) very properly points out:

an attempt to group plants on a systematic basis can only succeed if their varying ecological requirements are met, and in any case can hardly ever produce an aesthetically satisfactory result.

And to continue quoting Compton:

The idea of 'landscaping' Kirstenbosch was always rendered futile by the grandeur and diversity of its setting, making any sort of 'improvement' seem foolish. No botanic garden in the world has a more magnificent site, with its hills, slopes, streams and forests, and its superbly bold mountain background and distant views. The landscape was already there and the main thing was to ensure that it should not be spoilt by the uses to which it would be put. The Central theme of the National Botanic Gardens has always been the indigenous flora of South Africa, its study, cultivation and display, and in developing Kirstenbosch this ideal has always been borne in mind and the natural features of the Gardens have been used and adapted to these ends.

KIRSTENBOSCH AS ONE SEES IT

Kirstenbosch is one of the most fascinating of botanic gardens and it would be so even if the site, the topography, were less advantageous, because of the very remarkable and particular quality of the Cape flora. Consider for a moment only these: the spring bulb flora including over a score of species of *Gladiolus, Watsonia, Ornithogalum, Ixia, Dierama, Schizostylis* and other genera of the *Iridaceae*; the six hundred species of *Erica*, with flowers so richly coloured and which make the Cape the world centre for that genus; the extraordinary genus *Protea*, so very much more spectacular and diverse than those of any other part of Africa. Even such spectacularly beautiful species as *Erythrina abyssinica* and *E. kaffra*, the Rubiaceous *Alberta magna*, and *Burchellia bubelina* lose their interest where there are *Proteas* in flower to look at.

But to take the setting first. Picture a vast amphitheatre formed by very steep, in places almost sheer, mountain slope, rocky, wooded, dotted with glaucous-leaved *Proteas*, rising to the flat top of Table Mountain on the right, to a series of sharp peaks on the left, high enough to be impressive yet not so enormously high as to dwarf and diminish all else in their neighbourhood; now set on the lower slopes, high enough to dominate the foreground yet properly in the middle distance and following the curve of the amphitheatre, a deep screen of forest trees composed chiefly of natives such as *Podocarpus*, partly of *Eucalyptus*, chiefly the gloriously red-flowered *ficifolia*, and a few intruding conifers and even English oaks. In front of the trees, a little lower on the foot-slopes, picture a second and lesser screen, lengthening and emphasising the perspective, of silky-leaved 'Silver Trees', *Leucadendron argenteum*. Such is the scene, the backcloth, for Kirstenbosch. Between it and the foreground lies a vast, steep, rocky area of acid, gravelly soil in which thousands of Cape Heaths, hundreds of *Proteas*, and hundreds of thousands of the Cape bulb flowers flourish as in the wild. Embraced within this great wild, woodland garden are, at progressively lower levels, such features as the Cycad Amphitheatre and the Dell; and, by way of kernel, the comparatively formal rock garden and the flower gardens.

A flower calendar of Kirstenbosch—and I am sorry but cannot help it if it be meaningless to those who are ignorant of plant names—would be composed of the following; and all these are Cape natives or at least South African natives:

August, September and October

CAPE ANNUALS

Arctotis, Felicia, Dimorphoteca, Gazania, Ursinia, Venidium Charies.

CAPE BULBS

Babiana, Watsonia, Tritonia, Ixia, Lachenalia, Homeria, Moraea, Ornithogalum, Zantedeschia, Gladiolus, Bulbinella.

ALSO

Heliophila ('Blue Flax')
Greyia ('Bottle Brush')
Liparia ('Orange Nodding Head'),
Chironia baccifera ('Wild Gentian')
Podalyria ('Keurtjie')
Strelitzia reginae ('Crane Flower')
Sutherlandia ('Gansies')
Virgilia divaricata ('Keurboom')
Pelargonium ('Geranium')
Dorotheanus ('Bokbaai Vygie')
Phoenocoma prolifera ('Pink Everlasting')
Leucodendron argentum ('Silver Tree')
Lebeckia simsiana ('Dwarf Broom')

PROTEAS

Protea repens ('Sugar Bush')
Protea nana ('Mountain Rose')
Protea arborea ('Waboom')
Protea cynaroides ('Giant Protea')
Serruria florida ('Blushing Bride')
Leucospermums ('Pincushions')

November, December, January

Agapanthus ('Blue Lily')
Aristea ('Blousuurknol')
Gloriosa ('Superb Lily')
Begonia sutherlandia
Dicrama pendulum ('Fairy Bells')
Aspalathus and *Cyclopia* (Tea)
Carissa ('Amatungulu'), *Erica*: various (Heath)
Dais Continifolia ('Kannabast', 'Cape Daphne')
Gardenia thunbergia ('Katjiepiering')
Gerbera jamesonii ('Barberton Daisy')
Geranium incanum ('Berg Tree')
Leonotis leonurus ('Wild Dagga')
Podranea ricasoliana ('Port St John Creeper')
Barleria obtusa, Rochea coccinea ('Red Crassula')
Thunbergia alata ('Black-eyed Susan')

Wachendorfia thyrsiflora ('Bloodroot')
Disa uniflora ('Pride of Table Mountain')
February, March and April
Amaryllis belladonna ('March Lily')
Barosma, *Agathosma* and *Diosma* ('Buchu')
Bauhinia galpinii ('Pride of de Kaap')
Cissus juttae (fruit) ('Desert Grape')
Dissotis incana
Dovyalis caffra (fruit) ('Kei Apple')
Erica: various (Heath)
Eucomis undulata ('Pineapple flower')
Podranea ricasoliana ('Port St John's Creeper')
Salvia africana ('Wild Blue Sage')
Sutera grandiflora
Virgilia oroboides ('Keurboom')
May, June and July
Mesembryanthemum and allied plants (Mesems, Vygies, Sour figs)
Chironia baccifera (fruit) ('Wild Gentian')
Proteas: various, *Leucadendron*: various, *Aloes*: various.

The Dell is centred about Colonel Bird's bird-shaped pool known as Bird's Bath or, absurdly, 'Lady Barnard's Bath'. Overlooking this small pool, contained in fine brick-work, is a stone bridge in the deep shade of trees and of fine tree-ferns; the little waterfall and stream out of the pool are lined with ferns. This is a cool, damp, pleasant place, a refuge from South Africa's burning sun and a fine collection of ferns and mosses in considerable variety.

From the Dell you climb—this Garden is all climbs—into another feature unique to Kirstenbosch, the Cycad Amphitheatre. This is the heart of the Garden and the first part which Pearson laid out and planted. The Cycads are primitive seed-plants, living fossils, related to and perhaps lying between Ferns and Conifers. Occurring in the tropics and sub-tropics of both hemispheres, is their genus *Encephalartos*. These have short trunk-like stocks, magnificently architectural, pinnate, frond-like leaves, in some species hard and with the pinnae ending in spines; the leaf colour varies from deep green to a green so glaucous as to be nearly blue. The gigantic cones produced at the crown are very beautiful. At Kirstenbosch you can see many specimens of twenty-four out of the twenty-five known species. One big specimen of *Ecephalartos woodii* is the parent of the only six individuals of the species known to exist—it was found in the wild and no other individual has ever been found; it is a male plant, (its 'offspring' were shoots from the root). The exquisitely regular cones of *E. transversos* weigh as much as 75 lb apiece. The Garden also has the largest plant of the very slow-growing *E. latifolius* known to botany. Very spectacular are the great cones of *E. longifolius* which, when broken open by birds, have orange pulp and red seeds, the latter poisonous, the former wholesome, and often called 'Kaffir Bread'. The paths and stairways up through the Cycad Amphitheatre are of fine masonry, one of the pleasing details of Kirstenbosch.

The grown-out remains of the hedge planted by Governor Van Riebeck to keep the natives out of the settlement. Preserved as a national monument in the Botanic Garden.

The Garden has another unique feature, the only hedge in the world which has been declared a National Monument. In 1660 Van Riebeck decided that a good stout hedge would help to keep marauding Hottentots out of the territory of the Colony. For this purpose he chose the native *Braebium stellatifolium*, and planted his hedge the entire length of his frontier. That part of it which crosses the Botanic Garden is about 20 feet high and densely branched, and is in excellent shape after three centuries. It follows a stream bed and it helps to protect one of the lawns from winds. Another feature of Kirstenbosch which I have seen in no other garden is the school: a small but fully equipped school in a pretty cottage-like building, it is at present manned by two permanent teachers and will shortly have more, possibly as many as six. Every day it receives classes seconded from schools all over the city and its environs. The children spend half a day in school learning some botany and particularly about the conservation of their native flora; and the other half day being taught out of doors in the Garden itself.

Of the flowering shrub gardens the two most remarkable are the *Protea* garden and the *Heath* garden. The most spectacular of the *Proteas* is *P. cynaroides*, the 'Giant Protea' or 'King Protea', with cerise and silver flowers sometimes a foot in diameter. These can be seen in flower in October and again in February. The very rare *P. aristata* is, to my mind, even more beautiful, with flowers a kind of silvery raspberry-red. The species is probably extinct in the wild but it is well established as a cultivar. Other very striking species of this peculiarly South African genus are *P. repens*, a ground-covering, free-flowering shrub; *P. compacta*, and *P. nerifolia*. The proteaceous *Serruria florida*, known as 'Blushing Bride' has relatively small, blush-pink flowers.

The Heaths at Kirstenbosch seem legion; I do not know how many of the 600 species are to be found in this Garden. Flower colour ranges from white through all the yellows to orange; and through rose-pink to deep crimson and some other reds. Among the best I starred *Erica bauera, E. coccinea* and *E. regia*.

DISA UNIFLORA

Special mention must be made of the most beautiful of the Cape Orchids: the province has a considerable number which have been the object of an excellent monograph illustrated in colour and with fine line drawings, by Professor Schelpe of Cape Town University. *Disa uniflora* is by far the most spectacular and is, in fact, one of the most beautiful orchids in the world. It grows in a limited number of sites, mostly somewhere on Table Mountain but also elsewhere in the Peninsula. It is a hardy ground orchid but not easy to cultivate because it needs rather special conditions—a very acid soil, peaty and grown over with sphagnum or some other moss, over granite, absolutely perfect drainage, and a cool atmosphere constantly saturated with water. This orchid, with from three to six or more flowers per spike varying in colour from pale pink to either orange-red or crimson-red or even scarlet, grows at altitudes where its site is almost continually shrouded in cloud; or in the spray from waterfalls. Most growers have failed with it, but there have been notable exceptions: for many years it has been well grown in quantity at the Botanic Gardens, Göteborg, Sweden; it was once superbly grown at Chatsworth, where up to twelve flowers per spike were obtained. It is now grown well at Kirstenbosch and still better by a member of the staff at the Stellenbosch Botanic Garden (see below). Although, like any other plant, this *Disa* flowers seasonally in the wild, in the Kirstenbosch greenhouses it can be seen in flower at any time of the year.

Stellenbosch

The University of Stellenbosch, one of the prettiest small towns in the world and an almost unspoilt example of seventeenth- to eighteenth-century Cape Dutch architecture, has an interesting little botanic garden. The University botanical department is an important one, so that the Garden receives proper attention; although primarily a teaching garden, it is prettily laid out and full of pleasure for the eye.

The Succulent Houses are the most interesting feature of this Garden from the botanical point of view, and in one of them are grown the *Welwitschias*, for which this botanic garden is world-famous among botanists. Stellenbosch Botanic Garden was the first to succeed in cultivating these very strange and rather grotesque plants from Southwest Africa's arid regions. From a burly, corky trunk a few inches tall grow two leaves, never more, in the same axis. These get longer and longer and wider and wider year after year. The plants flower after twenty years, male and female flowers on different plants, very distinct and peculiar. The older plants at Stellenbosch are in their forties. The oldest yet dated, in the wild, is an enormous, lumpish object carbon-dated to plus or minus 850 years.

Another extraordinary family of plants cultivated at both Stellenbosch and Kirstenbosch are the *Asclepidaceae*: in the genera *Stapelia* and *Ceropegia* the long, scandent stems are often entirely leafless, though some species sometimes have leaves. The flowers are indescribably complex in shape, in some like five-pointed hats of some Tibetan abbot as imagined by a romantic artist, in others like rococo parachutes. Then there are the *Dioscoreas* at Stellenbosch, great lumps of corky stuff out of which grow slender green stems bearing small leaves and weird flowers. *Fockea* is another grotesque genus cultivated in this botanic garden; the Curator, incidentally, says that Vienna's claim to have, in the Schönbrunn Botanic Garden, specimens of a *Fockea* species several centuries old and which are the only ones of the kind in the world, cannot be allowed.

BUENOS AIRES

Ficus species, cultivated by Carlos Thays, became important street trees in Argentina.

The Botanical Garden of Buenos Aires can no longer be called 'great'; but in the past it did important work in the botany of South American plants; its taxonomists have added to our sum of knowledge of plants; its publications have been valuable; in the hands of its newest Director it may well enjoy a new period of prosperity. For at least thirty years it has been badly neglected; Argentina, like most countries in the modern world but much more than most, has suffered grievously at the hands of politicians. But things may be slowly on the change; even in Argentina control is slowly passing from the politician to the technologist; and the Buenos Aires garden, or rather gardens, for they are all over the city, may have a future as useful as their past.

First it should be explained that the 8-hectare Carlos Thays Garden (named for its founder) at the far end of the Avenida Santa Fe which connects it directly with the Plaza San Martin in the heart of Buenos Aires, is not and never has been the whole Botanical Garden of this city. Owing to the nature of the founder and first director's appointment by the Municipality of Buenos Aires in February 1892, and to the wording of the instructions which he received from, but had himself suggested to, the Intendente Municipal, he was required to plant not only the Jardin Botanico itself with flowers and trees representative of the flora of Argentina, but to plant all the parks and open spaces of Buenos Aires in the same fashion—parks and open spaces of which Carlos Thays was also made Director General. Thus we may say that the Jardin Botanico of Buenos Aires is planted all over the city, and the actual 'Carlos Thays' garden is merely its concentrated centre.

But as such it is the most important part of the whole; its site was chosen by Thays himself; he could not be expected to foresee the consequences of air pollution by motor-car exhaust fumes. It is very pleasant to have one's Botanical Garden in the centre of the city, as they do here, in Copenhagen, and in some other great centres. But it is apt to kill plants. At all events, the land which Thays (he was, by the way, a Frenchman) wanted, belonged to the State; however, at the intercession of the Intendente (Lord Mayor) of Buenos Aires, it was ceded to the Municipality.

Joined with Carlos Thays in the original planning and planting of the Jardin Botanico were a number of distinguished citizens, and of these the most useful was Eugenio Antran, professor of the *Facultad de Ciencias Naturales,* who undertook the classification of the specimens. The principal purpose of the garden was scientific; but from the beginning it was also conceived as an ornamental garden, to demonstrate the art of horticulture as well as the science of botany. In this context it is much to the point that Carlos Thays was Argentina's greatest designer of parks and gardens, being responsible for the creating of both private and municipal gardens all over the country; he was, in fact, Argentina's 'Capability' Brown. Thus the garden contains a little *Jardin Romano* made with Mediterranean plant material, not a very good example of an Italian garden, though it is centred on a statue of the she-wolf feeding Romulus and Remus. In the same spirit there is a pretty little 'French' garden centred on a formal pool with a nymph amongst the nymphaeas.

But the greater part of the Jardin Botanico is planted in the English manner of the late nineteenth century, that is, as a park of trees, shrubs and lawns, with some borders for herbaceous plants. Only, of course, the groupings are according to a system, geobotanical, so that each section, separated by broad, shaded walks from the next, represents the flora of a particular region.

In his original planting in the central garden Carlos Thays managed to include 721 Argentinian species; more if one takes in his plantings in the city parks. In a monograph published on the species actually in the garden in 1959, *Ingenieros Agronomos* E. L. Ratera and R. G. Montani list only 316 Argentinian species; that is a sad falling-off but by no means all of it has been due to neglect: Thays seems to have hoped that species from the tropical north of the country, from the arid *pampa secca*, and from the cold, wet south, would all survive, even if they did not exactly flourish, in the alkaline soil, cool winters and very hot humid summers of the capital. He was hoping for the impossible: Argentina has three million square kilometres of territory between 21° 30′ and 54° S—that is, it extends over 33 degrees of latitude. The extreme south has approximately the climate of Northern Scotland, the extreme North that of tropical Brazil. Moreover, in altitude range it varies from sea level to Aconcagua's 22,200 feet. The climate of Buenos Aires is, no doubt, 'temperate', but the extreme northern plants do not like its midwinter frosts, slight as a rule but occasionally as much as 10°F; on the other hand the southern plants do not survive the intense midsummer heats, nor those of the *pampa secca* the very high humidity. So that many of Thays' original plantings have failed for climatic reasons.

Since the Carlos Thays Garden is a teaching garden a certain area of its 15 acres is given over to demonstrating the systematic classification of plants. There are collections of *Gramineae* forming part of a planting of *Monocotyledons* including bananas which survive the Buenos Aires winters, and *Strelitzias*. Several other major groups are similarly represented. But here again the Buenos Aires gardeners are up against difficult conditions: their garden, like Kew, but with not a tenth of the extent, is a favourite recreation park for the citizens and their children. Children are destructive pests in a garden and mischievous small boys often indulge in the game of label-swapping. It is surprising that the Carlos Thays garden is as well ordered as it is.

TREES

Argentina is rich in fine trees: from the warm north and the mountains of the central tropical zone come the glorious *Araucarias*; there are pure stands of *Araucaria imbricata* which are very beautiful, and it is only a pity that the value of the timber has led to far too much felling. From the north again come the gigantic *Enterolobiums*—'Timbos' in the vernacular—and the violet-flowered *Jacaran-*

das, whose beauty and excellence as street trees have led to their introduction to so many tropical countries; and the gorgeous *Erythrina crista-galli*, which Carlos Thays planted on such a scale in Buenos Aires that in their season the parks and gardens are on fire with their orange-scarlet flowers. From the central hinterland come the stately 'Lapochos'—*Tabebuia avellana* or *T. ipe*—tall, graceful trees which, in spring, are completely covered with their rose-mauve trumpet flowers. There are gigantic 'Ivirapita' trees, *Pettophorum vogelianum*, and the 'Cedro', *Cedrela odorata*. As well as scores of broad-leaved and coniferous native trees Argentina has its native palms—*Cocos australis* and *Cocos yatay* among others.

Many of the native arboreal species are represented by some well-grown specimens in the Botanical Garden; and if by no means all of them are to be seen there it is because there is not room for them, nor does the climate of the city suit all of them. But other species have been planted in the satellite gardens, the city's parks and open spaces, and Buenos Aires has some of the most magnificent old *Ficus elastica* trees in the world.

As well as *Erythrina crista-galli*, the garden has specimens of the other 'Ceibos', *E. splendens* and *E. speciosa*; *splendens* has even more vividly coloured racemes of giant pea-flowers than the other two. Native shrubs are represented by such fragrant beauties as the *Brunfelsias* with their white and blue flowers, and the magnificent native privet, whose heavily scented flower-spikes, like those of a white lilac, make the great flats of the La Plata delta look as if snow had fallen in mid-spring. The vernacular name for *Brunfelsia* is 'Yasmin del Paraguay', but the plant is a native of north-eastern Argentina and yet hardy, reaching about 10 feet in Buenos Aires. Other fine flowering shrubs are *Caesalpina gilliesii*, the lovely *Calliandras* with their finely divided leaves and powder-puff flowers of crimson or rose; *Gleditchias, Indigoferas, Mimosas* (Argentina has eight species), *Sophoras*, and many more. Of them all, perhaps the most pleasing to the gardener if not to the botanist are the 'Ceibos', the Erythrinas so beloved of the *Bonaerense*.

Argentina is as rich in herbaceous as in woody species and many are represented in the 'Carlos Thays' garden, although once again not as many as one would wish. There are the native *Calceolarias*, scarlet *Mimulus, Nicoteanas* in variety, *Physalis mirabilis, Schizanthus*, various lovely *Tropaeolums, Eryngiums*—these are but a few of the names which will suggest to gardeners how much they owe to Argentine botanical gardeners. House-plant enthusiasts owe almost as great a debt, for the country is rich in terrestrial epiphytic and even parasitic *Bromeliads* which have been cultivated in the Botanical Garden; they include the spectacular *Puyas* and *Pitcairnias*, and the troublesome but decorative *Tillandsias*, which bear such poetical or evocative vernacular names as *Flor del aire* or *Barbe de viejo*.

I have already mentioned some of Argentina's best

Carlos Thays was much influenced by French and Italian styles in designing Buenos Aires' Botanic Garden.

leguminous species, notably the *Erythrinas* and the *Calliandras*. Many more of this family are named by Ratera and Montani (*op. cit.*) as still surviving in 1959 not all of which I was able to identify as still there in 1967: three *Acacias,* three *Mimosas*, four *Piptadanias*, *Bauhinia candicans*; the *Caesalpinas* and *Cassias*, the spectacular *Sesbania punicea,* the massive and stately *Tipuana tipu*, *Gleditsias*, *Dalbergias* and *Lonchocarpus* in three species, these are only a few of them. But from 1898 to 1959 there was, in this as in other families a serious falling-off, for Carlos Thays originally planted no less than eighty-nine species of leguminous plants native to Argentina. By 1959 forty-nine of these had disappeared although happily some of the most interesting and curious, for example, *Parkinsonia aculeata*, were among the survivors.

ECONOMIC BOTANY

The one great work in the field of economic botany carried out in the Carlos Thays garden, comparable, although on a lesser scale, with Kew's work on tea and rubber, or that of the Rio de Janeiro Jardim Botanico on tea and coffee, is sufficiently interesting to be worth a brief account.

The Spanish and Portuguese colonists of South America had taken over from the natives the practice, soon to become a habit, of drinking a refreshing and stimulating infusion of a holly, *Ilex paraguayensis*, or, in the vernacular, *Yerba-maté*. The Jesuits of Paraguay popularized this excellent herb, so that it became widely known as 'Jesuits' Tea'. By 1890 or thereabouts Argentinos were drinking so much of this *yerba-maté* (which smells of China tea), it had become to such an extent the drink of the people that, since the plant is not native to Argentina, they had to import over 40,000 tons of it a year from Brazil and Paraguay at an annual cost of £1m. (gold). Carlos Thays was of the opinion that the country's requirements in *maté* could perfectly well be grown at home; in 1895 he obtained seeds of *Ilex paraguayensis* from an expert in its cultivation, an Argentino working in Paraguay, Dr Honorio Leguizamon. The seeds proved very difficult to germinate. The Jesuits had come up against this difficulty a hundred years earlier, had overcome it, and taken their secret with them when they were driven from the country. Acutally this so-called secret was not a profound one; it seems likely that Jesuit botanists had observed that the *testa* of *yerba-maté* seed is so hard that it must be partially destroyed by some agency before germination can take place; when berries were eaten by birds the seed was softened and, being passed out in the droppings, germinated. Probably the Jesuits fed the berries ot their chickens and planted the dung. Thays did not at once adopt the same solution; he tried various ways of softening the *testa* and got good results after soaking the seeds in hot water. He obtained three seedlings from the first batch of seed, but was getting 60 per cent germination from later batches. He published his results and when he sent out seeds for trial in possible commercial plantations, sent his special instructions with them. In 1898 he started raising large numbers of plants in the Botanical Garden and they proved hardy in the Buenos Aires winters. By 1906 some of the original seedlings were 15 feet tall and producing a vast number of seeds for propagation. Seedlings were distributed to suitable growers, and Argentina's *yerba-maté* industry began.

At present Argentina has, in a typical year, about 147,000 hectares (368,000 acres) of *yerba-maté* plantation in production; Argentina's exports are worth about £2.75m. a year for the 3,000,000 kg of prepared leaf exported not only to Uruguay, Chile and Bolivia, but to the Lebanon (500 metric tons) and Syria (250 metric tons). This trade, then, is a debt which Argentina owes to Carlos Thays and his Botanical Garden.

Today the Garden's socially important work consists chiefly in running a very good training school for the professional gardeners which the country badly needs for both market gardening and ornamental horticulture. The children adopting this trade enter from school at fourteen years of age and graduate from it at eighteen; the four years' course includes both theory and practice.

In the Carlos Thays Garden and its satellites in the city's parks and open spaces Argentina possesses the past and traditions of a great botanical garden. With these by way of foundation, gardens representative of a much greater part of the country's wonderful flora could be made. Any plant-lover who has, as I have, seen the flora of the *pampa secca,* so rich in arid soil shrubs and sub-shrubs and notably in succulents; seen that of Patagonia, too, which shares, with Chile, a rich and lovely range of trees, shrubs and herbaceous plants; and seen the sub-tropical and tropical flora of the north—will agree that in the present state of botany as a science the sensible policy for the Buenos Aires botanical gardeners would be to give up trying to display exotic species and to concentrate on endemics. The flaming 'Ceibos' are there, but where are the *Gevuinas*, the *Embothriums*, the *Notofagi* and the rest? In the changes which Argentina is trying to make in her institutions and way of life the Botanical Garden and the work of Carlos Thays should not be overlooked.

JAPAN

JAPAN

Left Tree-training demonstrated in the Kyoto Botanic Garden.

Top right Botanic gardening on European lines is a very late development in Japan, and Japanese botanic gardens are not in the traditional styles of Japanese garden art. But *below right* they are rich in examples of traditional Japanese garden craftsmanship, such as this one of Chrysanthemum training.

The art of gardens has been carried in Japan to heights which have never been achieved elsewhere. From the seventh century to the middle of the nineteenth century garden designing was a fine art on a par with, and in a sense, a part of landscape painting; it was practised by painters, poets and by several of the Shoguns; as well as being a fine art, it was a religious act, and rich or powerful men could acquire merit by garden-making, as by church-building in Europe. The art reached its zenith of refinement under the influence of Sung (Chinese) monochrome landscape painting, and in the hands of Zen Buddhist monks practising as landscape-gardening artists who created 'abstract' gardens of rocks and raked sand which are known as Dry Landscape gardens.

But for all that, and also despite the brilliance of the Japanese in floriculture (which has a much longer history as a scientific craft in Japan than in Europe), pure botanical gardening has not long been practised in Japan, for although there was an ancient Japanese descriptive botany, based on a still more ancient Chinese science, botany as we understand it was a Western science adopted, with other of our sciences, by the Far Eastern civilizations. It is true that there were Physic Gardens in Japan in the fourteenth century and perhaps earlier, and that from the seventeenth century the Tokugawa Shogunate was establishing more such gardens, the first of them being the Oyakuen at Yedo (Tokyo). But all these gardens were for the cultivation of medicinal herbs which had long been imported from China and which the Government preferred to have growing at home; they were not botanical gardens in the proper sense of the term.*

The movement for the making of real botanic gardens did not, in fact, begin in Japan until the end of World War II, but it then started with such impetus that in only twenty years some sixty such gardens have been planted, the northernmost being that of Hokkaido University at Sapporo, the most southerly the so-called Nakamura Nursery at Miyazaki. However, a few of these gardens are of older foundation; that of Hokkaido University was, for example, started in 1886; and the oldest true botanic garden in Japan, the Koishikawa, dates from 1684.

The Koishikawa is now the Botanic Garden of Tokyo University's Faculty of Science, and has been so since 1877. The garden is divided into four principal areas—Arboretum, Systematic Garden, Fern Garden and Physic Garden. About 6,000 species are in cultivation in an area of 40 acres. The Arboretum has 600 of these—260 genera, 88 families—from East Asia, and some hundreds of exotics: this part of the garden includes a classical Japanese landscape garden, good collections of *Camellia* and *Prunus* (chiefly Cherry), and an Alpine garden. The Fernery is a very good one, and so is the collection of evergreen oaks and other trees, both broad-leaved and coniferous. The Garden also has excellent collections of *Primula, Azalea* and Bonsai trees, and some pleasing avenues—one of *Chaenomeles,* which is unusual, others of Maples which are glorious in autumn, and of limes.

Associated with this garden as a satellite is the Nikko Botanic garden in the Tochigi Prefecture, established in 1902 and specializing in Alpine plants. The area is 26 acres. Although both the big rock garden, and the bog garden, are planted chiefly with native Japanese alpines, there are some alpine exotics; the sphagnum bog garden is probably the best of its kind in the world. But although alpines are the *raison d'etre* of this Nikko garden, they are not all that is to be seen there; the garden has a very complete collection of native *Prunus*, and above all of Japanese *Rhododendron* (chiefly *Azalea* series) which make it very beautiful in spring and early summer; finally, there is a remarkable collection of Japanese Maples (*Acer*).

Kyoto was, and in some respects still is, the greatest centre of garden art in Japan, with scores of historically and aesthetically famous gardens in its neighbourhood. At a moment in Japan's history when the aristocracy was

Botanical Gardens in Japan, Fumio Maekawa (1964).

Chrysanthemum breeding is an important part of the work done in some Japanese botanic gardens and one such garden has 800 different cultivars. (See text.) Some of the flower forms first developed in Japanese gardens.

Below The pretty domed conservatory in the Kyoto Botanic Garden.

in decline and the warrior caste rising to power, Kyoto was a centre of conservatism in all the arts, where romantic landscape gardens were still being made while elsewhere, taste in garden art was becomingly increasingly austere and simple. But Kyoto's botanic garden was only started in 1917, beautifully sited between a view of Mount Hiei to the east, the River Kamo on the west, the Kitayama Hills to the north, and the old city to the south. It was replanted and its buildings reconstructed in 1955, and now covers about 57 acres. This garden has a pleasing topography and an attempt has been made to do the planting in ecologically natural association. The garden is remarkable for its domed greenhouse, unlike anything we have in the West.

The old Sapporo garden of Hokkaido University, 32 acres in the heart of the city, is particularly interesting for its conservation of an area of primeval lowland forest which has not been modified by human interference. Around this some trees from other parts of Japan have been planted into the natural landscape in order to enhance its beauty, notably the Japanese elm, *Ulmus japonica*; *Acer mono*; *A. palmatum* var. *matsumurae*; *Quercus mongolica* cult. 'Mizunara' and some others. There is, as in most Japanese botanic gardens, a Physic Garden. There are also special areas, devoted to Roses and to flowering shrubs, and an alpine garden which is very large. A permanent experimental garden is devoted to the genus *Trillium* and another to *Aconitum*. The greenhouses have adequately representative collections of tropical plants.

An interesting alpine and bog garden is maintained at a high altitude (3,000 feet) on Mount Hakkoda by the Botanical Faculty of Tohoku University. A considerable part of it is unmodified natural vegetation—alpine plants at the higher levels, sphagnum bog plants below. Some of the common species are *Rhododendron fauriei, Pinus pumila, Sorbus commixta, Acer tsonoskii* and *Ilex sugeroki*, and there are native *Gaultherias* and *Vaccineums*; *Drosera, Oxycoccus* and *Eriophorum* are to be found in the sphagnum. As well as the natural rock and bog gardens, there are man-made ones into which a few exotics, but all Japanese (for example from Mount Ozeghara) have been planted. Tohoku University also maintains a much larger botany 'garden'—125 acres—in the old Aobayama Castle grounds on the outskirts of the city of Sendai and here again a considerable area is unmodified primeval vegetation carefully preserved, much of it *Abies firma* forest. This is, in fact, much more of a nature reserve than a garden; and so, although man-made and planted solely with local endemics, is the 16-acre Sendai Municipal Wild

If, in the modern Japanese botanic (Imperial Gardens, Tokyo) one is reminded of English rather than Japanese landscape gardening *below*, still, distinctively Japanese elements are preserved *right* even in gardens whose primary purpose may be scientific.

Plants garden which overlooks the city and which is remarkable for its *Lespedezas*—a favourite shrub with Japanese gardeners for over a thousand years. The Japanese are fond of such small reserves for wild plants, natural or man-made, and small arboreta of native trees like the Aritaki Arboretum with its 200 genera of woody plants and which developed out of the Shokaken private garden.

Apart from botanic gardens for purely scientific or commercial purposes, Japan has a number of teaching gardens. A fair example is the Yatsu Yuen garden on Tokyo Bay, where 500 species each representative of a family of plants, and Japan's finest rose garden (700 cultivars, 3,000 plants) are maintained for the instruction of high school children. A much larger teaching 'garden' is the 50-acre National Park for Nature Study maintained by the Natural Science Museum at Minato-Ku, Tokyo. It is chiefly, again, a nature reserve of native species, part of it being a vestige of the ancient virgin forest of Mushashino. Its water garden has plants collected from a wider area.

The Shinjuku Gyoen National Garden, although classed as 'botanical' by the Japan Association of Botanical Gardens, is very much more horticultural than botanical, and is a good example of how, following the Meiji Restoration, Japan temporarily gave up her lovely native styles in garden art to adopt those of the West. This garden was designed by Professor Henri Martin of the Ecole Nationale d'Horticulture de Versailles. It has Le Nôtrian vistas composed of avenues of plane trees. Into it were introduced, for the first time into Japan, such Western species as cedars, tulip-trees and swamp cypress, as well as planes. The garden has been a centre of floricultural technology since 1893. It has a vast greenhouse with the largest collection of tropical plants in the Far East. And it is famous for its chrysanthemums, and its cherries; it grows about 800 different chrysanthemum cultivars.

Japan still maintains a number of special botanic gardens which are wholly or primarily Physic Gardens. A good example of these is the Tsummura Botanical Garden of Medicinal Plants. Although one of its purposes is to preserve species threatened with extinction in the Mushashino virgin forest, its principal work consists in cultivating both foreign and the old Japanese and Chinese medicinal herbs, studying their chemistry, and also cultivating plants which, not in the old pharmaceutical books, seem likely to be valuable as medicinal herbs.

Some botanic gardens have been planted exclusively with tropical plants. Their sites, of course, have to be chosen carefully; a case in point is the Enoshima Tropical Plants Garden, on the island of Enoshima. This island (lat. 35° 17′ N; long. 139° 29′ E) is in the middle of the warm Kuroshio current which maintains a warm and humid microclimate throughout the year. Enoshima has become the northern limit of outdoor cultivation of several species of *Ficus,* several *Strelitzias* and *Erithrinas,* numerous *Cacti* and some other tropical species, including *Butia capitata.* A very different tropical garden is maintained in the Hakone mountains under glass by making use of the local hot springs. The plants grown are principally economic ones—esculents, fibre plants, tropical, medicinals. There are also some insectivorous plants.

Bamboo is of great economic importance in Japan as it is all over the Far East, and it has, in the Fuji Bamboo Garden, its own botanic garden. Here more than 600 species are represented, but the garden's most important work is the conservation of rare or new cultivars. There are bamboos with black or red, as well as yellow canes; bamboos with variegated foliage, swollen internodes, wrinkled culms and many other singularities; and bamboos varying in height from 5 inches at maturity to as much as 25 feet (only the tropical bamboos exceed this).

The Izu peninsula in the Shizuoka Prefecture has a subtropical climate and its Useful Plants Garden takes advantage of it and its freedom from frost in winter to grow subtropical and even some tropical fruits, fibre and food plants, and an increasing range of ornamentals. Its scientific department, however, is concerned chiefly with

Japan

The stone lanterns which Westerners associate with Japanese gardens and which turn up even in botanic gardens, were originally added in the sixteenth century to Japanese garden art by Oribe, a disciple of the Tea Master Rikuyu. Oribe was a Christian convert and the lantern originally represented the Cross.

genetical work on chrysanthemums, amaryllis, carnation and gladiolus, as well as on tropical plants. There, scientific work and commercial research in floriculture are on the whole more important than either teaching or recreation, but the reverse is true of another class of Japanese botanical gardens, the Higashlyama in Nagoya being fairly typical, for although some educational work is done there, this and similar gardens are principally recreational. It is about 80 acres in extent and centred upon a large, conventional conservation of tropical plants arranged 'after nature', one section of this being given over to an instructively representative collection of xerophytic plants. In the open garden, planted after the English landscape style, there are quite good collections of native *Rhododendron, Prunus* and *Camellia*, and there is a native Pinetum. There is also a garden planted to American ornamental cultivars, and a very good rose garden.

Japanese botanists and gardeners are keenly interested in alpine florae, especially their own, and there are many good rock gardens. One of the best of them is the Rokko Alpine Garden between Kobe and Osaka, 2,830 feet above sea-level. Pleasingly laid out as a natural alpine-type landscape about a central pool, it is planted with about 800 species of Japanese alpine plants, and about 500 alien species, chiefly from other parts of the Far East. Particularly good is the Himalayan section, which contains the 200 mountain species collected during the course of a plant-hunting expedition sponsored by Kyoto University. The garden includes an area of bog and marsh where the *Primula japonica* are outstanding.

Although botanical gardening on European lines is relatively new to Japan, the country has an active and very intelligent approach to it. As for scientific work, Japanese biologists seem, like the Russians we met, to believe that mankind is still far from having discovered and domesticated all the useful plants in the world.

INDEX

INDEX

Page numbers in bold type indicate a colour plate; those in italic refer to a black and white illustration.

D

E

Index

M

N

O

P

Q

R

S

THE WORLD'S BOTANICAL GARDENS

This list has been compiled from the *International Directory of Botanical Gardens* (Utrecht, 1963).

CANADA

1. Edmonton: Botanic Garden and Field Laboratory, Department of Botany, University of Alberta, Edmonton.
2. Vancouver: Botanic Garden, University of British Columbia, Vancouver 8, British Columbia.
3. Indian Head: Forest Nursery Station, Indian Head, Saskatchewan.
4. Morden: Experimental Farm, Research Branch, Canada Department of Agriculture, Morden, Manitoba.
5. Macdonald College: Morgan Arboretum, Box 240, Macdonald College, Quebec.
6. Montreal: Montreal Botanic Gardens, 4101 Sherbrooke Street E., Montreal, Quebec.
7. Ottawa: Dominion Arboretum and Botanic Garden, Plant Research Institute, Central Exp. Station, Ottawa.
8. Hamilton: Royal Botanic Gardens, Hamilton, Ontario.

USA

9. Seattle: Drug Plant Gardens, College of Pharmacy, University of Washington, Seattle 5, Washington.
10. Seattle: University of Washington Arboretum, Seattle 5, Washington.
11. Carson: Wind River Arboretum, Carson, Washington.
12. Portland: Hoyt Arboretum, 4000 S.W. Fairview Boulevard, Portland 1, Oregon.
13. Moscow: Charles Houston Shattuck Arboretum, College of Forestry, University of Idaho, Moscow, Idaho.
14. Placerville: Eddy Arboretum, Institute of Forest Genetics, PO Box 552, Placerville, California.
15. Sacramento: C. M. Goethe Arboretum, Sacramento State College, Sacramento, California.
16. Arcadia: Los Angeles State and County Arboretum, Arcadia, California.
17. Davis: University of California Arboretum, University of California, Davis, California.
18. San Francisco: The Strybing Arboretum and Botanical Garden, Golden Gate Park, San Francisco 17, California.
19. Provo: Brigham Young University Botanical Garden, Provo, Utah.
20. Saratoga: Saratoga Horticultural Foundation, PO Box 308, Saratoga, California.
21. La Canada: Descanso Gardens, Department of Arboreta and Botanic Gardens, 1418 Descanso Drive, La Canada, California.
22. San Marino: Huntingdon Botanical Gardens, 1151 Oxford Road, San Marino, California.
23. Santa Barbara: Santa Barbara Botanic Garden, Santa Barbara, California.
24. Berkeley: Regional Parks Botanic Garden Tilden Regional Park, Berkeley 8, California.
25. Berkeley: University of California Botanical Garden, Berkeley, California.
26. Los Angeles: Botanical Garden, University of California, Los Angeles 24, California.
27. Claremont: Rancho Santa Ana Botanic Garden, 1500 North College Avenue, Claremont, California.
28. Tempe: Desert Botanical Garden of Arizona, Phoenix, Arizona (PO Box 547, Tempe, Arizona).
29. Honolulu: Foster Botanical Garden, 50 N. Vineyard Street, Honolulu 17, Hawaii.
30. Honolulu: Kapiolani Hibiscus Garden, 3620 Leahi Avenue, Honolulu, Hawaii.
31. Honolulu: Lyon Botanical Garden, C/- Foster Botanical Garden, 50 N. Vineyard Street, Honolulu 17, Hawaii.
32. Wahiawa: Wahiawa Botanical Garden, 1396 California Avenue, Wahiawa, Oahu, Hawaii.
33. Nebraska City: Arbor Lodge State Historical Park, Nebraska City, Nebraska.
34. Denver: Denver Botanic Gardens, 909 York Street, Denver 6, Colorado.

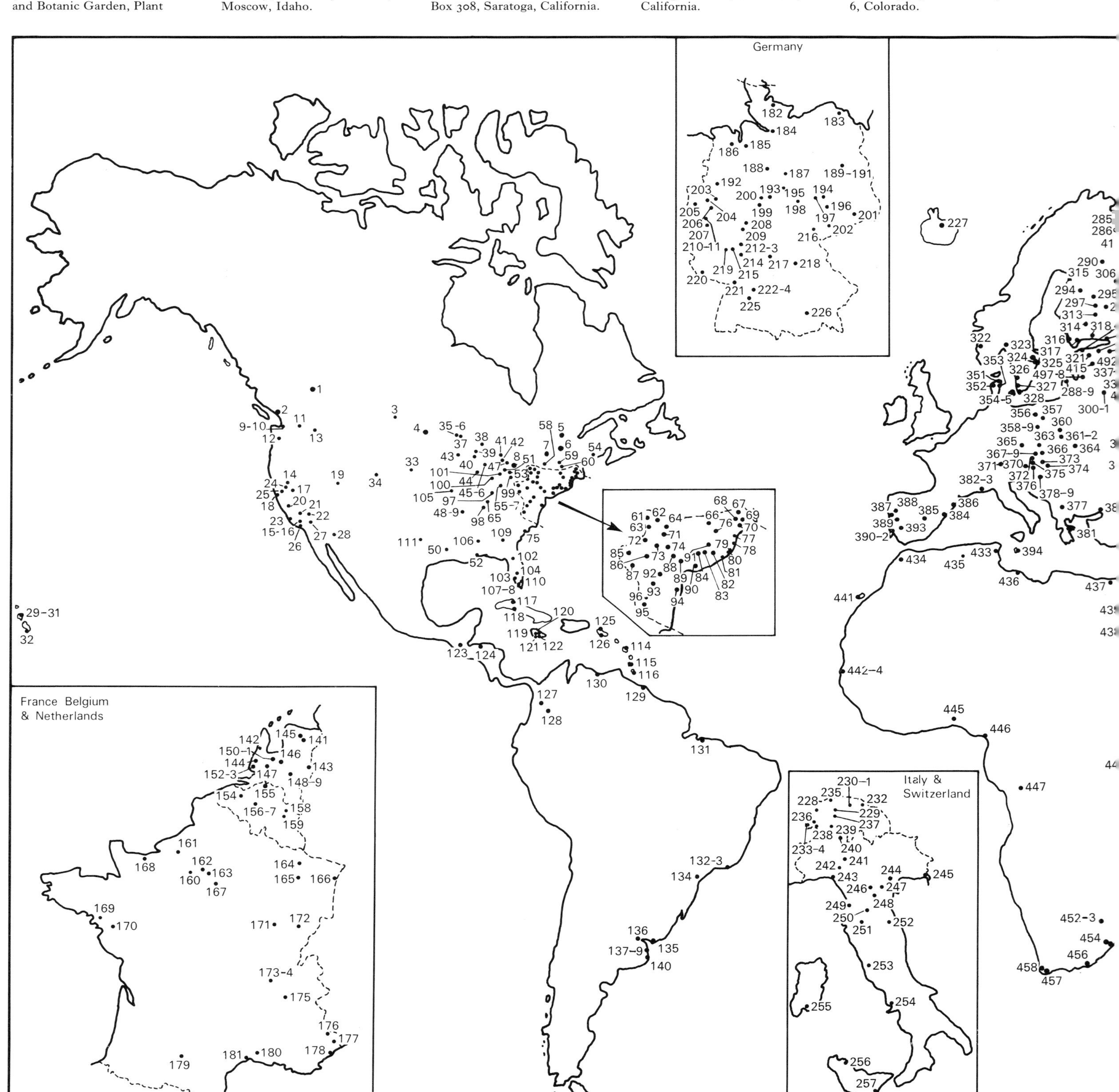

35. Minneapolis: Eloise Butler Wild Flower Garden, Minneapolis Park Board, City Hall, Minneapolis, Minnesota.
36. Minneapolis: Greenhouse of the College of Pharmacy, University of Minnesota, Minneapolis, Minnesota.
37. St Paul: University of Minnesota Landscape Arboretum, University of Minnesota, St Paul, Minnesota.
38. Oshkosh: Paine Art Center and Arboretum, 1410 Algoma Boulevard, Oshkosh, Wisconsin.
39. Madison: University of Wisconsin Arboretum, Birge Hall, University of Wisconsin, Madison 6, Wisconsin.
40. Hales Corners: Alfred L. Boerner Botanical Gardens, 5879 South 92 Street, Hales Corners, Wisconsin.
41. East Lansing: Beal-Garfield Botanic Garden, Building A-1, South Campus, Michigan State University, East Lansing, Michigan.
42. Ann Arbor: Nichols Arboretum, University of Michigan, Ann Arbor, Michigan.
43. Cedar Falls: Iowa State Teachers College Gardens, Cedar Falls, Iowa.
44. Lisle: The Morton Arboretum, Lisle, Illinois.
45. Chicago: Department of Botany, University of Chicago, Chicago 37, Illinois.
46. Chicago: Garfield Park Conservatory, 300 N. Central Park Avenue, Chicago 24, Illinois.
47. Hillsdale: Slayton Arboretum of Hillsdale College, Hillsdale, Michigan.
48. Memphis: Southwestern Arboretum, Memphis Campus, Southwestern at Memphis, Memphis 12, Tennessee.
49. Memphis: The W. C. Paul Arboretum, 800 South Cherry Road, Memphis 17, Tennessee.
50. Many: Hodges Gardens, PO Box 921, Many, Louisiana.
51. Mansfield: Kingwood Center, 900 Park Avenue West, Mansfield, Ohio.
52. Theodore: The Camellia Arboretum, Bellingrath Gardens, Theodore, Alabama.
53. Wooster: Secrest Arboretum, Ohio Agricultural Experiment Station, Wooster, Ohio.
54. Orono: Botanical Plantations of the University of Maine, 203 Deering Hall, University of Maine, Orono, Maine.
55. Cincinnati: Eden Park (Krohn) Conservatory, Cincinnati Park Board, 950 Eden Park Drive, Cincinnati 2, Ohio.
56. Cincinnati: Mt. Airy Arboretum, 2083 Colerian Avenue, Cincinnati, Ohio.
57. Cincinnati: The Stanley M. Rowe Arboretum, 4500 Muchmore Road, Cincinnati 43, Ohio.
58. Rochester: Highland and Durand-Eastman Park, 5 Castle Park, Rochester 20, New York.
59. Esperance: George Landis Arboretum, Esperance, New York.
60. Schenectady: Jackson's Garden, Union College Campus, Schenectady, New York.
61. Washington Crossing: Bowman's Hill State Wild Flower Preserve, Washington Crossing State Park, Washington Crossing, Pennsylvania.
62. Swarthmore: Arthur Hoyt Scott Horticultural Foundation, Swarthmore College, Swarthmore, Pennsylvania.
63. Newtown Square: Ellis School Arboretum, Newtown Square, Pennsylvania.
64. Berwick: Elan Memorial Park, 116 E. Front Street, Berwick, Pennsylvania.
65. Clermont: Bernheim Forest Arboretum, The Isaac W. Bernheim Foundation, Clermont, Kentucky.
66. Oakdale: Bayard Cutting Arboretum, Oakdale, Long Island, New York.
67. Wellesley: Walter Hunnewell Arboretum, 845 Washington Street, Wellesley, Massachusetts.
68. Jamaica Plain: Arnold Arboretum of Harvard University, Jamaica Plain 30, Massachusetts.
69. Northampton: The Botanic Garden of Smith College, Northampton, Massachusetts.
70. South Sudbury: Garden in the Woods, South Sudbury, Massachusetts.
71. Haverford: Haverford College Arboretum, Haverford College, Haverford, Pennsylvania.
72. Pittsburgh: Phipps Conservatory, Schenley Park, Pittsburgh 13, Pennsylvania.
73. Kennett Square: Longwood Gardens, Kennett Square, Pennsylvania.
74. Merion Station: Arboretum of the Barnes Foundation, 300 Latch's Lane, Merion Station, Pennsylvania.
75. Pawleys Island: Brookgreen Gardens, Society for Southeastern Flora and Fauna, Pawleys Island, South Carolina.
76. Newburgh: Thomas C. Desmond Arboretum, R.D. 1, Newburgh, New York.
77. New Haven: Marsh Botanical Garden, 227 Mansfield Street, New Haven 11, Connecticut.
78. Oyster Bay: Planting Fields Arboretum, Planting Fields, Oyster Bay, Long Island, New York.
79. Brooklyn: Brooklyn Botanic Garden and Arboretum, Administration Building, 1000 Washington Avenue, Brooklyn 25, New York.
80. New York: The New York Botanical Garden, Bronx Park, New York 58, New York.
81. Newark: The Dawes Arboretum, R.F.D. 5, Newark, Ohio.
82. New Brunswick: Rutgers Display Gardens, Department of Horticulture, College of Agriculture, Rutgers University, New Brunswick, New Jersey.
83. Philadelphia: Morris Arboretum, 9414 Meadowbrook Avenue, Philadelphia 18, Pennsylvania.
84. Chester: Taylor Memorial Arboretum, 10 Ridley Drive, Garden City, Chester, Pennsylvania.
85. Boyce: The Orland E. White Research Arboretum, Blandy Experimental Farm, University of Virginia, Boyce, Virginia.
86. Mont Alto: Mont Alto Arboretum, Pennsylvania State University, College of Agriculture, School of Forestry, Mont Alto Branch, Mont Alto, Pennsylvania.
87. Morgantown: West Virginia University Arboretum, Department of Biology, West Virginia University, Morgantown, West Virginia.
88. Lima: John J. Tyler Arboretum, Lima, Delaware County, Pennsylvania.
89. Washington: United States National Arboretum, Washington 25, DC.
90. Winterthur: Winterthur Gardens, Winterthur, Delaware.
91. Reading: Reading Public Museum and Art Gallery, Reading, Pennsylvania.
92. Westtown: Westtown School Arboretum, Westtown, Pennsylvania.
93. Blacksburg: Virginia Polytechnic Institute Arboretum, Blacksburg, Virginia.
94. Norfolk: Norfolk Botanical Garden, Airport Road, Norfolk 13, Virginia.
95. Chapel Hill: Coker Arboretum, University of North Carolina, Chapel Hill, North Carolina.
96. Durham: Sarah P. Duke Gardens, North Carolina.
97. Louisville: General Electric Appliance Park, Louisville, Kentucky.
98. Nashville: The Tennessee Botanical Gardens and Fine Arts Center, Inc., Cheekwood, Nashville 5, Tennessee.
99. Mentor: The Holden Arboretum, Sperry Road, Mentor, Ohio.
100. Michigan City: International Friendship Gardens, Michigan City, Indiana.
101. Muncie: Christy Woods (Arboretum), Ball State Teachers College, Muncie, Indiana.
102. Gainesville: Wilmot Memorial Garden, University of Florida, Gainesville, Florida.
103. Fort Meyers: Edison Botanical Gardens, Edison Winter Home, Fort Meyers, Florida.
104. Naples: Caribbean Gardens, Inc., PO Box 623, Naples, Florida.
105. St Louis: Missouri Botanical Garden, 2315 Tower Grove Avenue, St Louis 10, Missouri.

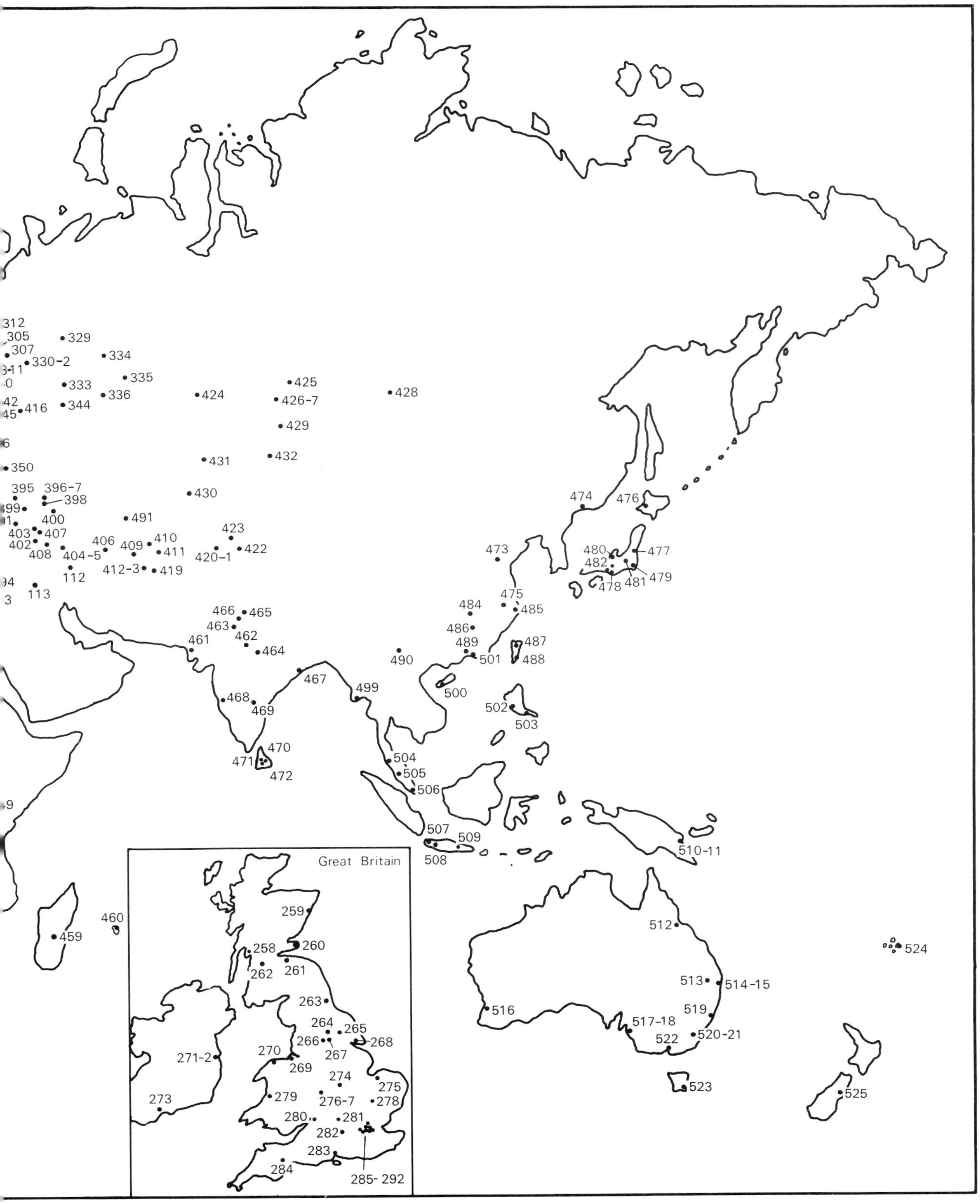

106. University: Arboretum of the University of Alabama, PO Box 1927, University (Tuscaloosa), Alabama.
107. Homestead: Orchid Jungle, Fennel Orchid Company, 26715 S.W. 157th Avenue, Homestead, Florida.
108. Homestead: Sub-Tropical Experiment Station, Route 1, Box 560, Homestead, Florida.
109. Pine Mt: Ida Cason Callaway Gardens, Pine Mt, Georgia.
110. Miami: Fairchild Tropical Garden, 10901 Old Cutler Road, Miami 56, Florida.
111. Fort Worth: Fort Worth Botanic Garden, 3220 Botanic Garden Drive, Fort Worth, Texas.

IRAN

112. Karadj: Botanical Garden, Agricultural College, University of Tehran, Karadj.

IRAQ

113. Baghdad: Zaafaraniyah Arboretum, Zaafaraniyah Horticultural Experiment Station, Baghdad.

BRITISH WEST INDIES

114. Roseau: Botanic Gardens, Roseau, Dominica.
115. Kingstown: St Vincent Botanic Gardens, Kingstown, St Vincent.
116. St George's: Botanic Gardens, Department of Agriculture, St George's, Grenada.

CUBA

117. Havana: Departmento de Botánica Economica, Estación Agronomica, Santiago de las Vegas, Havana.
118. Cienfuegos: Atkins Garden & Research Laboratory, Apartado 414, Cienfueges.

JAMAICA

119. Castleton: Castleton Gardens, St Mary.
120. Cinchona: The Hill Gardens, Cinchona, Hall's Delight, St Andrew.
121. Bath: Bath Garden, St Thomas.
122. Kingston: Royal Botanic Gardens, (Hope), Kingston 6.

GUATEMALA

123. Guatemala City: Botanic Garden of Guatemala, Avenida de la Reforma 0–42, Zona 10, Guatemala City.

HONDURAS

124. San Pedro Sula: Jardín Botánico, Perez Estrada, San Pedro Sula.

PUERTO RICO

125. Mayaguez: Plant Collection, Federal Experiment Station, Mayaguez.
126. San German: Arboretum and Casa Maria Gardens of the Inter-American University, San German.

COLOMBIA

127. Manizales: Jardín de la Facultad de Agronomía, Universidad de Caldas, Manizales.
128. Bogotá: Jardín Botánico José C. Mutis, Calle 19A, No. 7-A-23, Bogotá.

GUYANA

129. Georgetown: Botanic Garden, Department of Agriculture, Georgetown.

VENEZUELA

130. Caracas: Jardín Botánico de la Universidad Central, Caracas.

BRAZIL

131. Belem: Museu Paraense 'Emilio Goeldi', Caiza Postal 399, Belem, Para.
132. Rio de Janeiro: Divisao de Botânica do Museu Nacional-Horto Botânico, Quinta da Bõa Vista, Rio de Janeiro, Guanabara.
133. Rio de Janeiro: Jardim Botânico do Rio de Janeiro, Rio de Janeiro, Guanabara.
134. Sao Paulo: Jardim Botânico de São Paulo, Instituto de Botânico, Caixa postal 4005, São Paulo.

URUGUAY

135. Montevideo: Jardín Botánico, Facultad de Agronomía, Av. Garzón 780, Montevideo.

ARGENTINA

136. San Isidro: Instituto de Botánica Darwinion Lavardén y Del Campo, San Isidro, Provincia de Buenos Aires.
137. Buenos Aires: Instituto de Botánica Agricola—INTA—Araoz 2875, Buenos Aires.
138. Buenos Aires: Jardín Botánico 'Carlos Thays', Dependiente del Instituto Municipal de Botánica, Santa Fe 3951, Buenos Aires.
139. Buenos Aires: Jardín Botánico de la Facultad de Agronomia y Veterinaria, Av. San Martín 4453, Buenos Aires.
140. Llavallol: Jardín Agrobotánico de Santa Catalina, Instituto Fitotécnico de Santa Catalina, Llavallol, FNGR.

NETHERLANDS

141. Haren: Hortus 'De Wolf', Rijksstraatweg 74, Haren.
142. Leyden: Botanic Garden of the University, Nonnensteeg 3, Leyden.
143. De Lutte: Arboretum 'Poort-Bulten', De Lutte bij Oldenzaal.
144. Delft: Cultuurtuin voor Technische Gewassen, 1 Julianalaan 67, Delft.
145. Groningen: Hortus Botanicus, Grote Rozenstraat 31, Groningen.
146. Doorn: von Gimborn Arboretum, Dorpsstraat 75, Doorn.
147. Lunteren: 'De Dennenhorst', Lunteren.
148. Wageningen: Botanische Tuinen en Belmonte Arboretum der Landbouwhogeschool, Gen. Foulkesweg 37, Wageningen.
149. Wageningen: Institute of Horticultural Plant Breeding, 15 Dr S. L. Mansholtlaan, Wageningen.
150. Utrecht: Botanical Museum and Herbarium, 'Cantonspark', Lange Nieuwstraat 106, Utrecht.
151. Utrecht: Hortus Botanicus, Lange Nieuwstraat 106, Utrecht.
152. Rotterdam: Koninklijke Rotterdamse Diergaarde, Botanical Section, Rotterdam.
153. Rotterdam: Stichting 'Arboretum Trompenburg', Groene Wetering 46, Rotterdam.
154. Ghent: Plantentuin der Rijksuniversiteit Ghent, Ledeganckstraat 33, Ghent.
155. Kalmthout: Arboretum, Kalmthout.
156. Brussels: Arboretum Géographique de Tervuren, Dreve du Duc, 106, Brussels 17.
157. Brussels: Jardin Botanique de l'Etat, 236 Rue Royale, Brussels 3.
158. Liège: Jardin Botanique de l'Institut de Botanique de l'Université de Liège, 3 rue Fusch, Liège.
159. Boncelles: Domaine du Sart Tilman, Université de Liège, 134 Route du Condroz, Boncelles.

FRANCE

160. Alford: Jardin Botanique de l'Ecole Vétérinaire, Alford, Seine.
161. Rouen: Jardin Botanique de la Ville de Rouen, 114 avenue des Martyrs de la Résistance, Rouen, Seine-Mar.
162. Grignon: Jardin Botanique de l'Ecole Nationale d'Agriculture de Grignon, Grignon, Seine et Oise.
163. Paris: Jardin Botanique de la Faculté de Pharmacie, 4 avenue de l'Observatoire, Paris 6e.
164. Montigny les Metz: Jardin Botanique, rue de Pont-A-Mousson, Montigny les Metz, Moselle.
165. Nancy: Jardin Botanique de la Ville, 30 bis, rue Sainte Catherine, Nancy, Meurthe et Moselle.
166. Strasbourg: Jardin Botanique, 7 rue de l'Université, Strasbourg, Bas-Rhin.
167. Verrières-le-Buisson: Arboretum et Alpinium Vilmorin, 2 rue d'Estienne d'Orves, Vérrières-le-Buisson, Seine-et-Oise.
168. Caen: Jardin Botanique de la Ville et de l'Université de Caen, 39 rue Desmoueux, Caen, Calvados.
169. Nogent-sur-Vernisson: Arboretum des Barres et Fruticetum Vilmorinianum, Nogent-sur-Vernisson, Loiret.
170. Nantes: Jardin Botanique de Nantes, Place de la Monnaie, Nantes, Loire-Inf.
171. Dijon: Jardin Botanique de la Ville, l'avenue Albert Premier, Dijon, Cote d'Or.
172. Besançon: Jardin Botanique de l'Université et de la Ville, 1 Place du Maréchal Leclerc, Besançon, Doubs.
173. Lyon: Jardin Botanique de la Ville, Parc de la Tête d'Or, Lyon, Rhône.
174. Lyon: Jardin Botanique—Faculté de Médecine et Pharmacie, 8 avenue Rockefeller, Lyon, Rhône.
175. Col du Lautaret: Institut Alpin du Lautaret, 9 Place Bir-Hakeim, Grenoble, Isère.
176. La Colle sur Loup: Station d'Acclimatation Botanique du Château de la Tour de Montmuilhe, La Colle dur Loup, Alpes Maritimes.
177. Saint Jean-Cap Ferrat: Jardin Botanique 'Les Cèdres', Saint Jean-Cap Ferrat, Alpes Maritimes.
178. Antibes: Station de Botanique et de Pathologie Végétale, Villa Thuret, Antibes, Alpes Maritimes.
179. Toulouse: Jardin Botanique, Université de Toulouse, 2 rue Lamarck, Toulouse, Haute-Garonne.
180. Générargues: Maurice Nègre, Domaine de Prafrance, Générargues, Gard.
181. Montpellier: Jardin de Botanique de l'Université de Montpellier, Institut de Botanique, rue Auguste de Boutonnet, Montpellier, Hérault.

GERMANY

182. Kiel: Botanischer Garten der Universität Kiel, Düsterbrooker Weg 17–19, Kiel.
183. Greifswald: Botanischer Garten der Ernst-Moritz-Arndt-Universität, Greifswald, DDR.
184. Hamburg: Staatsinstitut for Allgemeine Botanik und Botanischer Garten, Hamburg 36, Jungiusstr. 6.
185. Bremen: Botanischer Garten, Marcus-Allee 60, Bremen 17.
186. Oldenburg: Staatlicher Botanischer Garten in Oldenburg, Oldenburg/Oldb., Philosophenweg.
187. Brunswick: Botanischer Garten der Technischen Hochschule, Humboltstrasse 1, Brunswick.
188. Hannover: Stadt. Botanischer Schulgarten Burg, Hannover, Burweg 1.
189. Berlin: Botanischer Garten und Museum, Berlin-Dahlem, Königin-Luise-Strasse 6–8.
190. Berlin: Forstbotanischer Garten Eberswalde der Humboldt—Universität Berlin, Berlin, DDR.
191. Berlin: Institut fur Kulturpflanzenforschung, Gatersleben, Kreis Aschersleben, Berlin.
192. Munster: Botanischer Garten der Universität Munster, Munster/Westf., Schlossgarten 3.
193. Göttingen: Botanischer Garten der Universität Göttingen, Untere Karspule.
194. Wörlitz: Staatliche Schlosser und Garten Wörlitz, Oranienbaum und Luisium in Wörlitz, Wörlitz, DDR.
195. Brocken/Oberharz: Schau-u. Versuchsgarten der Martin-Luther- Universität, Halle-Wittenberg, Brocken/Oberharz, DDR.
196. Leipzig: Botanischer Garten der Karl-Marx-Universität, Leipzig C-1, Linnestr. 1, DDR.
197. Halle/Saale: Botanischer Garten der Martin-Luther-Universität, Halle-Wittenberg, Halle/Salle, Am Kirchtor 3, DDR.
198. Sangerhausen: Rosarium Sangerhausen, Sangerhausen, Steinberger Weg 3, DDR.
199. Kassel: Botanischer Garten der Stadt Kassel, (16) Kassel, Bosestrasse 15.
200. Munden: Forstbotanischer Garten der Universität Göttingen, (20A) Hann. Munden, Werraweg 1.
201. Tharandt: Forstbotanischer Garten, Tharandt bei Dresden, Dippoldiswalder Str. 7.
202. Altenburg: Botanischer Garten der Stadt Altenburg, Heinrich-Zille-Strasse, Altenburg, DDR.
203. Dortmund: Botanischer Garten, Dortmund-Brunninghausen.
204. Essen: Botanischer Garten der Stadt Essen, Essen, Kulshammerweg 10.
205. Krefeld: Botanischer Garten der Stadt Krefeld, Krefeld.
206. Wuppertal: Botanischer Garten der Stadt Wuppertal, Wuppertal-Elberfeld, Elisenhohe 5.
207. Bonn: Botanischer Garten der Universität Bonn, Meckenheimer Allee 171, Bonn.
208. Marburg: Botanischer Garten der Philipps-Universität Marburg, Pilgrimstein 4, Marburg.
209. Giessen: Botanischer Garten der Universität Giessen, Giessen, Senckenbergstr. 6.
210. Koln-Merheim: Gartenbau-Abteilung, Dr. Madaus & Co., Koln-Merheim.
211. Koln-Riehl: Botanischer Garten, Koln-Riehl, Amsterdamer Str. 36.
212. Frankfurt am Main: Botanischer Garten der Johann Wolfgang Goethe-Universität, Siesmayerstr. 70, Frankfurt am Main.
213. Frankfurt am Main: Palmengarten, Frankfurt am Main.
214. Darmstadt: Botanischer Garten der Technischen Hochschule, Darmstadt.
215. Mainz: Botanischer Garten der Universität Mainz, Mainz.
216. Jena: Botanischer Garten der Friedrick-Schiller-Universität, Jena, Goetheallee 26, DDR.
217. Würzburg: Botanischer Garten der Universität Würzburg, Würzburg.
218. Erlangen: Botanischer Garten der Universität Erlangen, Schlossgarten 4, Erlangen.
219. Geisenheim: Institut für Gärtnerischen Pflanzenbau der Hess, Lehr- und Forschungs-anstalt für Wein-, Obst- und Gartenbau, Geisenheim (Rheingau).
220. Saarbrucken: Botanischer Garten, Saarbrucken 15.
221. Karlsruhe: Botanischer Garten der Technischen Hochschule, Karlsruhe, Am Fasenengarten.
222. Stuttgart: Botanisches Institut und Botanischer Garten der Technischen Hochschule, 212 Cannstatier Strasse, Stuttgart.
223. Stuttgart–Bad Cannstatt: Botanischer Garten Schloss Wilhelma, Stuttgart-Bad Cannstatt.
224. Stuttgart–Hohenheim: Botanischer Garten d. Landwirtschaftlichen Hochschule, Stuttgart–Hohenheim.
225. Tübingen: Botanischer Garten der Universität Tübingen, Wilhelmstr. 5, Tübingen.
226. Munich: Botanischer Garten, Menzingerstr. 63, Munich 19.

ICELAND

227. Akureyri: Lystigardur Akureyrar, Public Park and Botanic Garden, Akureyri.

SWITZERLAND

228. Porrentruy: Jardin Botanique, École Cantonale, Porrentruy, Neuchâtel.
229. Koppigen: Alpengarten Schynige Platte, Bern.
230. Zurich: Botanischer Garten und Botanisches Museum der Universität Zurich, Postfach Zurich 39.
231. Zurich: Stadtische Sukkulentensammlung Zurich, Mythenquai 88, Zurich 2.
232. St Gallen: Botanischer Garten St Gallen, Brauerstrasse 69, St Gallen.
233. Geneva: Conservatoire et Jardin Botaniques, 192 route de Lausanne, Geneva.
234. Geneva: Jardin Alpin d'Acclimatation 'Floraire', 50 Ave. Petit-Senn, Chêne-Bourg, Geneva.
235. Basel: Botanischer Garten der Universität, Schonbeinstrasse 6, Basel.
236. Montreux: Jardin Alpin 'Rambertia' Montreux, aux Rochers de Naye, Vaud.
237. Berne: Institut et Jardin Botanique de l'Université de Berne, Berne, Altenbergrain 21.
238. Lausanne: Jardin Botanique de la Ville et de l'Université de Lausanne, Palais de Rumine, Lausanne.
239. Champex-Lac: Florealpe, J. M. Aubert, Champex-Lac, Valais.

ITALY

240. Pallanza: Villa Taranto Botanic Gardens, Pallanza, Lake Maggiore.
241. Milan: Giardini Botanici dell'Isola Madre, Amministrazione Borromeo, Via Borromei 1/A, Milan (317).
242. Pavia: Istituto ed Orto Botanico dell'Università di Pavia, Via S. Epifanio 14, Pavia.
243. Genoa: Istituto ed Orto Botanico 'Hanbury' dell'Università, Via Balbi 5, Genoa.
244. Padua: Orto Botanica dell'Università di Padova, Via Orto Botanico 15, Padua.
245. Trieste: Civico Orto Botanico, Via Carlo de Marchesetti N.2, Trieste.
246. Modena: Istituto ed Orto Botanico dell'Università, Viale Caduti in guerra 127, Modena.
247. Ferrara: Istituto ed Orto Botanico dell'Università di Ferrara, Via del Paradiso 7, Ferrara.
248. Bologna: Orto Botanico dell'Università, Via Irnerio 42, Bologna.
249. Pisa: Istituto Botanico dell'Università di Pisa, Via Luca Ghini No. 2, Pisa.
250. Florence: Orto dell'Istituto Botanico dell'Università, Via Lamarmora 4, Florence.
251. Siena: Istituto Botanico dell'Università, Via P. A. Mattioli 4 (S. Agostino), Siena.
252. Urbino: Orto Botanico, Via Bramsute 28, Urbino.
253. Rome: Istituto ed Orto Botanico della Università, Citta Universitaria, Rome.
254. Naples: Istituto ed Orto Botanico della Facoltà di Agraria Portici, Naples.
255. Cagliari: Orto Botanico dell'Università degli Studi, Viale Fra Ignazio da Laconi 13, Cagliari, Sardinia.
256. Palermo: Istituto Botanico e Giardino Coloniale, Via Lincoln, Palermo, Sicily.
257. Catania: Istituto ed Orto Botanico dell'Università, Via Antonino Longo 19, Catania.

SCOTLAND

258. Benmore: Younger Botanic Garden, Benmore, by Dunoon, Argyll.
259. Aberdeen: Cruickshank Botanic Garden, St Machar Drive, Old Aberdeen.
260. St Andrews: Botanic Garden, The University of St Andrews, St Andrews, Fife.
261. Edinburgh: Royal Botanic Gardens, Edinburgh 3.
262. Glasgow: Botanic Gardens, Glasgow, W2.

ENGLAND

263. Durham: Botanical Garden, University Science Laboratories, South Road, Durham.
264. Harrowgate: Northern Horticultural Society Gardens, Harlow Car, Harrogate, Yorkshire.
265. York: Museum Gardens, Yorkshire Museum, York.
266. Bradford: Bradford Botanic Garden, Lister Park, Bradford 9, Yorkshire.
267. Leeds: Botany Department Experimental Gardens, The University, Leeds 2.
268. Hull: University of Hull, Botanic Garden and Experimental Garden, Thwaite Street, Cottingham, Yorkshire.
269. Ness: The University of Liverpool Botanic Garden, Ness, Wirral, Cheshire.

WALES

270. Llandudno: Haulfre Gardens, Town Hall, Llandudno.

EIRE

271. Dublin: National Botanic Gardens, Glasnevin, Dublin 9.
272. Dublin: Trinity College Botanic Garden, Shelbourne Road, Dublin.
273. Cork: Botanic Gardens, Botany Department, University College, Cork.

ENGLAND

274. Leicester: Beaumont Hall, Stoughton Drive South, Leicester.
275. King's Lynn: Garden of Leonard Maurice Mason, Talbot Manor, Fincham, King's Lynn, Norfolk.
276. Birmingham: Birmingham Botanical and Horticultural Society Ltd, Edgbaston, Birmingham.
277. Birmingham: The University Botanic Garden, Department of Botany, University of Birmingham, Birmingham.
278. Cambridge: University Botanic Garden, Cambridge.

WALES (cont.)

279. Aberystwyth: University College of Wales, Botany Garden, Aberystwyth.

ENGLAND (cont.)

280. Westonbirt: Westonbirt Arboretum, Westonbirt, near Tetbury, Gloucestershire.
281. Oxford: University Botanic Garden, Rose Lane, Oxford.
282. Reading: Agricultural Botany Garden, Cutbush Lane, Shinfield, Reading, C/- Department of Agricultural Botany, The University, Reading, Berkshire.
283. Southampton: University Botanic Garden, Southampton.
284. Exeter: Garden of the Department of Botany, University of Exeter, Exeter, Devon.
285. Enfield: Myddelton House, Enfield, Middlesex.
286. Englefield Green: University of London Botanical Supply Unit, Elm Lodge, Englefield Green, Surrey.
287. Chelsea: Chelsea Physic Garden, Royal Hospital Road, Chelsea, London, SW3.
288. London: South London Botanical Institute, 323 Norwood Road, London, SE24.
289. Kew: Royal Botanic Gardens, Kew, Richmond, Surrey.
290. Wisley: The Royal Horticultural Society's Gardens, Wisley, Ripley, Woking, Surrey.
291. Godalming: Winkworth Arboretum, Hascombe Road, Godalming, Surrey.
292. Hawkhurst: Bedgebury National Pinetum, Hawkhurst, Kent.

USSR

293. Kaunas: Kaunas Botanical Garden of the Botanical Institute of the Lithuanian Academy of Sciences, Botanikos Prospect 1, Kaunas 19, Lithuania.
294. Uzhgorod: Botanical Garden of the Uzhgorod State University, Oktyabrskaya Street 60, Zakarpatskaya Region, Uzhgorod, Ukraine.
295. Lvov: Botanical Garden of the Lvov State University, Marc Cheremshina Street 44 and Lomonosov Street 4, Lvov, Ukraine.
296. Lvov: Arboretum of the Lvov Forest Institute, Pushkin Street 103, Lvov, Ukraine.
297. Kremenets: Kremenets Botanical Garden, Krupskaya Street 1, Kremenets, Ternopol Region, Ukraine.
298. Uman: Dendrological Park 'Sofievka', Uman, Chercassy Region, Ukraine.
299. Petrozavodsk: Botanical Garden of the Petrozavodsk State University, Petrozavodsk, Solomennoje.
300. Minsk: Central Botanical Garden of the Academy of Sciences of the Byelorussian SSR, Academicheskaya Street 31, Minsk 13, Byelorussian SSR.
301. Minsk: Botanical Garden of the Byelorussian State University, Minsk, Byelorussian SSR.
302. Leningrad: Botanical Garden of the Komarov Botanical Institute of the USSR Academy of Sciences, Professor Popov Street 2, Leningrad 2.
303. Leningrad: Botanical Garden of Leningrad State University, Universitetskaya quay 7–9, Leningrad V-164.
304. Leningrad: Arboretum of the S. M. Kirov Forest Academy, Institutsky Lane 5, Leningrad K-18.
305. Vitebsk: Botanical Garden of the Vitebsk Medical Institute, Frunze Avenue 7, Vitebsk, Byelorussian SSR.
306. Vilnius: Botanical Garden of the Vilnius State University, Churlionio 110, Vilnius, Lithuania.
307. Yaroslavl: Botanical Garden of the Yaroslavl Pedagogical Institute, Republikanskaya Street 108, Yaroslavl.
308. Moscow: Botanical Garden of Officinal Plants of the First Moscow Medical Institute, 4th Krasnogvardeisky proezd 20, Moscow D-100.
309. Moscow: Main Botanical Garden, Botanicheskaya Street 4, Moscow I-276.
310. Moscow: Botanical Garden and Schreder's Dendrarium of the Moscow Timirjazev Agricultural Academy, Prjanishnikov Street, Moscow A-8.
311. Moscow: Botanical Garden of the Moscow Lomonosov State University, Lenin Hills, Moscow V-234.
312. Vilar: Botanical Garden of the All-Union Research Institute of Medicinal and Aromatic Plants, Post Office, Vilar, Moscow Region.
313. Kishinev: Botanical Garden of the Academy of Sciences of the Moldavian SSR, Dunaevsky Street 5, Kishinev, Moldavian SSR.
314. Odessa: Botanical Garden of the Mechnikov State University, Proletarsky Boulevard 48–50, Odessa, Ukraine.

FINLAND

315. Oulu: Botanic Garden of the University of Oulu, Department of Botany, Oulu.
316. Turku: Botanical Garden, University of Turku, Turku.
317. Helsinki: Helsingin Yliopiston Kasvitieteellinen Puutarha, Unioninkatu 44, Helsinki.
318. Elimäki: Arboretum Mustila, Elimäki.

USSR (cont.)

319. Zhitomir: Botanical Garden of the Zhitomir Agricultural Institute, Stalin Street 9, Zhitomir, Ukraine.
320. Kamenets-Podolsky: Kamenets-Podolsky Botanical Garden, Leningradskaya Street 78, Kamenets-Podolsky, Khmelnitsky Region, Ukraine.
321. Chernovtsky: Botanical Garden of Chernovtsky State University, Fedkovich Street 11, Chernovtsky, Ukraine.

NORWAY

322. Bergen: Botanisk Hage, J. Frieles gt. 1, Bergen.
323. Oslo: Botanic Garden, University of Oslo, Trondheimsv. 23B, Oslo.

SWEDEN

324. Stockholm: Hortus Botanicus Bergianus, Stockholm 50.
325. Uppsala: The Botanical Garden, PO Box 123, Uppsala 1.
326. Göteborg: Göteborgs Botaniska Trädgården, Frölundagatan 22, Göteborg SV.
327. Hälsingborg: Botanic Garden, Hälsingborg.
328. Lund: The Botanical Garden, Lund.

USSR (cont.)

329. Kirov: Botanical Garden of the Kirov Pedagogical Institute, Karl Marx Street 95, Kirov.
330. Gorki: Botanical Garden of the Byelorussian Agriculture Academy, Gorki, Mogilev Region, Byelorussian SSR.
331. Gorky: Agrobotanical Garden of the Gorky Agricultural Institute, Teaching and Experimental Farm 'Shcherbinki', Gorky 9.
332. Gorky: Botanical Garden of Gorky State University, Gorky 62.
333. Kazan: Kazan Zoobotanic Garden, Khadi Taktash Street 112, Kazan, Tartar Autonomous SSR.
334. Perm: A. G. Genkel Botanical Garden of the Perm State University, Genkel Street 7, Perm 5, Zaimka.
335. Sverdlovsk: Botanical Garden of Biological Institute of the Ural Branch of the Academy of Sciences of the USSR, 8 March Street 102, Sverdlovsk-8.
336. Ufa: Botanical Garden of the Bashkir Branch of Academy of Sciences of USSR, Polnarnaya Street 8, Ufa, Bashkir.
337. Kiev: Central Republic Botanical Garden of the Ukrainian Academy of Sciences, Timirjazevskaya Street 1, Kiev 14, Ukraine.
338. Kiev: Botanical Garden of the Kiev State University, Komintern Street 1, Kiev, Ukraine.
339. Belaya Tserkov: Dendrological Park 'Alexandria' of the Ukrainian Academy of Sciences, Belaya Tserkov, Kiev Region, Ukraine.
340. Chernigov: Chernigov Botanical Garden, Projectnaya Street 4, Chernigov, Ukraine.
341. Ichnya: Dendrological Park 'Trostyanets', Post Office Ichnya, Chernigov Region, Ukraine.
342. Lipetsk: Forest Steppe Selection Experimental Station of Decorative Cultures, Post Office, Meshcherskoe, Lipetsk Province.
343. Poltava: Botanical Garden of Agronomy-Biological Station of the Poltava Pedagogical Institute, Ostrogradsky Street 2, Poltava, Ukraine.
344. Kuibyshev: Kuibyshev Botanical Garden, Kuibyshev, PO 24, Ovrag Podpolshikov.
345. Voronezh: Prof. B. M. Kozopolansky Botanical Garden of the Voronezh State University, Voronezh 12.
346. Kharkov: Botanical Garden of the Kharkov State University, Klochkovskaya 52, Kharkov.
347. Dniepropetrovsk: Dniepropetrovsk Botanical Garden of the State University, Dniepropetrovsk 10, Ukraine.
348. Yalta: Nikita Botanical Garden, Yalta, Crimea, Ukraine.
349. Askania Nova: Botanical Park of the Ukrainian Scientific Research Institute of Cattle Breeding, Askania Nova, Khersonsky Region, Novo-Troitsky District, Ukraine.
350. Rostov: Botanical Garden of the Rostov State University, Rostov Don, PO 2.

DENMARK

351. Hørsholm: Arboretum Hørsholm.
352. Hvedde: The Desert Arboretum, Hvedde, Kibaek.
353. Charlottenlund: Forest Botanic Garden, Charlottenlund.
354. Copenhagen: Botanic Garden of the University of Copenhagen, Oster Farimagsgade 2B, Copenhagen K.
355. Copenhagen: Den Kgl. Veterinaer—og Landbohojskøles Haver, Bulowsvej 13, Copenhagen V.

POLAND

356. Kórnik: Polish Academy of Sciences, Kórnik Arboretum of the Institute of Dendrology, Kórnik, near Poznán.
357. Goluchów: Wyższa Szkola Rolniczaw Poznaniu, Arboretum Goluchów, Poczta Goluchów, Powiat Pleszew.
358. Wroclaw: Ogród Botaniczny Uniwersytetu Wroclawskiego, Wroclaw, ul. Kanonia 6/8.
359. Wroclaw: Ogród Roslin Leczniczych Akademii Medycznej we Wroclawiu, ul. Kochanowskiego 10.
360. Rogow: Arboretum, Lasy Doświadczalne, Szkoly Glównej Gospodarstwa Wiejskiego, Rogow K/Koluszek.
361. Kraków: Polish Academy of Science, Pharmacological Institute, Division of the Medical Plants, 16 Grzegorzecka Str. Kraków.
362. Kraków: Hortus Botanicus Universitatis Jacellonicae, Kraków, ul. Kopernika 27.

CZECHOSLOVAKIA

363. Opavy: Krajské Arboretum Novy Dvur, p. Steborice u Opavy.
364. Kosice: Botanical Garden of the Pedagogical Institute, Kosice.
365. Prague: The Botanic Garden of the Charles University, Benátská 2, Prague.
366. Bratislava: Botanic Garden of Komensky University, Bratislava.

AUSTRIA

367. Vienna: Alpengarten im Belvedere, Vienna III, Prinz-Eugen Strasse 27.
368. Vienna: Botanischer Garten der Universität Wien, Vienna III, Rennweg 14.
369. Vienna: Verwaltung der Bundesgarten, Vienna XIII, Schönbrunn.
370. St Veit: Alpengarten Rannach, St Veit bei Graz, Steiermark.
371. Innsbruck: Botanischer Garten der Universität Innsbruck, Botanikerstrasse 10, Innsbruck.
372. Graz: Botanischer Garten der Universität Graz, Holteigasse 6, Graz.

HUNGARY

373. Vácrátót: Botanical Garden and Geobotanical Laboratorium, Hungarian Academy of Sciences, Vácrátót.
374. Tápiószele: Orszagos Agrobotanikai Intezet, Tápiószele.

YUGOSLAVIA

375. Ljubljana: Botanicki vrt univerze v. Ljubljani, Izanska c. 15, Ljubljana.
376. Radomlje: Arboretum Volcji potok, p. Radomlje, Slovenia.
377. Skopje-Gazibaba: Botanic Garden of the Botanic Institute, University of Skopje, Skopje-Gazibaba, Box 439, Makedonija.
378. Zagreb: Botanicki vrt Farmaceutskog fakulteta, Alagoviceva 43, Zagreb II.
379. Zagreb: Botanicki vrt Prirodoslovno-matematickog fakulteta u Zagrebu, Marulicev trg 9a, Zagreb.

TURKEY

380. Istanbul: Istanbul Üniversitesi Botanik Bahçesi, Suleymaniye, Istanbul.

GREECE

381. Athens: Botanical Garden of the University, 58 Odos Spiros Patsis, Athens 3.

MONACO

382. Monte Carlo: Service de Botanique de la Principauté de Monaco, 5 impasse de la Fontaine, Monte Carlo.
383. Monte Carlo: Jardin Exotique, BP 105, Monte Carlo.

SPAIN

384. Barcelona: Jardín Botánico de Barcelona, Barcelona.
385. Madrid: Jardín Botánico de Madrid, Plaza de Murillo 2, Madrid.
386. Blanes: Jardín de Aclimatación 'Pinya de Rosa', Blanes, Costa Brava.

PORTUGAL

387. Porto: Instituto de Botânica 'Dr Gonçalo Sampaio', R. do Campo Alegre, 1191, Porto.
388. Sacavém: Estação Agronómica Nacional, Sacavém.
389. Coimbra: Museu, Laboratório e Jardim Botânico da Faculdade de Ciências da Universidade de Coimbra.
390. Lisbon: Jardim Botânico da Faculdade de Ciências, Universidade de Lisboa, Rua da Escola Politécnica, Lisbon 2.
391. Lisbon: Jardim Botânico da Ajuda, Calçada da Ajuda, Lisbon.
392. Belém: Jardim e Museu Agrícola do Ultramar, Calçada do Galvão, Belém, Lisbon.
393. Elvas: Estação de Melhoramento de Plantas, Elvas.

MALTA

394. Floriana: Argotti Botanic Gardens, Floriana.

USSR (cont.)

395. Sochy: Arboretum of the Sochy Scientific Experimental Station of Subtropical Forest and Forest-Parks Economy, Kurortny Avenue 74, Sochy.
396. Stavropol: Botanical Garden of the Stavropol Pedagogical Institute, Pushkin Street 1, Stavropol-Caucasia.
397. Stavropol: Stavropol Botanical Garden, Post Box 22, Stavropol-Caucasia.
398. Pyatigorsk: Botanical Garden of the Pyatigorsk Pharmacy Institute, Kirov Avenue 33, Pyatigorsk.
399. Sukhumi: Sukhumi Botanical Garden of the Georgian Academy of Sciences, Chavchavadze Street 18, Sukhumi, Georgia.
400. Nalchik-Dolinsk: Kabardino-Balkar Botanical Garden, Nalchik-Dolinsk, Kabardino-Balkar Autonomous SSR.
401. Batumi: Batumi Botanical Garden of the Georgian Academy of Sciences, Batumi, Zeleny mis., Georgian SSR.
402. Yerevan: Botanical Garden of Armenian Academy of Sciences, Yerevan, Kanaker, Armenian SSR.
403. Tbilisi: Tbilisi Central Botanical Garden, Tbilisi, PO 5, Georgian SSR.
404. Baku: Dendropark of the Azerbaijan Scientific Research Institute of Horticulture, Viticulture and Subtropical Cultures, Street of 26 Comissarov 89, Baku 44, Mardakyany, Azerbaijan SSR.
405. Baku: Botanical Garden of Komarov Botanical Institute of the Azerbaijan Academy of Sciences PO 1, Lokbatanskoe chaussee, Baku, Azerbaijan SSR.
406. Ashkhabad: Botanical Garden of the Academy of Sciences of Turkman SSR, Ashkhabad 12, Keshi, Turkman SSR.
407. Kirovakan: Kirovakan Branch of Botanical Garden of the Armenian Academy of Sciences, Kirovakan, Vanadzorskoye Ushchelis, Armenian SSR.
408. Sevan: Sevan Branch of Botanical Garden of the Armenian Academy of Sciences, Sevan, Armenian SSR.
409. Samarkand: Botanical Garden of the Samarkland State University, Gorky Boulevard 3, Samarkand, Uzbek SSR.
410. Tashkent: Botanical Garden of the Uzbek Academy of Sciences, Karamurtskaya Street 272, Tashkent 53, Uzbek SSR.
411. Leninabad: Leninabad Botanic Garden, Pravyi bereg, PO 6, Leninabad, Tadjik SSR.

412. Dushanbe: Botanical Garden of the Botanical Institute of the Academy of Sciences of Tadjik SSR, Karamov Street 19, Dushanbe-17, Tadjik SSR.
413. Dushanbe: Botanical Garden of the Tadjik State Universitym Lenin Street 17, Dushanbe, Tadjik SSR.
414. Tallin: Tallin Botanical Garden, Kloostrimetsa Street 44, Tallin, Estonia.
415. Tartu: Botanical Garden of the Tartu State University, Michurin Street 40, Tartu, Estonia.
416. Penza: Penza Botanical Garden of the Belinsky Pedagogical Institute, Carl Marx Street 4, Penza.
417. Kirovsk: Polar-Alpine Botanical Gardens of the Kolsky Branch of Academy of Sciences USSR, Kirov Street 9a, Kirovsk, Murmansk Region.
418. Cherkassy: Botanical Garden of the Cherkassy Pedagogical Institute, Gromov Street 26, Cherkassy, Ukraine.
419. Khorog: Pamir Botanical Garden of the Tadjik Academy of Sciences, Khorog, Tadjik SSR.
420. Frunze: Agrobotanical Garden of the Kirghiz State University, Belinsky Street 100, Frunze 14, Kirghiz SSR.
421. Frunze: Botanical Garden of the Kirghiz Academy of Sciences, Volgogradskaya Street 100, Frunze, Kirghiz SSR.
422. Przhevalsk: Przhevalsk Zonal Experimental Station of the All-Union Research Institute of Medicinal and Aromatical Plants, Przhevalsk, Kirghiz SSR.
423. Alma-Ata: Alma-Ata Botanical Garden of the Kazakh Academy of Sciences, PO 10, Alma-Ata, Kazakh SSR.
424. Omsk: Botanical Garden of the Omsk Agricultural Institute, Post Office 8, Omsk.
425. Tomsk: Siberian Botanical Garden of the Tomsk State University, Lenin Avenue 36, Tomsk.
426. Novosibirsk: Botanical Garden of the Novosibirsk Agricultural Institute, Dobrolubov Street 289, Novosibirsk.
427. Novosibirsk: Central Siberian Botanical Garden of the Siberian Division of the USSR Academy of Sciences, Zaeltsovsky District, Novosibirsk.
428. Irkutsk: Botanical Garden of the Irkutsk State University, Vuzovskaya quay 20, Irkutsk.
429. Barnaul: Dendropark of Altai Fruit-Berry Experiment Station, Smeinogorsky Tract 25, Barnaul 20.
430. Dzhezkazgan: Dzhezkazgan Botanical Garden, Dzhezkazgan 1, Karaganda Region, Kazakh SSR.
431. Karaganda: Karaganda Botanical Garden, Karaganda, Kazakh SSR.
432. Leninogorsk: Altai Botanical Garden, Leninogorsk, East-Kazakhstan Region, Kazakh SSR.

TUNISIA

433. Ariana: Service Botanique et Agronomique de Tunisie, Service Botanique, Ariana.

MOROCCO

434. Rabat: Institut Scientifique Chérifien, Laboratoire de Phanérogamie, Avenue Biarnay, Rabat.

ALGERIA

435. Hamma: Jardin d'Essais du Hamma, Rue de Lyon, Hamma.

LIBYA

436. Tripoli: Sidi Mesri Experiment Station, Tripoli.

EGYPT

437. Alexandria: Botanic Garden, Botany Department, Faculty of Science, University of Alexandria, Moharram Bey, Alexandria.
438. Aswan: The Plant Garden, Aswan.
439. Cairo: Orman Botanic Garden, Giza-Orman, Cairo.
440. Cairo: Zohariya Trial Gardens, Gezira, Cairo.

SPAIN (cont.)

441. Tenerife: Jardin de Aclimatación de la Orotava, Puerto de la Cruz, Tenerife, Canary Islands.

SENEGAL

442. Dakar: Institut Francais d'Afrique Noire, BP 206, Dakar.
443. Dakar: Jardin Botanique de la Faculté des Sciences, BP 6049, Dakar.
444. Dakar: Parc Forestier et Zoologique de Hann, BP 1831, Dakar.

NIGERIA

445. Ibadan: Botanical Garden, University College, Ibadan.

CAMEROONS

446. Victoria: Victoria Botanic Gardens, Victoria.

CONGO REPUBLIC

447. Kisantu: Jardin Gillet, Kisantu.

UGANDA

448. Entebbe: Entebbe Botanic Gardens, PO Box 40, Entebbe.

KENYA

449. Nairobi: Nairobi Arboretum Forest Reserve, The Chief Conservator of Forests, Forest Department, PO 30027, Nairobi.
450. Kikuyu: East Africa Agriculture and Forestry Research Organization, PO Box 21, Kikuyu.

RHODESIA

451. Salisbury: Federal Botanic Garden, PO Box 8100, Causeway, Salisbury.

SOUTH AFRICA

452. Pretoria: Botanic Garden of the University of Pretoria, Pretoria.
453. Pretoria: Pretoria National Botanic Garden, National Herbarium, Division of Botany, Department of Agriculture, PO Box 994, Pretoria.
454. Pietermaritzburg: Botanic Gardens of Pietermaritzburg, PO Box 99, Pietermaritzburg, Natal.
455. Durban: Botanical Gardens, St Thomas Road, Durban.
456. Grahamstown: Grahamstown Municipal Botanical Gardens, PO Box 176, Grahamstown.
457. Stellenbosch: Botanical Garden, University of Stellenbosch, Stellenbosch.
458. Kirstenbosch: National Botanic Gardens of South Africa, Kirstenbosch, Newlands, Cape Province.

MALAGASY REPUBLIC

459. Tananarive: Jardin Botanique de Tsimbazaza, Boîte postale 434, Tsimbazaza, Tananarive.

MAURITIUS

460. Pamplemousses: Royal Botanic Gardens, Pamplemousses.

PAKISTAN

461. Karachi: M. Gandhi Garden, Burns Garden & Frere Hall Garden, Karachi.

INDIA

462. Lucknow: National Botanic Gardens, Lucknow.
463. New Delhi: Division of Botany, Indian Agricultural Research Institute, New Delhi.
464. Allahabad: Experimental Garden, Central Botanical Laboratory, 10 Chatham Lines, Allahabad.
465. Dehra Dun: Botanical Garden of the Forest Research Institute, New Forest, Dehra Dun.
466. Saharanpur: Governmental Horticultural Research Institute, Saharanpur, Uttar Pradesh.
467. Calcutta: Indian Botanic Garden, PO Botanic Garden, Calcutta, Howrah.
468. Poona: Experimental Garden, Botanical Survey of India, Western Circle, 7 Koregaon Road, Poona-1.
469. Hyderabad-Deccan: Botanic Garden. Osmania University, Botany Department, Hyderabad-Deccan, AP.

CEYLON

470. Peradeniya: Royal Botanic Gardens, Peradeniya.
471. Gampaha: Botanic Garden, Heneratgoda, Gampaha.
472. Hakgala: Botanic Gardens, Hakgala.

CHINA

473. Peking: Peking Botanical Garden, Peking, Hopei.

USSR (cont.)

474. Vladivostok: Botanical Garden of the Far East Branch of the Siberian Division of the Academy of Sciences of USSR, Makovsky Street 27, PO Oceanskaya, Vladivostok.

CHINA (cont.)

475. Nanking: Nanking San Yat-Sen Botanical Garden, Nanking, Kiangsu.

JAPAN

476. Sapporo: The Botanic Garden, Faculty of Agriculture, Hokkaido University, Sapporo.
477. Sendai: Botanical Garden, Tohoku University, Sendai.
478. Osaka: The Botanical Garden of Osaka City University, Kisaich, Katano-cho, Kitakawochigun, Osaka Pref.
479. Tokyo: Botanic Garden, Koishikawa, Bunkyoku, Tokyo.
480. Kanazawa: Botanic Garden of the Faculty of Science, University of Kanazawa, Isikawa-Ken.
481. Kasukabe-Shi: Kasukabe Experimental Station for Medicinal Plants, National Institute of Hygienic Sciences, 30 Kasukabe-Shi, Saitama-Ken.
482. Kyoto: Kyoto Takeda Herbal Garden, Ichijoji, Sakyo-ku, Kyoto.
483. Kobe: Kobe Municipal Arboretum, 4–1 Nakaichiri-ya, Shimotanigami, Yammada-cho, Hyogo-Ku, Kobe.

CHINA (cont.)

484. Wuhan: Wuhan Botanical Garden, Wuhan, Hupei.
485. Hangchow: Hangchow Botanical Garden, Hangchow, Chekiang.
486. Lushan: Lushan Botanical Garden, Lushan, Kiangsi.
487. Taipei: Taiwan Forestry Research Institute Botanical Garden, Po-Ai Road, Taipei, Taiwan.
488. Ping-Tung: Heng-Chun Tropical Forest Botanical Garden, Heng-Chun, Ping-tung, Taiwan.
489. Canton: Canton Botanical Garden, Canton, Kwantung.
490. Kunming: Kunming Botanical Garden, Kunming, Yunnan.

USSR (cont.)

491. Nukus: Botanic Garden of the Karakalpak Branch of the Uzbek Academy of Sciences, K. Marx Street 2, Nukus, Uzbek SSR.
492. Kaliningrad: Kaliningrad Botanical Garden, Belomorskaya Street 20, Kaliningrad.

ISRAEL

493. Jerusalem: The Botanic Garden of the Hebrew University, Jerusalem.
494. Nathanya: Ilanoth, PO Box 88, Nathanya.
495. Holon: Botanical Garden 'Mikveh-Israel', Holon.
496. Tel-Aviv: Botanic Garden of Tel-Aviv University, 155 Herzl Street, Tel-Aviv.

USSR (cont.)

497. Riga: Botanical Garden of the Latvian Academy of Sciences, Riga, PO Salaspils, Latvia.
498. Riga: Botanical Garden of the Latvian State University, L. Kandavas Street 2, Riga 7, Latvia.

BURMA

499. Rangoon: The Agri-Horticultural Society of Burma (Kandawgalay), Rangoon.

CHINA (cont.)

500. Hainan: Hainan Botanical Garden, Hainan, Island of Hainan.

HONG KONG

501. Hong Kong: Botanic Gardens, Park and Playground Section, Urban Services Department, Central Government Offices, West Wing.

PHILIPPINES

502. College: Makiling National Park, College of Forestry, University of the Philippines, College, Laguna.
503. Los Banos: Los Banos College of Agriculture Arboretum, Laguna.

MALAYSIA

504. Penang: Botanic Gardens, Penang.
505. Kepong: Forest Research Institute, Kepong, Selangor.

SINGAPORE

506. Singapore: Botanic Gardens, Singapore 10.

INDONESIA

507. Bogor: Kebun Raya Indonesia, Bogor.
508. Sindanglaja: Mountain Garden Tjibodas, Sindanglaja, Java.
509. Lawang: Purwodadi Botanic Garden, Lawang, Java.

NEW GUINEA

510. Lae: Botanic Gardens, Division of Botany, Department of Forests, Lae.
511. Lae: Botanical Garden, Lae.

AUSTRALIA

512. Townsville: Botanic Gardens, Townsville, North Queensland.
513. Toowoomba: Toowoomba Botanical Gardens, Toowoomba, Queensland.
514. Brisbane: Botanic Gardens, Brisbane, Queensland.
515. Brisbane: Sherwood Arboretum, Sherwood, Brisbane, Queensland.
516. Perth: Botanic Gardens of Western Australia, King's Park, Perth, Western Australia.
517. Adelaide: Botanic Garden, North Terrace, Adelaide, South Australia.
518. Adelaide: Waite Agricultural Research Institute, Private Mail Bag 1, Adelaide, South Australia.
519. Sydney: Royal Botanic Gardens, Sydney, New South Wales.
520. Canberra: Canberra Botanic Garden, Parks and Gardens Section, Department of the Interior, Canberra, ACT.
521. Canberra: Plant Introduction Section, Division of Plant Industry, C.S.I.R.O., PO Box 109, Canberra, ACT.
522 Melbourne: Royal Botanic Gardens and National Herbarium, South Yarra, SE1, Melbourne, Victoria.
523. Hobart: Tasmanian Botanical Gardens, Queen's Domain, Hobart, Tasmania.

FIJI ISLANDS

524. Suva: Suva Gardens, Queen Elizabeth Drive, Suva.

NEW ZEALAND

525. Christchurch: Christchurch Botanic Gardens, Christchurch, C1.